The Haraway Reader

The Haraway Reader

Donna Haraway

ROUTLEDGE
NEW YORK AND LONDON

Published in 2004 by
Routledge
29 West 35th Street
New York, NY 10001
www.routledge-ny.com

Published in Great Britain by
Routledge
11 New Fetter Lane
London EC4P 4EE
www.routledge.co.uk

Library of Congress Cataloging in Publication Data

Haraway, Donna Jeanne.
    The Haraway reader / Donna Haraway.
        p.   cm.
Includes bibliographical references and index.
    ISBN 0-415-96688-4 (HB : alk. paper)—ISBN 0-415-96689-2 (PB. : alk. paper)
    1. Feminist theory.   2. Feminist criticism.   3. Technology—Social aspects.
4. Science—Social aspects.   I. Title.
HQ1190 .H364   2003
305.42′01—dc21   2003005861

To all my companion species

CONTENTS

INTRODUCTION
A KINSHIP OF FEMINIST FIGURATIONS

I learned to read and write inside worlds at war. I was born near the end of World War II, grew up in the Cold War, attended graduate school during the Viet Nam War, and I am preparing this *Reader* for publication during my country's invasion of Iraq. And that's the short list. These wars are personal. They make me who I am; they throw me into inherited obligations, whether I like it or not. These worlds at war are the belly of the monster from which I have tried to write into a more vivid reality a kin group of feminist figures. My hope is that these marked figures might guide us to a more livable place, one that in the spirit of science fiction I have called "elsewhere."

Figures collect up hopes and fears and show possibilities and dangers. Both imaginary and material, figures root peoples in stories and link them to histories. Stories are always more generous, more capacious, than ideologies; in that fact is one of my strongest hopes. I want to know how to inhabit histories and stories rather than deny them. I want to know how critically to live both inherited and novel kinships, in a spirit neither of condemnation nor celebration. I want to know how to help build ongoing stories rather than histories that end. In that sense, my kinships are about keeping the lineages going, even while defamiliarizing their members and turning lines into webs, trees into esplanades, and pedigrees into affinity groups.

My kinships are made up of the florid machinic, organic, and textual entities with which we share the earth and our flesh. These entities are full of bumptious life, and it would be a serious mistake to figure them mainly anthropomorphically or anthropocentrically. All of the agencies, all of the actors, are not human, to say the least. Indeed, if in his potent little book Bruno Latour convinced me that *We Have Never Been Modern,* I firmly believe that we have never been human, much less man. That's one reason I like to explore figurations that do not resolve into the lineaments of man, even when they seem born to do so.

Nonetheless, in my view, people are human in at least one important sense. We are members of a biological species, *Homo sapiens.* That puts us solidly inside science, history, and nature, right at the heart of things. Furthermore, I am in love with biology—the discourse and the beings, the way of knowing and the world known through those practices. Biology is relentlessly historical, all the way down. There is no border where evolution ends and history begins, where genes stop and environment takes up, where culture rules and nature submits, or vice versa. Instead, there are turtles upon turtles of naturecultures all the way down. Every being that matters is a congeries of its formative histories—all of them— even as any genome worth the salt to precipitate it is a convention of all the infectious events cobbled together into the provisional, permanently emerging things Westerners call individuals, but Melanesians, perhaps more presciently, call dividuals. Perhaps all of this is why, before the end of this introduction, I will go to the dogs for a life sentence.

Sometimes, re-reading the essays that make up this volume, I feel that I have written the same paper twenty times. All of these papers take up one or another aspect of inherited dualisms that run deep in Western cultures. All of these dualisms escape philosophical confinement or religious ritual to find themselves built into weapons, states, economies, taxonomies, national parks, museum displays, intimate bodily practices, and much else. All of my writing is committed to swerving and tripping over these bipartite, dualist traps rather than trying to reverse them or resolve them into supposedly larger wholes. These papers are full of tropes. That is surely because I have a perverse love of words, which have always seemed like tart physical beings to me. But tropes do more than please the palate of the effete of the twenty-first century, c.e. Tropes swerve; they defer the literal, forever, if we are lucky; they make plain that to make sense we must always be ready to trip. Tropes are a way of swerving around a death-defying and death-worshipping culture bent on total war, in order to re-member—in material-semiotic reality—the fragile, mortal, and juicy beings we really are.

Metaplasm is my favorite trope these days. It means remolding or remodeling. I want my writing to be read as an orthopedic practice for learning how to remold kin links to help make a kinder and unfamiliar world. It was Shakespeare who taught me about the sometimes violent play between kin and kind at the dawn of "modernity." It is my queer family of feminists, anti-racists, scientists, scholars, genetically engineered lab rodents, cyborgs, dogs, dog people, vampires, modest witnesses, writers, molecules, and both living and stuffed apes who teach me how to locate kin and kind now, when all of the cosmic correspondences, which Shakespeare understood not to be legible to moderns, might be traceable in non-Euclidean geometries for those who have never been either human or modern.

The papers in the *Reader* often rage against what I also love. All of them insist that science and feminism, anti-racism and science studies, biology and cultural theory, fiction and fact closely cohabit and should do so. Rage is not relativism in the sense that either facts or fictions are matters of "personal" opinion or "multicultural" difference. Quite the opposite. The colonialist epistemological dualisms of relativism and realism require tropic swerving in a spirit of love and rage. Anarchists knew that kind of thing; and anarchists made strong knowledge claims, not vapid truces. In the face of many established disorders we need to practice saying "none of the above." There can be an elsewhere, not as a utopian fantasy or relativist escape, but an elsewhere born out of the hard (and sometimes joyful) work of getting on together in a kin group that includes cyborgs and goddesses working for earthly survival.

Many of the entities that command my attention in this *Reader* were birthed through the reproductive apparatuses of war. Perhaps chief among them is the cyborg. The "Cyborg Manifesto" was not only the first paper I wrote on a computer; it was also a somewhat desperate effort in the early Reagan years to hold together impossible things that all seemed true and necessary simultaneously. Laughing at and crying over cybernetics, the "Manifesto" was an effort at a kind of systems-run-wild triage in dangerous times. Too many people, forgetting the discipline of love and rage, have read the "Manifesto" as the ramblings of a blissed-out, technobunny, fembot. For me, the "Cyborg Manifesto" was a nearly sober socialist-feminist statement written for the *Socialist Review* to try to think through how to do critique, remember war and its offspring, keep ecofeminism and technoscience joined in the flesh, and generally honor possibilities that escape unkind origins. Many readers have put the mutated and contradictory cyborg of the "Manifesto" to work in their own performance art, science studies, and feminist theory.

"Ecce Homo" is my affirmation of a sort of humanism, in spite of myself. The figure of the suffering servant is a troubling inheritance, especially in a race- and gender-saturated culture in which religious and secular masochism and mortification of the flesh take such deadly forms. In this essay, I tried to come to terms with figures of suffering and prophesy, Jesus and Sojourner Truth, whose words and bodies are fetishized and turned into tests of literal authenticity in contemporary feminist and anti-feminist "fundamentalisms." The point of the essay could not be simpler: All of the philology, hermeneutics, and textual criticism that lie behind Western humanism teach a very important truth; to wit, we have nothing but non-innocent translations, all the way down. And that is more than enough for our most lively figures.

"The Promises of Monsters" wreaks havoc with some of the finest technology of structuralist and poststructuralist analysis—the semiotic square. Reeling through real space, outer space, inner space, and finally into virtual space, the traveler in this essay learns, in the words of John Varley's SF short story, that to have a chance for a world that does not end, for stories that are ongoing, we have to "Press Enter." We have to get inside all the excessive connections and unruly categories in order to make sense at all. Better paranoid through too much connection than dead through none at all.

"Otherworldly Conversations" is a confessional piece, one that names the roots of my erotic desire in the fusions promised by molecular cell biology and finds analytical tools for considering the nature of individuality in the life ways of *Mixotricha paradoxa,* an undecidably multiple parasite in the hindgut of a south Australian termite. My situated knowledges tend to be recursively biological, and not even I am sure what is a metaphor and what is not.

"Teddy Bear Patriarchy" and "Morphing in the Order" both meditate on scientific practices for knowing other primates, especially the great apes. In field sites, museums, and labs, primate studies have been crucial means for constructing and contesting popular and professional categories of nature and culture. I believe it is impossible to study primate sciences without appreciating the fact that because of these knowledge practices, humans on this planet know much more about our near biological kin than we ever could before. Progress is not a swear word; new knowledge is a fragile, precious achievement. Furthermore, at every level of the onion, scientific knowledge, like all other kinds, remains constitutively historical and non-innocent. All of the actors, human and nonhuman alike, in these knowledge practices are situated in dense, worldly webs. Primate sciences—both what is known and how knowledge gets crafted—are naturalcultural all the way down. No more than any other kind of knowing, progressive knowledge does not ever precipitate out of the viscid brew of worldly configurations. That means that sometimes raciallyinflected, class-based, and gender-saturated discourses are not pseudo-sciences, but also that categories like race, gender, and class might be precisely the wrong ones for getting at the odd alliances and agencies that come together in power-saturated and power-producing ways of knowing.

"Modest_Witness@Second_Millenium" remembers how important cumulative, secular, revisable, stuttering, reliable knowledge is to everything I care about and how important "unbiased" but "situated" accounts of the world really are. Thus, this essay takes up the apparatuses of witness production in technoscience to argue for mutated modest witnesses who can be more alert to matters like gender- and race-in-the-making in the potent technologies of the experimental way of life. Then, "Race: Universal Donors in a Vampire Culture" tries to show how the figure of modern humanity is produced and reproduced through the biological categories of race, population, and genome. My chief actors in this piece are vampires—those toothy monsters who pollute the blood of the normal on their wedding night and who remind us of the infectious association of Jews, lesbians, intellectuals, foreigners, and other deviants in Western culture. Vampires remind us why alliances to defeat normalization remain crucial in the age of the genome, enterprised up.

The *Reader* ends in the kennel, where the old slogan "Cyborgs for Earthly Survival!" lives side-by-side with two new mottos born of my love affair with dogs, "Run fast; bite hard!" and "Shut up and train!" These mottos might serve better for getting through the aggressive Bush years. But my real concerns in the last essays of the *Reader* are to explore the layered meanings of historically cohabiting companion species of many ontological kinds, organic and not. Companion species give me another way not just to think about kin groups of feminist figurations, but also to live them. Feminists, anti-racists, and socialists have always argued for collective action if we are to have any hope for more livable worlds. In that sense, like the "Cyborg Manifesto," the move to a "Companion Species Manifesto"[1] is an effort to do socialist feminist anti-war work once again. Perhaps the same paper needs to be written again and again.

Katie King taught me in how many ways writings are always technologies for world building. Her current book project, *Feminism and Writing Technologies* (see http://www.womensstudies.umd.edu/wmstfac/kking/research/bookbit.html), explores the layers of locals and globals that reconfigure the possible knowledges and bodies that worldly feminists must care about. Her eye is on the interdigitating communications technologies and the webs of conversation that constitute the field of mobile, contentious practice called feminism. King was also one of the graduate students who welcomed me when I first arrived as a new faculty member in feminist theory in the History of Consciousness program at the University of California at Santa Cruz in 1980. That program—its institutional situation, its people, its history, and its wider scholarly and activist networks—has enabled every line I have written for twenty-five years. The conversation between King and me is one cherished strand in the many ramifying webs of reading and writing that I try to signal in the excessive footnotes and citations of my papers, books, and lectures. I am not, in fact, so much writing the same paper again and again as writing in the embrace of a complex, collective practice, in which the many writers loop through each other, tracing together the barely discernible figure of an elsewhere.

NOTE

1. See Donna Haraway, *The Companion Species Manifesto: Dogs, People, and Significant Otherness* (Chicago: Prickly Paradigm Press, 2003), distributed by the University of Chicago Press.

1

A MANIFESTO FOR CYBORGS: SCIENCE, TECHNOLOGY, AND SOCIALIST FEMINISM IN THE 1980s

AN IRONIC DREAM OF A COMMON LANGUAGE FOR WOMEN IN THE INTEGRATED CIRCUIT

This essay is an effort to build an ironic political myth faithful to feminism, socialism, and materialism. Perhaps more faithful as blasphemy is faithful, than as reverent worship and identification. Blasphemy has always seemed to require taking things very seriously. I know no better stance to adopt from within the secular-religious, evangelical traditions of United States politics, including the politics of socialist-feminism. Blasphemy protects one from the moral majority within, while still insisting on the need for community. Blasphemy is not apostasy. Irony is about contradictions that do not resolve into larger wholes, even dialectically, about the tension of holding incompatible things together because both or all are necessary and true. Irony is about humor and serious play. It is also a rhetorical strategy and a political method, one I would like to see more honored within socialist feminism. At the center of my ironic faith, my blasphemy, is the image of the cyborg.

A cyborg is a cybernetic organism, a hybrid of machine and organism, a creature of social reality as well as a creature of fiction. Social reality is lived social relations, our most important political construction, a world-changing fiction. The international women's movements have constructed "women's experience," as well as uncovered or discovered this crucial collective object. This experience is a fiction and fact of the most crucial, political kind. Liberation rests on the construction of the consciousness, the imaginative apprehension, of oppression, and so of possibility. The cyborg is a matter of fiction and lived experience that changes what counts as women's experience in the late twentieth century. This is a struggle over life and death, but the boundary between science fiction and social reality is an optical illusion.

Contemporary science fiction is full of cyborgs—creatures simultaneously animal and machine, who populate worlds ambiguously natural and crafted. Modern medicine is also full of cyborgs, of couplings between organism and machine, each conceived as coded devices, in an intimacy and with a power that was not generated in the history of sexuality. Cyborg "sex" restores some of the lovely replicative baroque of ferns and invertebrates (such nice organic prophylactics against heterosexism). Cyborg replication is uncoupled from organic reproduction. Modern production seems like a dream of cyborg colonization of work, a dream that makes the nightmare of Taylorism seem idyllic. And modern war is a cyborg orgy, coded by C^3I, command-control-communication-intelligence, an \$84 billion item in 1984's U.S. defense budget. I am making an argument for the cyborg as a fiction mapping our social and bodily reality and as an imaginative resource suggesting some very fruitful couplings. Foucault's biopolitics is a flaccid premonition of cyborg politics, a very open field.

By the late twentieth century, our time, a mythic time, we are all chimeras, theorized and fabricated hybrids of machine and organism; in short, we are cyborgs. The cyborg is our ontology; it gives us our politics. The cyborg is a condensed image of both imagination and material reality, the two joined centers structuring any possibility of historical transformation. In the traditions of "Western" science and politics—the tradition of racist, male-dominant capitalism; the tradition of progress; the tradition of the appropriation of nature as resource for the productions of culture; the tradition of reproduction of the self from the reflections of the other—the relation between organism and machine has been a border war. The stakes in the border war have been the territories of production, reproduction, and imagination. This essay is an argument for *pleasure* in the confusion of boundaries and for *responsibility* in their construction. It is also an effort to contribute to socialist-feminist culture and theory in a post-modernist, non-naturalist mode and in the utopian tradition of imagining a world without gender, which is perhaps a world without genesis, but maybe also a world without end. The cyborg incarnation is outside salvation history. Nor does the cyborg mark time on an Oedipal calendar, attempting to heal the terrible cleavages of gender in an oral-symbiotic utopia or post-oedipal apocalypse. As Zoe Sofoulis argues in her unpublished manuscript on Jacques Lacan and Melanie Klein, *Lacklein,* the most terrible and perhaps the most promising monsters in cyborg worlds are embodied in non-oedipal narratives with a different unconscious and a different logic of repression, which we need to understand for our survival.

The cyborg is a creature in a post-gender world; it has no truck with bisexuality, pre-Oedipal symbiosis, unalienated labor, or other seductions to organic wholeness through a final appropriation of all the powers of the parts into a higher unity. In a sense, the cyborg has no origin story in the Western sense; a "final" irony since the cyborg is also the awful apocalyptic *telos* of the "West's" escalating dominations of abstract individuation, an ultimate self untied at last from all dependency, a man in space. An origin story in the "Western," humanist sense depends on the myth of original unity, fullness, bliss and terror, represented by the phallic mother from whom all humans must separate, the task of individual development and of history, the twin potent myths inscribed most powerfully for us in psychoanalysis and Marxism. Hilary Klein has argued that both Marxism and psychoanalysis, in their concepts of labor and of individuation and gender formation, depend on the plot of original unity out of which difference must be produced and enlisted in a drama of escalating domination of woman/nature. The cyborg skips the step of original unity, of identification with nature in the Western sense. This is its illegitimate promise that might lead to subversion of its teleology as star wars.

The cyborg is resolutely committed to partiality, irony, intimacy, and perversity. It is oppositional, utopian, and completely without innocence. No longer structured by the polarity of public and private, the cyborg defines a technological polis based partly on a revolution of social relations in the *oikos,* the household. Nature and culture are reworked; the one can no longer be the resource for appropriation or incorporation by the other. The relationships for forming wholes from parts, including those of polarity and hierarchical domination, are at issue in the cyborg world. Unlike the hopes of Frankenstein's monster, the cyborg does not expect its father to save it through a restoration of the garden; i.e., through the fabrication of a heterosexual mate, through its completion in a finished whole, a city and cosmos. The cyborg does not dream of community on the model of the organic family, this time without the Oedipal project. The cyborg would not recognize the Garden of Eden; it is not made of mud and cannot dream of returning to dust. Perhaps that is why I want to see if cyborgs can subvert the apocalypse of returning to nuclear dust in the manic compulsion to name the Enemy. Cyborgs are not reverent; they do not remember the cosmos. They are wary of holism, but needy for connection—they seem to have a natural feel for united front politics, but without the vanguard party. The main trouble with cyborgs, of course, is that they are the illegitimate offspring of militarism and patriarchal capitalism, not to mention state socialism. But illegitimate offspring are often exceedingly unfaithful to their origins. Their fathers, after all, are inessential.

I will return to the science fiction of cyborgs at the end of this essay, but now I want to signal three crucial boundary breakdowns that make the following political fictional (political scientific) analysis possible. By the late twentieth century in United States scientific culture, the boundary between human and animal is thoroughly breached. The last beachheads of uniqueness have been polluted if not turned into amusement parks—language, tool use, social behavior, mental events, nothing really convincingly settles the separation of human and animal. And many people no longer feel the need of such a separation; indeed, many branches of feminist culture affirm the pleasure of connection of human and other living creatures. Movements for animal rights are not irrational denials of human uniqueness; they are clear-sighted recognition of connection across the discredited breach of nature and culture. Biology and evolutionary theory over the last two centuries have simultaneously produced modern organisms as objects of knowledge and reduced the line between humans and animals to a faint trace re-etched in ideological struggle or professional disputes between life and social sciences. Within this framework, teaching modern Christian creationism should be fought as a form of child abuse.

Biological-determinist ideology is only one position opened up in scientific culture for arguing the meanings of human animality. There is much room for radical political people to contest for the meanings of the breached boundary.[1] The cyborg appears in myth precisely where the boundary between human and animal is transgressed. Far from signaling a walling off of people from other living beings, cyborgs signal disturbingly and pleasurably tight coupling. Bestiality has a new status in this cycle of marriage exchange.

The second leaky distinction is between animal-human (organism) and machine. Pre-cybernetic machines could be haunted; there was always the specter of the ghost in the machine. This dualism structured the dialogue between materialism and idealism that was settled by a dialectical progeny, called spirit or history, according to taste. But basically machines were not self-moving, self-designing, autonomous. They could not achieve man's dream, only mock it. They were not man, an author to himself, but only a caricature of that masculinist reproductive dream.

To think they were otherwise was paranoid. Now we are not so sure. Late-twentieth-century machines have made thoroughly ambiguous the difference between natural and artificial, mind and body, self-developing and externally designed, and many other distinctions that used to apply to organisms and machines. Our machines are disturbingly lively, and we ourselves frighteningly inert.

Technological determinism is only one ideological space opened up by the reconceptions of machine and organism as coded texts through which we engage in the play of writing and reading the world.[2] "Textualization" of everything in post-structuralist, post-modernist theory has been damned by Marxists and socialist feminists for its utopian disregard for lived relations of domination that ground the "play" of arbitrary reading.[3]* It is certainly true that post-modernist strategies, like my cyborg myth, subvert myriad organic wholes (e.g., the poem, the primitive culture, the biological organism). In short, the certainty of what counts as nature—a source of insight and a promise of innocence—is undermined, probably fatally. The transcendent authorization of interpretation is lost, and with it the ontology grounding "Western" epistemology. But the alternative is not cynicism or faithlessness, i.e., some version of abstract existence, like the accounts of technological determinism destroying "man" by the "machine" or "meaningful political action" by the "text." Who cyborgs will be is a radical question; the answers are a matter of survival. Both chimpanzees and artifacts have politics, so why shouldn't we?[4]

The third distinction is a subset of the second: the boundary between physical and non-physical is very imprecise for us. Pop physics books on

---

* A provocative, comprehensive argument about the politics and theories of "post-modernism" is made by Frederick Jameson, who argues that post-modernism is not an option, a style among others, but a cultural dominant requiring radical reinvention of left politics from within; there is no longer any place from without that gives meaning to the comforting fiction of critical distance. Jameson also makes clear why one cannot be for or against post-modernism, an essentially moralist move. My position is that feminists (and others) need continuous cultural reinvention, post-modernist critique, and historical materialism; only a cyborg would have a chance. The old dominations of white capitalist patriarchy seem nostalgically innocent now: they normalized heterogeneity, e.g., into man and woman, white and black. "Advanced capitalism" and post-modernism release heterogeneity without a norm, and we are flattened, without subjectivity, which requires depth, even unfriendly and drowning depths. It is time to write *The Death of the Clinic*. The clinic's methods required bodies and works; we have texts and surfaces. Our dominations don't work by medicalization and normalization anymore; they work by networking, communications redesign, stress management. Normalization gives way to automation, utter redundancy. Michel Foucault's *Birth of the Clinic, History of Sexuality*, and *Discipline and Punish* name a form of power at its moment of implosion. The discourse of biopolitics gives way to technobabble, the language of the spliced substantive; no noun is left whole by the multinationals. These are their names, listed from one issue of *Science*: Tech-Knowledge, Genentech, Allergen, Hybritech, Compupto, Genen-cor, Syntex, Allelix, Agrigenetics Corp., Syntro, Codon, Repligen, Micro-Angelo from Scion Corp., Percom Data, Inter Systems, Cyborg Corp., Statcom Corp., Intertec. If we are imprisoned by language, then escape from that prison house requires language poets, a kind of cultural restriction enzyme to cut the code; cyborg heteroglossia is one form of radical culture politics.

the consequences of quantum theory and the indeterminacy principle are a kind of popular scientific equivalent to the Harlequin romances as a marker of radical change in American white heterosexuality: they get it wrong, but they are on the right subject. Modern machines are quintessentially microelectronic devices: they are everywhere and they are invisible. Modern machinery is an irreverent upstart god, mocking the Father's ubiquity and spirituality. The silicon chip is a surface for writing; it is etched in molecular scales disturbed only by atomic noise, the ultimate interference for nuclear scores. Writing, power, and technology are old partners in Western stories of the origin of civilization, but miniaturization has changed our experience of mechanism. Miniaturization has turned out to be about power; small is not so much beautiful as pre-eminently dangerous, as in cruise missiles. Contrast the TV sets of the 1950s or the news cameras of the 1970s with the TV wrist bands or hand-sized video cameras now advertised. Our best machines are made of sunshine; they are all light and clean because they are nothing but signals, electromagnetic waves, a section of a spectrum. And these machines are eminently portable, mobile—a matter of immense human pain in Detroit and Singapore. People are nowhere near so fluid, being both material and opaque. Cyborgs are ether, quintessence.

The ubiquity and invisibility of cyborgs is precisely why these sunshine-belt machines are so deadly. They are as hard to see politically as materially. They are about consciousness—or its simulation.[5] They are floating signifiers moving in pickup trucks across Europe, blocked more effectively by the witch-weavings of the displaced and so unnatural Greenham women, who read the cyborg webs of power very well, than by the militant labor of older masculinist politics, whose natural constituency needs defense jobs. Ultimately the "hardest" science is about the realm of greatest boundary confusion, the realm of pure number, pure spirit, C^3I, cryptography, and the preservation of potent secrets. The new machines are so clean and light. Their engineers are sun-worshipers mediating a new scientific revolution associated with the night dream of post-industrial society. The diseases evoked by these clean machines are "no more" than the miniscule coding changes of an antigen in the immune system, "no more" than the experience of stress. The nimble little fingers of "Oriental" women, the old fascination of little Anglo-Saxon Victorian girls with doll houses, women's enforced attention to the small take on quite new dimensions in this world. There might be a cyborg Alice taking account of these new dimensions. Ironically, it might be the unnatural cyborg women making chips in Asia and spiral dancing in Santa Rita whose constructed unities will guide effective oppositional strategies.

So my cyborg myth is about transgressed boundaries, potent fusions, and dangerous possibilities which progressive people might explore as one part of needed political work. One of my premises is that most American socialists and feminists see deepened dualisms of mind and body, animal and machine, idealism and materialism in the social practices, symbolic formulations, and physical artifacts associated with "high technology" and scientific culture. From *One-Dimensional Man* to *The Death of Nature*,[6] the analytic resources developed by progressives have insisted on the necessary domination of technics and recalled us to an imagined organic body to integrate our resistance. Another of my premises is that the need for unity of people trying to resist worldwide intensification of domination has never been more acute. But a slightly perverse shift of perspective might better enable us to contest for meanings, as well as for other forms of power and pleasure in technologically mediated societies.

From one perspective, a cyborg world is about the final imposition of a grid of control on the planet, about the final abstraction embodied in a Star War apocalypse waged in the name of defense, about the final appropriation of women's bodies in a masculinist orgy of war.[7] From another perspective, a cyborg world might be about lived social and bodily realities in which people are not afraid of their joint kinship with animals and machines, not afraid of permanently partial identities and contradictory standpoints. The political struggle is to see from both perspectives at once because each reveals both dominations and possibilities unimaginable from the other vantage point. Single vision produces worse illusions than double vision or many-headed monsters. Cyborg unities are monstrous and illegitimate; in our present political circumstances, we could hardly hope for more potent myths for resistance and recoupling. I like to imagine LAG, the Livermore Action Group, as a kind of cyborg society dedicated to realistically converting the laboratories that most fiercely embody and spew out the tools of technological apocalypse, and committed to building a political form that actually manages to hold together witches, engineers, elders, perverts, Christians, mothers, and Leninists long enough to disarm the state. Fission Impossible is the name of the affinity group in my town. (Affinity: related not by blood but by choice, the appeal of one chemical nuclear group for another, avidity.)

FRACTURED IDENTITIES

It has become difficult to name one's feminism by a single adjective— or even to insist in every circumstance upon the noun. Consciousness of exclusion through naming is acute. Identities seem contradictory, partial, and strategic. With the hard-won recognition of their social and historical constitution, gender, race, and class cannot provide the basis for belief in "essential" unity. There is nothing about being "female" that naturally binds women. There is not even such a state as "being" female, itself a highly complex category constructed in contested sexual scientific discourses and other social practices. Gender, race, or class consciousness is an achievement forced on us by the terrible historical experience of the contradictory social realities of patriarchy, colonialism, and capitalism. And who counts as "us" in my own rhetoric? Which identities are available to ground such a potent political myth called "us," and what could motivate enlistment in this collectivity? Painful fragmentation among feminists (not to mention among women) along every possible fault line has made the concept of *woman* elusive, an excuse for the matrix of women's dominations of each other. For me—and for many who share a similar historical location in white, professional middle class, female, radical, North American, mid-adult bodies—the sources of a crisis in political identity are legion. The recent history for much of the U.S. left and U.S. feminism has been a response to this kind of crisis by endless splitting and searches for a new essential unity. But there has also been a growing recognition of another response through coalition—affinity, not identity.[8]

Chela Sandoval, from a consideration of specific historical moments in the formation of the new political voice called women of color, has theorized a hopeful model of political identity called "oppositional consciousness," born of the skills for reading webs of power by those refused stable membership in the social categories of race, sex, or class.[9] "Women of color," a name contested at its origins by those whom it would incorporate, as well as a historical consciousness marking systematic breakdown of all the signs of Man in "Western" traditions, constructs a kind of post-modernist identity out of otherness and difference. This post-modernist identity is fully political, whatever might be said about other possible post-modernisms.

Sandoval emphasizes the lack of any essential criterion for identifying who is a woman of color. She notes that the definition of the group has been by conscious appropriation of negation. For example, a Chicana or U.S. black woman has not been able to speak as a woman or as a black person or as a Chicano. Thus, she was at the bottom of a cascade of negative identities, left out of even the privileged oppressed authorial categories called "women and blacks," who claimed to make the important revolutions. The category "woman" negated all non-white women; "black" negated all non-black people, as well as all black women. But there was also no "she," no singularity, but a sea of differences among U.S. women who have affirmed their historical identity as U.S. women of color. This identity marks out a self-consciously constructed space

that cannot affirm the capacity to act on the basis of natural identification, but only on the basis of conscious coalition, of affinity, of political
kinship.[10] Unlike the "woman" of some streams of the white women's
movement in the United States, there is no naturalization of the matrix,
or at least this is what Sandoval argues is uniquely available through the
power of oppositional consciousness.

Sandoval's argument has to be seen as one potent formulation for
feminists out of the worldwide development of anti-colonialist discourse,
i.e., discourse dissolving the "West" and its highest product—the one
who is not animal, barbarian, or woman; i.e., man, the author of a
cosmos called history. As orientalism is deconstructed politically and
semiotically, the identities of the occident destabilize, including those of
feminists.[11] Sandoval argues that "women of color" have a chance to build
an effective unity that does not replicate the imperializing, totalizing
revolutionary subjects of previous Marxisms and feminisms which had
not faced the consequences of the disorderly polyphony emerging from
decolonization.

Katie King has emphasized the limits of identification and the po-
litical/poetic mechanics of identification built into reading "the poem,"
that generative core of cultural feminism. King criticizes the persistent
tendency among contemporary feminists from different "moments" or
"conversations" in feminist practice to taxonomize the women's movement to make one's own political tendencies appear to be the *telos* of the
whole. These taxonomies tend to remake feminist history to appear to
be an ideological struggle among coherent types persisting over time, especially those typical units called radical, liberal, and socialist feminism.
Literally, all other feminisms are either incorporated or marginalized,
usually by building an explicit ontology and epistemology.[12] Taxonomies
of feminism produce epistemologies to police deviation from official
women's experience. And of course, "women's culture," like women of
color, is consciously created by mechanisms inducing affinity. The ritu-
als of poetry, music, and certain forms of academic practice have been
pre-eminent. The politics of race and culture in the U.S. women's movements are intimately interwoven. The common achievement of King and
Sandoval is learning how to craft a poetic/political unity without relying
on a logic of appropriation, incorporation, and taxonomic identification.

The theoretical and practical struggle against unity-through-domination
or unity-through-incorporation ironically not only undermines the justifications for patriarchy, colonialism, humanism, positivism, essentialism, scientism, and other unlamented -isms, but *all* claims for an
organic or natural standpoint. I think that radical and socialist/Marxist

feminisms have also undermined their/our own epistemological strategies and that this is a crucially valuable step in imagining possible unities. It remains to be seen whether all "epistemologies" as Western political people have known them fail us in the task to build effective affinities.

It is important to note that the effort to construct revolutionary standpoints, epistemologies as achievements of people committed to changing the world, has been part of the process showing the limits of identification. The acid tools of post-modernist theory and the constructive tools of ontological discourse about revolutionary subjects might be seen as ironic allies in dissolving Western selves in the interests of survival. We are excruciatingly conscious of what it means to have a historically constituted body. But with the loss of innocence in our origin, there is no expulsion from the Garden either. Our politics lose the indulgence of guilt with the naïveté of innocence. But what would another political myth for socialist feminism look like? What kind of politics could embrace partial, contradictory, permanently unclosed constructions of personal and collective selves and still be faithful, effective—and, ironically, socialist feminist?

I do not know of any other time in history when there was greater need for political unity to confront effectively the dominations of "race," "gender," "sexuality," and "class." I also do not know of any other time when the kind of unity we might help build could have been possible. None of "us" have any longer the symbolic or material capability of dictating the shape of reality to any of "them." Or at least "we" cannot claim innocence from practicing such dominations. White women, including socialist feminists, discovered (i.e., were forced kicking and screaming to notice) the non-innocence of the category "woman." That consciousness changes the geography of all previous categories; it denatures them as heat denatures a fragile protein. Cyborg feminists have to argue that "we" do not want any more natural matrix of unity and that no construction is whole. Innocence, and the corollary insistence on victimhood as the only ground for insight, has done enough damage. But the constructed revolutionary subject must give late-twentieth-century people pause as well. In the fraying of identities and in the reflexive strategies for constructing them, the possibility opens up for weaving something other than a shroud for the day after the apocalypse that so prophetically ends salvation history.

Both Marxist/socialist feminisms and radical feminisms have simultaneously naturalized and denatured the category "woman" and consciousness of the social lives of "women." Perhaps a schematic caricature can highlight both kinds of moves. Marxian socialism is rooted in an analysis of wage labor which reveals class structure. The consequence of the wage relationship is systematic alienation, as the worker is dissociated from his [*sic*] product. Abstraction and illusion rule in knowledge, domination rules in practice. Labor is the pre-eminently privileged category enabling the Marxist to overcome illusion and find that point of view which is necessary for changing the world. Labor is the humanizing activity that makes man; labor is an ontological category permitting the knowledge of a subject, and so the knowledge of subjugation and alienation.

In faithful filiation, socialist feminism advanced by allying itself with the basic analytic strategies of Marxism. The main achievement of both Marxist feminists and socialist feminists was to expand the category of labor to accommodate what (some) women did, even when the wage relation was subordinated to a more comprehensive view of labor under capitalist patriarchy. In particular, women's labor in the household and women's activity as mothers generally, i.e., reproduction in the socialist feminist sense, entered theory on the authority of analogy to the Marxian concept of labor. The unity of women here rests on an epistemology based on the ontological structure of "labor." Marxist/socialist feminism does not "naturalize" unity; it is a possible achievement based on a possible standpoint rooted in social relations. The essentializing move is in the ontological structure of labor or of its analogue, women's activity.[13*] The inheritance of Marxian humanism, with its pre-eminently Western self, is the difficulty for me. The contribution from these formulations has been the emphasis on the daily responsibility of real women to build unities, rather than to naturalize them.

Catharine MacKinnon's version of radical feminism is itself a caricature of the appropriating, incorporating, totalizing tendencies of Western theories of identity grounding action.[14] It is factually and politically wrong to assimilate all of the diverse "moments" or "conversations" in recent women's politics named radical feminism to MacKinnon's version. But the teleological logic of her theory shows how an epistemology and ontology—including their negations—erase or police difference. Only one of the effects of MacKinnon's theory is the rewriting of the history of the polymorphous field called radical feminism. The major effect

---

* The central role of object-relations versions of psychoanalysis and related strong universalizing moves in discussing reproduction, caring work, and mothering in many approaches to epistemology underline their authors' resistance to what I am calling post-modernism. For me, both the universalizing moves and the versions of psychoanalysis make analysis of "women's place in the integrated circuit" difficult and lead to systematic difficulties in accounting for or even seeing major aspects of the construction of gender and gendered social life.

is the production of a theory of experience, of women's identity, that is a kind of apocalypse for all revolutionary standpoints. That is, the totalization built into this tale of radical feminism achieves its end—the unity of women—by enforcing the experience of and testimony to radical non-being. As for the Marxist/socialist feminist, consciousness is an achievement, not a natural fact. And MacKinnon's theory eliminates some of the difficulties built into humanist revolutionary subjects, but at the cost of radical reductionism.

MacKinnon argues that radical feminism necessarily adopted a different analytical strategy from Marxism, looking first not at the structure of class, but at the structure of sex/gender and its generative relationship, men's constitution and appropriation of women sexually. Ironically, MacKinnon's "ontology" constructs a non-subject, a non-being. Another's desire, not the self's labor, is the origin of "woman." She therefore develops a theory of consciousness that enforces what can count as "women's" experience—anything that names sexual violation, indeed, sex itself as far as "women" can be concerned. Feminist practice is the construction of this form of consciousness; i.e., the self-knowledge of a self-who-is-not.

Perversely, sexual appropriation in this radical feminism still has the epistemological status of labor, i.e., the point from which analysis able to contribute to changing the world must flow. But sexual objectification, not alienation, is the consequence of the structure of sex/gender. In the realm of knowledge, the result of sexual objectification is illusion and abstraction. However, a woman is not simply alienated from her product, but in a deep sense does not exist as a subject, or even potential subject, since she owes her existence as a woman to sexual appropriation. To be constituted by another's desire is not the same thing as to be alienated in the violent separation of the laborer from his product.

MacKinnon's radical theory of experience is totalizing in the extreme; it does not so much marginalize as obliterate the authority of any other women's political speech and action. It is a totalization producing what Western patriarchy itself never succeeded in doing—feminists' consciousness of the non-existence of women, except as products of men's desire. I think MacKinnon correctly argues that no Marxian version of identity can firmly ground women's unity. But in solving the problem of the contradictions of any Western revolutionary subject for feminist purposes, she develops an even more authoritarian doctrine of experience. If my complaint about socialist/Marxian standpoints is their unintended erasure of polyvocal, unassimilable, radical difference made visible in anti-colonial discourse and practice, MacKinnon's intentional erasure of all difference through the device of the "essential" non-existence of women is not reassuring.

In my taxonomy, which like any other taxonomy is a reinscription of history, radical feminism can accommodate all the activities of women named by socialist feminists as forms of labor only if the activity can somehow be sexualized. Reproduction had different tones of meanings for the two tendencies, one rooted in labor, one in sex, both calling the consequences of domination and ignorance of social and personal reality "false consciousness."

Beyond either the difficulties or the contributions in the argument of any one author, neither Marxist nor radical feminist points of view have tended to embrace the status of a partial explanation; both were regularly constituted as totalities. Western explanation has demanded as much; how else could the "Western" author incorporate its others? Each tried to annex other forms of domination by expanding its basic categories through analogy, simple listing, or addition. Embarrassed silence about race among white radical and socialist feminists was one major, devastating political consequence. History and polyvocality disappear into political taxonomies that try to establish genealogies. There was no structural room for race (or for much else) in theory claiming to reveal the construction of the category woman and social group women as a unified or totalizable whole. The structure of my caricature looks like this:

Socialist Feminism—
    structure of class//wage labor//alienation
    labor, by analogy reproduction, by extension sex, by addition race
Radical Feminism—
    structure of gender//sexual appropriation//objectification
    sex, by analogy labor, by extension reproduction, by addition race

In another context, the French theorist Julia Kristeva claimed women appeared as a historical group after World War II, along with groups like youth. Her dates are doubtful; but we are now accustomed to remembering that as objects of knowledge and as historical actors, "race" did not always exist, "class" has a historical genesis, and "homosexuals" are quite junior. It is no accident that the symbolic system of the family of man—and so the essence of woman—breaks up at the same moment that networks of connection among people on the planet are unprecedentedly multiple, pregnant, and complex. "Advanced capitalism" is inadequate to convey the structure of this historical moment. In the "Western" sense, the end of man is at stake. It is no accident that woman disintegrates into women in our time. Perhaps socialist feminists were not substantially guilty of producing essentialist theory that suppressed women's particularity and contradictory interests. I think we have been, at least through unreflective participation in the logics, languages, and practices of white humanism and through searching for a single ground of domination to secure our revolutionary voice. Now we have less excuse. But in the consciousness of our failures, we risk lapsing into boundless difference and giving up on the confusing task of making partial, real connection. Some differences are playful; some are poles of world historical systems of domination. "Epistemology" is about knowing the difference.

THE INFORMATICS OF DOMINATION

In this attempt at an epistemological and political position, I would like to sketch a picture of possible unity, a picture indebted to socialist and feminist principles of design. The frame for my sketch is set by the extent and importance of rearrangements in worldwide social relations tied to science and technology. I argue for a politics rooted in claims about fundamental changes in the nature of class, race, and gender in an emerging system of world order analogous in its novelty and scope to that created by industrial capitalism; we are living through a movement from an organic, industrial society to a polymorphous, information system—from all work to all play, a deadly game. Simultaneously material and ideological, the dichotomies may be expressed in the following chart of transitions from the comfortable old hierarchical dominations to the scary new networks I have called the informatics of domination:

Representation	Simulation
Bourgeois novel, realism	Science fiction, post-modernism
Organism	Biotic component
Depth, integrity	Surface, boundary
Heat	Noise
Biology as clinical practice	Biology as inscription
Physiology	Communications engineering
Small group	Subsystem
Perfection	Optimization
Eugenics	Population control
Decadence, *Magic Mountain*	Obsolescence, *Future Shock*
Hygiene	Stress management
Microbiology, tuberculosis	Immunology, AIDS
Organic division of labor	Ergonomics/cybernetics of labor

Functional specialization	Modular construction
Reproduction	Replication
Organic sex role specialization	Optimal genetic strategies
Biological determinism	Evolutionary inertia, constraints
Community ecology	Ecosystem
Racial chain of being	Neo-imperialism, United Nations humanism
Scientific management in home/factory	Global factory/Electronic cottage
Family/Market/Factory	Women in the Integrated Circuit
Family wage	Comparable worth
Public/Private	Cyborg citizenship
Nature/Culture	Fields of difference
Cooperation	Communications enhancement
Freud	Lacan
Sex	Genetic engineering
Labor	Robotics
Mind	Artificial Intelligence
World War II	Star Wars
White Capitalist Patriarchy	Informatics of Domination

This list suggests several interesting things.[15] First, the objects on the right-hand side cannot be coded as "natural," a realization that subverts naturalistic coding for the left-hand side as well. We cannot go back ideologically or materially. It's not just that "god" is dead; so is the "goddess." In relation to objects like biotic components, one must think not in terms of essential properties, but in terms of strategies of design, boundary constraints, rates of flows, systems logics, costs of lowering constraints. Sexual reproduction is one kind of reproductive strategy among many, with costs and benefits as a function of the system environment. Ideologies of sexual reproduction can no longer reasonably call on the notions of sex and sex role as organic aspects in natural objects like organisms and families. Such reasoning will be unmasked as irrational, and ironically corporate executives reading *Playboy* and anti-porn radical feminists will make strange bedfellows in jointly unmasking the irrationalism.

Likewise for race, ideologies about human diversity have to be formulated in terms of frequencies of parameters, like blood groups or intelligence scores. It is "irrational" to invoke concepts like primitive and civilized. For liberals and radicals, the search for integrated social systems gives way to a new practice called "experimental ethnography" in which an organic object dissipates in attention to the play of writing. At the level of ideology, we see translations of racism and colonialism into languages of development and underdevelopment, rates and constraints of modernization. Any objects or persons can be reasonably thought of in terms of disassembly and reassembly; no "natural" architectures constrain system design. The financial districts in all the world's cities, as well as the export-processing and free-trade zones, proclaim this elementary fact of "late capitalism." The entire universe of objects that can be known scientifically must be formulated as problems in communications engineering (for the managers) or theories of the text (for those who would resist). Both are cyborg semiologies.

One should expect control strategies to concentrate on boundary conditions and interfaces, on rates of flow across boundaries—and not on the integrity of natural objects. "Integrity" or "sincerity" of the Western self gives way to decision procedures and expert systems. For example, control strategies applied to women's capacities to give birth to new human beings will be developed in the languages of population control and maximization of goal achievement for individual decision-makers. Control strategies will be formulated in terms of rates, costs of constraints, degrees of freedom. Human beings, like any other component or subsystem, must be localized in a system architecture whose basic modes of operation are probabilistic, statistical. No objects, spaces, or bodies are sacred in themselves; any component can be interfaced with any other if the proper standard, the proper code, can be constructed for processing signals in a common language. Exchange in this world transcends the universal translation effected by capitalist markets that Marx analyzed so well. The privileged pathology affecting all kinds of components in this universe is stress—communications breakdown.[16] The cyborg is not subject to Foucault's biopolitics; the cyborg simulates politics, a much more potent field of operations.

This kind of analysis of scientific and cultural objects of knowledge which have appeared historically since World War II prepares us to notice some important inadequacies in feminist analysis which has proceeded as if the organic, hierarchical dualisms ordering discourse in "the West" since Aristotle still ruled. They have been cannibalized, or as Zoe Sofia (Sofoulis) might put it, they have been "techno-digested." The dichotomies between mind and body, animal and human, organism and machine, public and private, nature and culture, men and women, primitive and civilized are all in question ideologically. The actual situation of women is their integration/exploitation into a world system of production/reproduction and communication called the informatics of domination. The home, workplace, market, public arena, the body itself—all can be dispersed and interfaced in nearly infinite, polymorphous ways, with large consequences for women and others—consequences that themselves are very different for different people and which make potent oppositional international movements difficult to imagine and essential for survival. One important route for reconstructing socialist-feminist politics is through theory and practice addressed to the social relations of science and technology, including crucially the systems of myth and meanings structuring our imaginations. The cyborg is a kind of disassembled and reassembled, post-modern collective and personal self. This is the self feminists must code.

Communications technologies and biotechnologies are the crucial tools recrafting our bodies. These tools embody and enforce new social relations for women worldwide. Technologies and scientific discourses can be partially understood as formalizations, i.e., as frozen moments, of the fluid social interactions constituting them, but they should also be viewed as instruments for enforcing meanings. The boundary is permeable between tool and myth, instrument and concept, historical systems of social relations and historical anatomies of possible bodies, including objects of knowledge. Indeed, myth and tool mutually constitute each other.

Furthermore, communications sciences and modern biologies are constructed by a common move—*the translation of the world into a problem of coding*, a search for a common language in which all resistance to instrumental control disappears and all heterogeneity can be submitted to disassembly, reassembly, investment, and exchange.

In communications sciences, the translation of the world into a problem in coding can be illustrated by looking at cybernetic (feedback controlled) systems theories applied to telephone technology, computer design, weapons deployment, or data base construction and maintenance. In each case, solution to the key questions rests on a theory of language and control; the key operation is determining the rates, directions, and probabilities of flow of a quantity called information. The world is subdivided by boundaries differentially permeable to information. Information is just that kind of quantifiable element (unit, basis of unity) which allows universal translation, and so unhindered instrumental power (called effective communication). The biggest threat to such power is interruption of communication. Any system breakdown is a function of stress. The fundamentals of this technology can be condensed into the metaphor C^3I, command-control-communication-intelligence, the military's symbol for its operations theory.

In modern biologies, the translation of the world into a problem in coding can be illustrated by molecular genetics, ecology, socio-biological evolutionary theory, and immunobiology. The organism has been translated into problems of genetic coding and read-out. Biotechnology, a writing technology, informs research broadly.[17] In a sense, organisms have ceased to exist as objects of knowledge, giving way to biotic components, i.e., special kinds of information processing devices. The analogous moves in ecology could be examined by probing the history and utility of the concept of the ecosystem. Immunobiology and associated medical practices are rich exemplars of the privilege of coding and recognition systems as objects of knowledge, as constructions of bodily reality for us. Biology is here a kind of cryptography. Research is necessarily a kind of intelligence activity. Ironies abound. A stressed system goes awry; its communication processes break down; it fails to recognize the difference between self and other. Human babies with baboon hearts evoke national ethical perplexity—for animal-rights activists at least as much as for guardians of human purity. Gay men, Haitian immigrants, and intravenous drug users are the "privileged" victims of an awful immune-system disease that marks (inscribes on the body) confusion of boundaries and moral pollution.

But these excursions into communications sciences and biology have been at a rarefied level; there is a mundane, largely economic reality to support my claim that these sciences and technologies indicate fundamental transformations in the structure of the world for us. Communications technologies depend on electronics. Modern states, multinational corporations, military power, welfare-state apparatuses, satellite systems, political processes, fabrication of our imaginations, labor-control systems, medical constructions of our bodies, commercial pornography, the international division of labor, and religious evangelism depend intimately upon electronics. Microelectronics is the technical basis of simulacra, i.e., of copies without originals.

Microelectronics mediates the translations of *labor* into robotics and word processing; *sex* into genetic engineering and reproductive technologies; and *mind* into artificial intelligence and decision procedures. The new biotechnologies concern more than human reproduction. Biology as a powerful engineering science for redesigning materials and processes has revolutionary implications for industry, perhaps most obvious today in areas of fermentation, agriculture, and energy. Communications sciences and biology are constructions of natural-technical objects of knowledge in which the difference between machine and organism is thoroughly blurred; mind, body, and tool are on very intimate terms. The "multinational" material organization of the production and reproduction of daily life and the symbolic organization of the production and reproduction of culture and imagination seem equally implicated. The boundary-maintaining images of base and superstructure, public and private, or material and ideal never seemed more feeble.

I have used Rachel Grossman's image of women in the integrated circuit to name the situation of women in a world so intimately restructured through the social relations of science and technology.[18] I use the odd circumlocution, "the social relations of science and technology," to indicate that we are not dealing with a technological determinism, but with a historical system depending upon structured relations among people. But the phrase should also indicate that science and technology provide fresh sources of power, that we need fresh sources of analysis and political action.[19] Some of the rearrangements of race, sex, and class rooted in high-tech-facilitated social relations can make socialist feminism more relevant to effective progressive politics.

THE HOMEWORK ECONOMY

The "new industrial revolution" is producing a new worldwide working class. The extreme mobility of capital and the emerging international division of labor are intertwined with the emergence of new collectivities, and the weakening of familiar groupings. These developments are neither gender- nor race-neutral. White men in advanced industrial societies have become newly vulnerable to permanent job loss, and women are not disappearing from the job rolls at the same rates as men. It is not simply that women in third-world countries are the preferred labor force for the science-based multinationals in the export-processing sectors, particularly in electronics. The picture is more systematic and involves reproduction, sexuality, culture, consumption, and production. In the prototypical Silicon Valley, many women's lives have been structured around employment in electronics-dependent jobs, and their intimate realities include serial heterosexual monogamy, negotiating childcare, distance from extended kin or most other forms of traditional community, a high likelihood of loneliness and extreme economic vulnerability as they age. The ethnic and racial diversity of women in Silicon Valley structures a microcosm of conflicting differences in culture, family, religion, education, language.

Richard Gordon has called this new situation the homework economy.[20] Although he includes the phenomenon of literal homework emerging in connection with electronics assembly, Gordon intends "homework economy" to name a restructuring of work that broadly has the characteristics formerly ascribed to female jobs, jobs literally done only by women. Work is being redefined as both literally female and feminized, whether performed by men or women. To be feminized means to be made extremely vulnerable; able to be disassembled, reassembled, exploited as a reserve labor force; seen less as workers than as servers; subjected to time arrangements on and off the paid job that make a mockery of a limited work day; leading an existence that always borders on being obscene, out of place, and reducible to sex. Deskilling is an old strategy newly applicable to formerly privileged workers. However, the homework economy does not refer only to large-scale deskilling, nor does it deny that new areas of high skill are emerging, even for women and men previously excluded from skilled employment. Rather, the concept indicates that factory, home, and market are integrated on a new scale and that the places of women are crucial—and need to be analyzed for differences among women and for meanings for relations between men and women in various situations.

The homework economy as a world capitalist organizational structure is made possible by (not caused by) the new technologies. The successs of the attack on relatively privileged, mostly white, men's unionized jobs is tied to the power of the new communications technologies to integrate and control labor despite extensive dispersion and decentralization. The consequences of the new technologies are felt by women both in the loss of the family (male) wage (if they ever had access to this white privilege) and in the character of their own jobs, which are becoming capital-intensive, e.g., office work and nursing.

The new economic and technological arrangements are also related to the collapsing welfare state and the ensuing intensification of demands on women to sustain daily life for themselves as well as for men, children, and old people. The feminization of poverty—generated by dismantling the welfare state, by the homework economy where stable jobs become the exception, and sustained by the expectation that women's wage will not be matched by a male income for the support of children—has become an urgent focus. The causes of various women-headed households are a function of race, class, or sexuality; but their increasing generality is a ground for coalitions of women on many issues. That women regularly sustain daily life partly as a function of their enforced status as mothers is hardly new; the kind of integration with the overall capitalist and progressively war-based economy is new. The particular pressure, for example, on U.S. black women, who have achieved an escape from (barely) paid domestic service and who now hold clerical and similar jobs in large numbers, has large implications for continued enforced black poverty *with* employment. Teenage women in industrializing areas of the third world increasingly find themselves the sole or major source of a cash wage for their families, while access to land is ever more problematic. These developments must have major consequences in the psychodynamics and politics of gender and race.

Within the framework of three major stages of capitalism (commercial/early industrial, monopoly, multinational)—tied to nationalism, imperialism, and multinationalism, and related to Jameson's three dominant aesthetic periods of realism, modernism, and postmodernism—I would argue that specific forms of families dialectically relate to forms of capital and to its political and cultural concomitants. Although lived problematically and unequally, ideal forms of these families might be schematized as (1) the patriarchal nuclear family, structured by the dichotomy between public and private and accompanied by the white bourgeois ideology of separate spheres and nineteenth-century Anglo-American bourgeois feminism; (2) the modern family mediated (or enforced) by the welfare state and institutions like the family wage, with a flowering of afeminist heterosexual ideologies, including their radical versions represented in Greenwich Village around World War I; and (3) the "family" of the homework economy with its oxymoronic structure of women-headed households and its explosion of feminisms and the paradoxical intensification and erosion of gender itself.

This is the context in which the projections for worldwide structural unemployment stemming from the new technologies are part of the picture of the homework economy. As robotics and related technologies put men out of work in "developed" countries and exacerbate failure to generate male jobs in third-world "development," and as the automated office becomes the rule even in labor-surplus countries, the feminization of work intensifies. Black women in the United States have long known what it looks like to face the structural underemployment ("feminization") of black men, as well as their own highly vulnerable position in the wage economy. It is no longer a secret that sexuality, reproduction, family, and community life are interwoven with this economic structure in myriad ways which have also differentiated the situations of white and black women. Many more women and men will contend with similar situations, which will make cross-gender and race alliances on issues of basic life support (with or without jobs) necessary, not just nice.

The new technologies also have a profound effect on hunger and on food production for subsistence worldwide. Rae Lessor Blumberg estimates that women produce about fifty per cent of the world's subsistence food.[21]* Women are excluded generally from benefiting from the increased high-tech commodification of food and energy crops, their days are made more arduous because their responsibilities to provide food do not diminish, and their reproductive situations are made more complex. Green Revolution technologies interact with other high-tech industrial production to alter gender divisions of labor and differential gender migration patterns:

The new technologies seem deeply involved in the forms of "privatization" that Ros Petchesky has analyzed, in which militarization, right-wing family ideologies and policies, and intensified definitions of corporate property as private synergistically interact.[22] The new communications technologies are fundamental to the eradication of "public life" for everyone. This facilitates the mushrooming of a permanent high-tech military establishment at the cultural and economic expense of most people, but especially of women. Technologies like video games and highly miniaturized television seem crucial to production of modern forms of "private life." The culture of video games is heavily oriented to individual competition and extraterrestrial warfare. High-tech, gendered imaginations are produced here, imaginations that can contemplate destruction of the planet and a sci-fi escape from its consequences. More than our imaginations is militarized, and the other realities of electronic and nuclear warfare are inescapable.

The new technologies affect the social relations of both sexuality and reproduction, and not always in the same ways. The close ties of sexuality and instrumentality, of views of the body as a kind of private satisfaction- and utility-maximizing machine, are described nicely in sociobiological origin stories that stress a genetic calculus and explain the inevitable dialectic of domination of male and female gender roles.[23] These sociobiological stories depend on a high-tech view of the body as a biotic component or cybernetic communications system. Among the many transformations of reproductive situations is the medical one, where women's bodies have boundaries newly permeable to both "visualization" and "intervention." Of course, who controls the interpretation of bodily boundaries in medical hermeneutics is a major feminist issue. The speculum served as an icon of women's claiming their bodies in the

---

* The conjunction of the Green Revolution's social relations with biotechnologies like plant genetic engineering makes the pressures on land in the third world increasingly intense. AID's estimates (*New York Times*, 14 October 1984) used at the 1984 World Food Day are that in Africa, women produce about 90 per cent of rural food supplies, about 60 to 80 per cent in Asia, and provide 40 per cent of agricultural labor in the Near East and Latin America. Blumberg charges that world organizations' agricultural politics, as well as those of multinationals and national governments in the third world, generally ignore fundamental issues in the sexual division of labor. The present tragedy of famine in Africa might owe as much to male supremacy as to capitalism, colonialism, and rain patterns. More accurately, capitalism and racism are usually structurally male dominant.

1970s; that hand-craft tool is inadequate to express our needed body politics in the negotiation of reality in the practices of cyborg reproduction. Self-help is not enough. The technologies of visualization recall the important cultural practice of hunting with the camera and the deeply predatory nature of a photographic consciousness.[24] Sex, sexuality, and reproduction are central actors in high-tech myth systems structuring our imaginations of personal and social possibility.

Another critical aspect of the social relations of the new technologies is the reformulation of expectations, culture, work, and reproduction for the large scientific and technical work force. A major social and political danger is the formation of a strongly bimodal social structure, with the masses of women and men of all ethnic groups, but especially people of color, confined to a homework economy, illiteracy of several varieties, and general redundancy and impotence, controlled by high-tech repressive apparatuses ranging from entertainment to surveillance and disappearance. An adequate socialist-feminist politics should address women in the privileged occupational categories, and particularly in the production of science and technology that constructs scientific-technical discourses, processes, and objects.[25]

This issue is only one aspect of inquiry into the possibility of a feminist science, but it is important. What kind of constitutive role in the production of knowledge, imagination, and practice can new groups doing science have? How can these groups be allied with progressive social and political movements? What kind of political accountability can be constructed to tie women together across the scientific-technical hierarchies separating us? Might there be ways of developing feminist science/technology politics in alliance with anti-military science facility conversion action groups? Many scientific and technical workers in Silicon Valley, the high-tech cowboys included, do not want to work on military science.[26] Can these personal preferences and cultural tendencies be welded into progressive politics among this professional middle class in which women, including women of color, are coming to be fairly numerous?

WOMEN IN THE INTEGRATED CIRCUIT

Let me summarize the picture of women's historical locations in advanced industrial societies, as these positions have been restructured partly through the social relations of science and technology. If it was ever possible ideologically to characterize women's lives by the distinction of public and private domains—suggested by images of the division of working-class life into factory and home, of bourgeois life into market and home, and of gender existence into personal and political realms—it is now a totally misleading ideology, even to show how both terms of these dichotomies construct each other in practice and in theory. I prefer a network ideological image, suggesting the profusion of spaces and identities and the permeability of boundaries in the personal body and in the body politic. "Networking" is both a feminist practice and a multinational corporate strategy—weaving is for oppositional cyborgs.

The only way to characterize the informatics of domination is as a massive intensification of insecurity and cultural impoverishment, with common failure of subsistence networks for the most vulnerable. Since much of this picture interweaves with the social relations of science and technology, the urgency of a socialist-feminist politics addressed to science and technology is plain. There is much now being done, and the grounds for political work are rich. For example, the efforts to develop forms of collective struggle for women in paid work, like SEIU's District 925, should be a high priority for all of us. These efforts are profoundly tied to technical restructuring of labor processes and reformations of working classes. These efforts also are providing understanding of a more comprehensive kind of labor organization, involving community, sexuality, and family issues never privileged in the largely white male industrial unions.

The structural rearrangements related to the social relations of science and technology evoke strong ambivalence. But it is not necessary to be ultimately depressed by the implications of late-twentieth-century women's relation to all aspects of work, culture, production of knowledge, sexuality, and reproduction. For excellent reasons, most Marxisms see domination best and have trouble understanding what can only look like false consciousness and people's complicity in their own domination in late capitalism. It is crucial to remember that what is lost, perhaps especially from women's points of view, is often virulent forms of oppression, nostalgically naturalized in the face of current violation. Ambivalence toward the disrupted unities mediated by high-tech culture requires not sorting consciousness into categories of "clear-sighted critique grounding a solid political epistemology" versus "manipulated false consciousness," but subtle understanding of emerging pleasures, experiences, and powers with serious potential for changing the rules of the game.

There are grounds for hope in the emerging bases for new kinds of unity across race, gender, and class, as these elementary units of socialist-feminist analysis themselves suffer protean transformations. Intensifications of hardship experienced worldwide in connection with the social relations of science and technology are severe. But what people are experiencing is not transparently clear, and we lack sufficiently subtle connections for collectively building effective theories of experience. Present efforts—Marxist, psychoanalytic, feminist, anthropological—to clarify even "our" experience are rudimentary.

I am conscious of the odd perspective provided by my historical position—a Ph.D. in biology for an Irish Catholic girl was made possible by Sputnik's impact on U.S. national science-education policy. I have a body and mind as much constructed by the post-World War II arms race and Cold War as by the women's movements. There are more grounds for hope by focusing on the contradictory effects of politics designed to produce loyal American technocrats, which as well produced large numbers of dissidents, rather than by focusing on the present defeats.

The permanent partiality of feminist points of view has consequences for our expectations of forms of political organization and participation. We do not need a totality in order to work well. The feminist dream of a common language, like all dreams for a perfectly true language, of perfectly faithful naming of experience, is a totalizing and imperialist one. In that sense, dialectics too is a dream language, longing to resolve contradiction. Perhaps, ironically, we can learn from our fusions with animals and machines how not to be Man, the embodiment of Western logos. From the point of view of pleasure in these potent and taboo fusions, made inevitable by the social relations of science and technology, there might indeed be a feminist science.

CYBORGS: A MYTH OF POLITICAL IDENTITY

I want to conclude with a myth about identity and boundaries which might inform late-twentieth-century political imaginations. I am indebted in this story to writers like Joanna Russ, Samuel R. Delany, John Varley, James Tiptree, Jr., Octavia Butler, Monique Wittig, and Vonda McIntyre.[27] These are our storytellers exploring what it means to be embodied in high-tech worlds. They are theorists for cyborgs. Exploring conceptions of bodily boundaries and social order, the anthropologist Mary Douglas should be credited with helping us to consciousness about how fundamental body imagery is to world view, and so to political language.[28] French feminists like Luce Irigaray and Monique Wittig, for all their differences, know how to write the body, how to weave eroticism, cosmology, and politics from imagery of embodiment, and especially for Wittig, from imagery of fragmentation and reconstitution of bodies.[29]

American radical feminists like Susan Griffin, Audre Lorde, and Adrienne Rich have profoundly affected our political imaginations—and perhaps restricted too much what we allow as a friendly body and political language.[30] They insist on the organic, opposing it to the technological.

But their symbolic systems and the related positions of ecofeminism and feminist paganism, replete with organicisms, can only be understood in Sandoval's terms as oppositional ideologies fitting the late twentieth century. They would simply bewilder anyone not preoccupied with the machines and consciousness of late capitalism. In that sense they are part of the cyborg world. But there are also great riches for feminists in explicitly embracing the possibilities inherent in the breakdown of clean distinctions between organism and machine and similar distinctions structuring the Western self. It is the simultaneity of breakdowns that cracks the matrices of domination and opens geometric possibilities. What might be learned from personal and political "technological" pollution? I will look briefly at two overlapping groups of texts for their insight into the construction of a potentially helpful cyborg myth: constructions of women of color and monstrous selves in feminist science fiction.

Earlier I suggested that "women of color" might be understood as a cyborg identity, a potent subjectivity synthesized from fusions of outsider identities. There are material and cultural grids mapping this potential. Audre Lorde captures the tone in the title of her *Sister Outsider*. In my political myth, Sister Outsider is the offshore woman, whom U.S. workers, female and feminized, are supposed to regard as the enemy preventing their solidarity, threatening their security. Onshore, inside the boundary of the United States, Sister Outsider is a potential amidst the races and ethnic identities of women manipulated for division, competition, and exploitation in the same industries. "Women of color" are the preferred labor force for the science-based industries, the real women for whom the worldwide sexual market, labor market, and politics of reproduction kaleidoscope into daily life. Young Korean women hired in the sex industry and in electronics assembly are recruited from high schools, educated for the integrated circuit. Literacy, especially in English, distinguishes the "cheap" female labor so attractive to the multinationals.

Contrary to orientalist stereotypes of the "oral primitive," literacy is a special mark of women of color, acquired by U.S. black women as well as men through a history of risking death to learn and to teach reading and writing. Writing has a special significance for all colonized groups. Writing has been crucial to the Western myth of the distinction of oral and written cultures, primitive and civilized mentalities, and more recently to the erosion of that distinction in "post-modernist" theories attacking the phallogocentrism of the West, with its worship of the monotheistic, phallic, authoritative, and singular word, the unique and perfect name.[31] Contests for the meanings of writing are a major form of contemporary political struggle. Releasing the play of writing is deadly serious. The poetry and stories of U.S. women of color are repeatedly about writing, about access to the power to signify; but this time that power must be neither phallic nor innocent. Cyborg writing must not be about the Fall, the imagination of a once-upon-a-time wholeness before language, before writing, before Man. Cyborg writing is about the power to survive, not on the basis of original innocence, but on the basis of seizing the tools to mark the world that marked them as other.

The tools are often stories, retold stories, versions that reverse and displace the hierarchical dualisms of naturalized identities. In retelling origin stories, cyborg authors subvert the central myths of origin of Western culture. We have all been colonized by those origin myths, with their longing for fulfillment in apocalypse. The phallogocentric origin stories most crucial for feminist cyborgs are built into the literal technologies— technologies that write the world, biotechnology and microelectronics— that have recently textualized our bodies as code problems on the grid of C^3I. Feminist cyborg stories have the task of recoding communication and intelligence to subvert command and control.

Figuratively and literally, language politics pervade the struggles of women of color; and stories about language have a special power in the rich contemporary writing by U.S. women of color. For example, retellings of the story of the indigenous woman Malinche, mother of the mestizo "bastard" race of the new world, master of languages, and mistress of Cortés, carry special meaning for Chicana constructions of identity. Cherríe Moraga in *Loving in the War Years* explores the themes of identity when one never possessed the original language, never told the original story, never resided in the harmony of legitimate hetero-sexuality in the garden of culture, and so cannot base identity on a myth or a fall from innocence and right to natural names, mother's or father's.[32] Moraga's writing, her superb literacy, is presented in her poetry as the same kind of violation as Malinche's mastery of the con-querer's language—a violation, an illegitimate production, that allows survival. Moraga's language is not "whole"; it is self-consciously spliced, a chimera of English and Spanish, both conqueror's languages. But it is this chimeric monster, without claim to an original language before violation, that crafts the erotic, competent, potent identities of women of color. Sister Outsider hints at the possibility of world survival not because of her innocence, but because of her ability to live on the boundaries, to write without the founding myth of original wholeness, with its inescapable apocalypse of final return to a deathly oneness that Man has imagined to be the innocent and all-powerful Mother, freed at the End from another spiral of appropriation by her son. Writing marks Moraga's body, affirms it as the body of a woman of color, against the possibility of passing into the unmarked category of the Anglo father or into the orientalist myth of "original illiteracy" of a mother that never was. Malinche was mother here, not Eve before eating the forbidden fruit. Writing affirms Sister Outsider, not the Woman-before-the-Fall-into-Writing needed by the phallogocentric Family of Man.

Writing is pre-eminently the technology of cyborgs, etched surfaces of the late twentieth century. Cyborg politics is the struggle for language and the struggle against perfect communication, against the one code that translates all meaning perfectly, the central dogma of phallogocentrism. That is why cyborg politics insist on noise and advocate pollution, rejoicing in the illegitimate fusions of animal and machine. These are the couplings which make Man and Woman so problematic, subverting the structure of desire, the force imagined to generate language and gender, and so subverting the structure and modes of reproduction of "Western" identity, of nature and culture, of mirror and eye, slave and master, body and mind. "We" did not originally choose to be cyborgs, but choice grounds a liberal politics and epistemology that imagines the reproduction of individuals before the wider replications of "texts."

From the perspective of cyborgs, freed of the need to ground politics in "our" privileged position of the oppression that incorporates all other dominations, the innocence of the merely violated, the ground of those closer to nature, we can see powerful possibilities. Feminisms and Marxisms have run aground on Western epistemological imperatives to construct a revolutionary subject from the perspective of a hierarchy of oppressions and/or a latent position of moral superiority, innocence, and greater closeness to nature. With no available original dream of a common language or original symbiosis promising protection from hostile "masculine" separation, but written into the play of a text that has no finally privileged reading or salvation history, to recognize "oneself" as fully implicated in the world, frees us of the need to root politics in identification, vanguard parties, purity, and mothering. Stripped of identity, the bastard race teaches about the power of the margins and the importance of a mother like Malinche. Women of color have transformed her from the evil mother of masculinist fear into the originally literate mother who teaches survival.

This is not just literary deconstruction, but liminal transformation. Every story that begins with original innocence and privileges the return to wholeness imagines the drama of life to be individuation, separation, the birth of the self, the tragedy of autonomy, the fall into writing, alienation; i.e., war, tempered by imaginary respite in the bosom of the

Other. These plots are ruled by a reproductive politics—rebirth without flaw, perfection, abstraction. In this plot women are imagined either better or worse off, but all agree they have less selfhood, weaker individuation, more fusion to the oral, to Mother, less at stake in masculine autonomy. But there is another route to having less at stake in masculine autonomy, a route that does not pass through Woman, Primitive, Zero, the Mirror Stage and its imaginary. It passes through women and other present-tense, illegitimate cyborgs, not of Woman born, who refuse the ideological resources of victimization so as to have a real life. These cyborgs are the people who refuse to disappear on cue, no matter how many times a "Western" commentator remarks on the sad passing of another primitive, another organic group done in by "Western" technology, by writing.[33] These real-life cyborgs, e.g., the Southeast Asian village women workers in Japanese and U.S. electronics firms described by Aiwa Ong, are actively rewriting the texts of their bodies and societies. Survival is the stakes in this play of readings.

To recapitulate, certain dualisms have been persistent in Western traditions; they have all been systemic to the logics and practices of domination of women, people of color, nature, workers, animals—in short, domination of all constituted as *others*, whose task is to mirror the self. Chief among these troubling dualisms are self/other, mind/body, culture/nature, male/female, civilized/primitive, reality/appearance, whole/part, agent/resource, maker/made, active/passive, right/wrong, truth/illusion, total/partial, God/man. The self is the One who is not dominated, who knows that by the service of the other; the other is the one who holds the future, who knows that by the experience of domination, which gives the lie to the autonomy of the self. To be One is to be autonomous, to be powerful, to be God; but to be One is to be an illusion, and so to be involved in a dialectic of apocalypse with the other. Yet to be other is to be multiple, without clear boundary, frayed, insubstantial. One is too few, but two are too many.

High-tech culture challenges these dualisms in intriguing ways. It is not clear who makes and who is made in the relation between human and machine. It is not clear what is mind and what body in machines that resolve into coding practices. Insofar as we know ourselves in both formal discourse (e.g., biology) and in daily practice (e.g., the homework economy in the integrated circuit), we find ourselves to be cyborgs, hybrids, mosaics, chimeras. Biological organisms have become biotic systems, communications devices like others. There is no fundamental, ontological separation in our formal knowledge of machine and organism, of technical and organic.

One consequence is that our sense of connection to our tools is heightened. The trance state experienced by many computer users has become a staple of science-fiction film and cultural jokes. Perhaps paraplegics and other severely handicapped people can (and sometimes do) have the most intense experiences of complex hybridization with other communication devices. Anne McCaffrey's *The Ship Who Sang* explored the consciousness of a cyborg, hybrid of girl's brain and complex machinery, formed after the birth of a severely handicapped child. Gender, sexuality, embodiment, skill: all were reconstituted in the story. Why should our bodies end at the skin, or include at best other beings encapsulated by skin? From the seventeenth century till now, machines could be animated—given ghostly souls to make them speak or move or to account for their orderly development and mental capacities. Or organisms could be mechanized—reduced to body understood as resource of mind. These machine/organism relationships are obsolete, unnecessary. For us, in imagination and in other practice, machines can be prosthetic devices; intimate components, friendly selves. We don't need organic holism to give impermeable wholeness, the total woman and her feminist variants (mutants?). Let me conclude this point by a very partial reading of the logic of the cyborg monsters of my second group of texts, feminist science fiction.

The cyborgs populating feminist science fiction make very problematic the statuses of man or woman, human, artifact, member of a race, individual identity, or body. Katie King clarifies how pleasure in reading these fictions is not largely based on identification. Students facing Joanna Russ for the first time, students who have learned to take modernist writers like James Joyce or Virginia Woolf without flinching, do not know what to make of *The Adventures of Alyx* or *The Female Man*, where characters refuse the reader's search for innocent wholeness while granting the wish for heroic quests, exuberant eroticism, and serious politics. *The Female Man* is the story of four versions of one genotype, all of whom meet, but even taken together do not make a whole, resolve the dilemmas of violent moral action, nor remove the growing scandal of gender. The feminist science fiction of Samuel Delany, especially *Tales of Nevèrÿon*, mocks stories of origin by redoing the neolithic revolution, replaying the founding moves of Western civilization to subvert their plausibility. James Tiptree, Jr., an author whose fiction was regarded as particularly manly until her "true" gender was revealed, tells tales of reproduction based on non-mammalian technologies like alternation of generations or male brood pouches and male nurturing. John Varley constructs a supreme cyborg in his arch-feminist exploration of

Gaea, a mad goddess-planet-trickster-old woman-technological device on whose surface an extraordinary array of post-cyborg symbioses are spawned. Octavia Butler writes of an African sorceress pitting her powers of transformation against the genetic manipulations of her rival (*Wild Seed*), of time warps that bring a modern U.S. black woman into slavery where her actions in relation to her white master-ancestor determine the possibility of her own birth (*Kindred*), and of the illegitimate insights into identity and community of an adopted cross-species child who came to know the enemy as self (*Survivor*).

Because it is particularly rich in boundary transgressions, Vonda McIntyre's *Superluminal* can close this truncated catalogue of promising monsters who help redefine the pleasures and politics of embodiment and feminist writing. In a fiction where no character is "simply" human, human status is highly problematic. Orca, a genetically altered diver, can speak with killer whales and survive deep ocean conditions, but she longs to explore space as a pilot, necessitating bionic implants jeopardizing her kinship with the divers and cetaceans. Transformations are effected by virus vectors carrying a new developmental code, by transplant surgery, by implants of microelectronic devices, by analogue doubles, and other means. Laenea becomes a pilot by accepting a heart implant and a host of other alterations allowing survival in transit at speeds exceeding that of light. Radu Dracul survives a virus-caused plague on his outerworld planet to find himself with a time sense that changes the boundaries of spatial perception for the whole species. All the characters explore the limits of language, the dream of communicating experience, and the necessity of limitation, partiality, and intimacy even in this world of protean transformation and connection.

Monsters have always defined the limits of community in Western imaginations. The Centaurs and Amazons of ancient Greece established the limits of the centered polis of the Greek male human by their disruption of marriage and boundary pollutions of the warrior with animality and woman. Unseparated twins and hermaphrodites were the confused human material in early modern France who grounded discourse on the natural and super-natural, medical and legal, portents and diseases—all crucial to establishing modern identity.[34] The evolutionary and behavioral sciences of monkeys and apes have marked the multiple boundaries of late-twentieth-century industrial identities. Cyborg monsters in feminist science fiction define quite different political possibilities and limits from those proposed by the mundane fiction of Man and Woman.

There are several consequences to taking seriously the imagery of cyborgs as other than our enemies. Our bodies, ourselves; bodies are maps of power and identity. Cyborgs are no exceptions. A cyborg body is not innocent; it was not born in a garden; it does not seek unitary identity and so generate antagonistic dualisms without end (or until the world ends); it takes irony for granted. One is too few, and two is only one possibility. Intense pleasure in skill, machine skill, ceases to be a sin, but an aspect of embodiment. The machine is not an *it* to be animated, worshiped and dominated. The machine is us, our processes, an aspect of our embodiment. We can be responsible for machines; *they* do not dominate or threaten us. We are responsible for boundaries; we are they. Up till now (once upon a time), female embodiment seemed to be given, organic, necessary; and female embodiment seemed to mean skill in mothering and its metaphoric extensions. Only by being out of place could we take intense pleasure in machines, and then with excuses that this was organic activity after all, appropriate to females. Cyborgs might consider more seriously the partial, fluid, sometimes aspect of sex and sexual embodiment. Gender might not be global identity after all.

The ideologically charged question of what counts as daily activity, as experience, can be approached by exploiting the cyborg image. Feminists have recently claimed that women are given to dailiness, that women more than men somehow sustain daily life, and so have a privileged epistemological position potentially. There is a compelling aspect to this claim, one that makes visible unvalued female activity and names it as the ground of life. But *the* ground of life? What about all the ignorance of women, all the exclusions and failures of knowledge and skill? What about men's access to daily competence, to knowing how to build things, to take them apart, to play? What about other embodiments? Cyborg gender is a local possibility taking a global vengeance. Race, gender, and capital require a cyborg theory of wholes and parts. There is no drive in cyborgs to produce total theory, but there is an intimate experience of boundaries, their construction and deconstruction. There is a myth system waiting to become a political language to ground one way of looking at science and technology and challenging the informatics of domination.

One last image: organisms and organismic, holistic politics depend on metaphors of rebirth and invariably call on the resources of reproductive sex. I would suggest that cyborgs have more to do with regeneration and are suspicious of the reproductive matrix and of most birthing. For sala-manders, regeneration after injury, such as the loss of a limb, involves regrowth of structure and restoration of function with the constant pos-sibility of twinning or other odd topographical productions at the site of former injury. The regrown limb can be monstrous, duplicated, po-tent. We have all been injured, profoundly. We require regeneration, not rebirth, and the possibilities for our reconstitution include the utopian dream of the hope for a monstrous world without gender.

Cyborg imagery can help express two crucial arguments in this essay: (1) the production of universal, totalizing theory is a major mistake that misses most of reality, probably always, but certainly now; (2) taking responsibility for the social relations of science and technology means refusing an anti-science metaphysics, a demonology of technology, and so means embracing the skillful task of reconstructing the boundaries of daily life, in partial connection with others, in communication with all of our parts. It is not just that science and technology are possible means of great human satisfaction, as well as a matrix of complex dominations. Cyborg imagery can suggest a way out of the maze of dualisms in which we have explained our bodies and our tools to ourselves. This is a dream not of a common language, but of a powerful infidel heteroglossia. It is an imagination of a feminist speaking in tongues to strike fear into the circuits of the super-savers of the new right. It means both building and destroying machines, identities, categories, relationships, spaces, stories. Though both are bound in the spiral dance, I would rather be a cyborg than a goddess.

ACKNOWLEDGMENTS

Research was funded by an Academic Senate Faculty Research Grant from the University of California, Santa Cruz. An earlier version of the paper on genetic engineering appeared as "Lieber Kyborg als Gottin: Für eine sozialistisch-feministische Unterwanderung der Gentechnologie," in Bernd-Peter Lange and Anna Marie Stuby, eds., *1984* (Berlin: Argument-Sonderband 105, 1984), pp. 66–84. The cyborg manifesto grew from "New Machines, New Bodies, New Communities: Political Dilemmas of a Cyborg Feminist," *The Scholar and the Feminist x: The Question of Technology*, Conference, April 1983.

The people associated with the History of Consciousness Board of UCSC have had an enormous influence on this paper, so that it feels collectively authored more than most, although those I cite may not recognize their ideas. In particular, members of graduate and under-graduate feminist theory, science and politics, and theory and methods courses have contributed to the cyborg manifesto. Particular debts here are due Hilary Klein ("Marxism, Psychoanalysis, and Mother Nature"); Paul Edwards ("Border Wars: The Science and Politics of Artificial Intelligence"); Lisa Lowe ("Julia Kristeva's *Des Chinoises:* Representing Cultural and Sexual Others"); Jim Clifford, "On Ethnographic Allegory: Essays," forthcoming.

Parts of the paper were my contribution to a collectively developed session, Poetic Tools and Political Bodies: Feminist Approaches to High Technology Culture, 1984 California American Studies Association, with History of Consciousness graduate students Zoe Sofoulis, "Jupiter Space"; Katie King, "The Pleasures of Repetition and the Limits of Identification in Feminist Science Fiction: Reimaginations of the Body after the Cyborg"; and Chela Sandoval, "The Construction of Subjectivity and Oppositional Consciousness in Feminist Film and Video." Sandoval's theory of oppositional consciousness was published as "Women Respond to Racism: A Report on the National Women's Studies Association Conference," Center for Third World Organizing, Oakland, California, n.d. For Sofoulis's semiotic-psychoanalytic readings of nuclear culture, see Z. Sofia, "Exterminating Fetuses: Abortion, Disarmament and the Sexo-Semiotics of Extraterrestrialism," Nuclear Criticism issue, *Diatritics*, vol. 14, no. 2 (1984), pp. 47–59. King's manuscripts. ("Questioning Tradition: Canon Formation and the Veiling of Power"; "Gender and Genre: Reading the Science Fiction of Joanna Russ"; "Varley's *Titan* and *Wizard*: Feminist Parodies of Nature, Culture, and Hardware") deeply inform the cyborg manifesto.

Barbara Epstein, Jeff Escoffier, Rusten Hogness, and Jaye Miller gave extensive discussion and editorial help. Members of the Silicon Valley Research Project of UCSC and participants in SVRP conferences and workshops have been very important, especially Rick Gordon, Linda Kimball, Nancy Snyder, Langdon Winner, Judith Stacey, Linda Lim, Patricia Fernandez-Kelly, and Judith Gregory. Finally, I want to thank Nancy Hartsock for years of friendship and discussion on feminist theory and feminist science fiction.

NOTES

1. Useful references to left and/or feminist radical science movements and theory and to biological/biotechnological issues include: Ruth Bleier, *Science and Gender: A Critique of Biology and Its Themes on Women* (New York: Pergamon, 1984); Elizabeth Fee, "Critiques of Modern Science: The Relationship of Feminist and Other Radical Epistemologies," and Evelyn Hammonds, "Women of Color, Feminism and Science," papers for Symposium on Feminist Perspectives on Science, University of Wisconsin, 11–12 April, 1985, in Ruth Bleier, ed., *Feminist Approaches to Science* (New York: Pergamon, 1986); Stephen J. Gould, *Mismeasure of Man* (New York: Norton, 1981); Ruth Hubbard, Mary Sue Henifin, Barbara Fried, eds., *Biological Woman, the Convenient Myth* (Cambridge, Mass.: Schenkman, 1982); Evelyn Fox Keller, *Reflections on Gender and Science* (New Haven: Yale University Press, 1985); R. C. Lewontin, Steve Rose, and Leon Kamin, *Not in Our Genes* (New York: Pantheon, 1984): *Radical Science Journal*, 26 Freegrove Road, London N7 9RQ; *Science for the People*, 897 Main St., Cambridge, MA 02139.
2. Starting points for left and/or feminist approaches to technology and politics include: Ruth Schwartz Cowan, *More Work for Mother: The Ironies of Household Technology from*

the *Open Hearth to the Microwave* (New York: Basic Books, 1983); Joan Rothschild, *Machina ex Dea: Feminist Perspectives on Technology* (New York: Pergamon, 1983); Sharon Traweek, "Up-time, Downtime, Spacetime, and Power: An Ethnography of U.S. and Japanese Particle Physics," Ph.D. thesis, UC Santa Cruz, History of Consciousness, 1982; R. M. Young and Les Levidov, eds., *Science, Technology, and the Labour Process*, vols. 1–3 (London: CSE Books); Joseph Weizenbaum, *Computer Power and Human Reason* (San Francisco: Freeman, 1976); Langdon Winner, *Autonomous Technology: Technics Out of Control as a Theme in Political Thought* (Cambridge, Mass.: MIT Press, 1977); Langdon Winner, *The Whale and the Reactor* (Chicago: University of Chicago Press, 1986); Jan Zimmerman, ed., *The Technological Woman: Interfacing with Tomorrow* (New York: Praeger, 1983); *Global Electronics Newsletter*, 867 West Dana St., #204, Mountain View, CA 94041; *Processed World*, 55 Sutter St., San Francisco, CA 94104; *ISES*, Women's International Information and Communication Service, P.O. Box 50 (Cornavin), 1211 Geneva 2, Switzerland, and Via Santa Maria dell'Anima 30,00186 Rome, Italy. Fundamental approaches to modern social studies of science that do not continue the liberal mystification that it all started with Thomas Kuhn, include: Karin Knorr-Cetina, *The Manufacture of Knowledge* (Oxford: Pergamon, 1981); K. D. Knorr-Cetina and Michael Mulkay, eds., *Science Observed: Perspectives on the Social Study of Science* (Beverly Hills, Calif.: Sage, 1983); Bruno Latour and Steve Woolgar, *Laboratory Life: The Social Construction of Scientific Facts* (Beverly Hills, Calif.: Sage, 1979); Robert M. Young, "Interpreting the Production of Science," *New Scientist*, vol. 29 (March 1979), pp. 1026–1028. More is claimed than is known about contesting productions of science in the mythic/material space of "the laboratory"; the 1984 Directory of the Network for the Ethnographic Study of Science, Technology, and Organizations lists a wide range of people and projects crucial to better radical analysis; available from NESSTO, P.O. Box 11442, Stanford, CA 94305.

3.   Frederic Jameson, "Post-modernism, or the Cultural Logic of Late Capitalism," *New Left Review*, July/August 1984, pp. 53–94. See Marjorie Perloff, "'Dirty' Language and Scramble Systems," *Sulfur* 11 (1984), pp. 178–183; Kathleen Fraser, *Something (Even Human Voices) in the Foreground, a Lake* (Berkeley, Calif.: Kelsey St. Press, 1984).

4.   Frans de Waal, *Chimpanzee Politics: Power and Sex among the Apes* (New York: Harper & Row, 1982); Langdon Winner, "Do Artifacts Have Politics?" *Daedalus*, Winter 1980, pp. 121–36.

5.   Jean Baudrillard, *Simulations*, trans. P. Foss, P. Patton, P. Beitchman (New York: Semiotext(e), 1983). Jameson ("Post modernism," p. 66) points out that Plato's definition of the simulacrum is the copy for which there is no original; i.e., the world of advanced capitalism, of pure exchange.

6.   Herbert Marcuse, *One-Dimensional Man* (Boston: Beacon, 1964); Carolyn Merchant, *Death of Nature* (San Francisco: Harper & Row, 1980).

7.   Zoe Sofia, "Exterminating Fetuses," *Diacritics*, vol. 14, no. 2 (Summer 1984), pp. 47–59, and "Jupiter Space" (Pomona, Calif.: American Studies Association, 1984).

8.   Powerful developments of coalition politics emerge from "third world" speakers, speaking from nowhere, the displaced center of the universe, earth: "We live on the third planet from the sun"—*Sun Poem* by Jamaican writer Edward Kamau Braithwaite, review by Nathaniel Mackey, *Sulfur*, 11 (1984), pp. 200–205. *Home Girls*, ed. Barbara Smith (New York: Kitchen Table, Women of Color Press, 1983), ironically subverts naturalized identities precisely while constructing a place from which to speak called home. See esp. Bernice Reagan, "Coalition Politics, Turning the Century," pp. 356–368, in Smith, 1983.

9.   Chela Sandoval, "Dis-Illusionment and the Poetry of the Future: The Making of Oppositional Consciousness," Ph.D. qualifying essay, UCSC, 1984.

10. bell hooks, *Ain't I a Woman?* (Boston: South End Press, 1981); Gloria Hull, Patricia Bell Scott, and Barbara Smith, eds., *All the Women Are White, All the Men Are Black, But Some of Us Are Brave: Black Women's Studies* (Old Westbury, Conn.: Feminist Press, 1982). Toni Cade Bambara, in *The Salt Eaters* (New York: Vintage/Random House, 1981), writes an extraordinary post-modernist novel, in which the women of color theater group, The Seven Sisters, explores a form of unity. Thanks to Elliott Evans's readings of Bambara, Ph.D. qualifying essay, UCSC, 1984.

11. On orientalism in feminist works and elsewhere, see Lisa Lowe, "Orientation: Representations of Cultural and Sexual 'Others,'" Ph.D. thesis, UCSC; Edward Said, *Orientalism* (New York: Pantheon, 1978).

12. Katie King has developed a theoretically sensitive treatment of the workings of feminist taxonomies as genealogies of power in feminist ideology and polemic: "Prospectus," *Gender and Genre: Academic Practice and the Making of Criticism* (Santa Cruz, Calif.: University of California, 1984). King examines an intelligent, problematic example of taxonomizing feminisms to make a little machine producing the desired final position: Alison Jaggar, *Feminist Politics and Human Nature* (Totowa, N.J.: Rowman & Allanheld, 1983). My caricature here of socialist and radical feminism is also an example.

13. The feminist standpoint argument is being developed by: Jane Flax, "Political Philosophy and the Patriarchal Unconsciousness," in Sandra Harding and Merill Hintikka, eds., *Discovering Reality* (Dordrecht: Reidel, 1983); Sandra Harding, *The Science Question in Feminism* (Ithaca: Cornell University Press, 1986); Harding and Hintikka, *Discovering Reality*; Nancy Hartsock, *Money, Sex, and Power* (New York: Longman, 1983) and "The Feminist Standpoint: Developing the Ground for a Specifically Feminist Historical Materialism," in Harding and Hintikka, *Discovering Reality*; Mary O'Brien, *The Politics of Reproduction* (New York: Routledge & Kegan Paul, 1981); Hilary Rose, "Hand, Brain, and Heart: A Feminist Epistemology for the Natural Sciences," *Signs*, vol. 9, no. 1 (1983), pp. 73–90; Dorothy Smith, "Women's Perspective as a Radical Critique of Sociology," *Sociological Inquiry* 44 (1974), and "A Sociology of Women," in J. Sherman and E. T. Beck, eds., *The Prism of Sex* (Madison: University of Wisconsin Press, 1979).

14. Catharine MacKinnon, "Feminism, Marxism, Method, and the State: An Agenda for Theory," *Signs*, vol. 7, no. 3 (Spring 1982), pp. 515–544. A critique indebted to MacKinnon, but without the reductionism and with an elegant feminist account of Foucault's paradoxical conservatism on sexual violence (rape), is Teresa de Lauretis, "The Violence of Rhetoric," *Semiotica*, special issue on "The Rhetoric of Violence," ed. Nancy Armstrong, vol. 54 (1985), pp. 11–31. A theoretically elegant feminist social-historical examination of family violence, that insists on women's, men's, children's complex agency without losing sight of the material structures of male domination, race, and class, is Linda Gordon, *Heroes of Their Own Lives* (New York: Viking Penguin, 1998).

15. My previous efforts to understand biology as a cybernetic command-control discourse and organisms as "natural-technical objects of knowledge" are: "The High Cost of Information in Post-World War II Evolutionary Biology," *Philosophical Forum*, vol. 13, nos. 2–3 (1979), pp. 206–237; "Signs of Dominance: From a Physiology to a Cybernetics of Primate Society," *Studies in History of Biology* 6 (1983), pp. 129–219; "Class, Race, Sex, Scientific Objects of Knowledge: A Socialist-Feminist Perspective on the Social Construction of Productive Knowledge and Some Political Consequences," in Violet Haas and Carolyn Perucci, eds., *Women in Scientific and Engineering Professions* (Ann Arbor: University of Michigan Press, 1984), pp. 212–229.

16. E. Rusten Hogness, "Why Stress? A Look at the Making of Stress, 1936–56," available from the author, 4437 Mill Creek Rd., Healdsburg, CA 95448.

17. A left entry to the biotechnology debate: *GeneWatch*, a Bulletin of the Committee for Responsible Genetics, 5 Doane St., 4th floor, Boston, MA 02109; Susan Wright "Recombinant DNA: The Status of Hazards and Controls," *Environment*, July/August

1982, pp. 12–20, 51–53; Edward Yoxen, *The Gene Business* (New York: Harper & Row, 1983).

18. Starting references for "women in the integrated circuit": Pamela D'Onofrio-Flores and Sheila M. Pfafflin, eds., *Scientific-Technological Change and the Role of Women in Development* (Boulder, Colo.: Westview Press, 1982); Maria Patricia Fernandez-Kelly, *For We Are Sold, I and My People* (Albany, N.Y.: SUNY Press, 1983); Annette Fuentes and Barbara Ehrenreich, *Women in the Global Factory* (Boston: South End Press, 1983), with an especially useful list of resources and organizations; Rachael Grossman, "Women's Place in the Integrated Circuit," *Radical America*, vol. 14, no. 1 (1980), pp. 29–50; June Nash and M. P. Fernandez-Kelly, eds., *Women and Men and the International Division of Labor* (Albany, N.Y.: SUNY Press, 1983); Aiwa Ong, *Spirits of Resistance and Capitalist Discipline* (Albany: State University of New York Press, 1987); Science Policy Research Unity, *Microelectronics and Women's Employment in Britain* (University of Sussex, 1982).

19. The best example is Bruno Latour, *Les Microbes: Guerre et Paix, suivi de Irreductions* (Paris: Metailie, 1984).

20. For the homework economy and some supporting arguments: Richard Gordon, "The Computerization of Daily Life, the Sexual Division of Labor, and the Homework Economy," in R. Gordon, ed., *Microelectronics in Transition* (Norwood, N.J.: Ablex, 1985); Patricia Hill Collins, "Third World Women in America," and Sara G. Burr, "Women and Work," in Barbara K. Haber, ed., *The Women's Annual, 1981* (Boston: G. K. Hall, 1982); Judith Gregory and Karen Nussbaum, "Race against Time: Automation of the Office," *Office: Technology and People* 1 (1982), pp. 197–236; Frances Fox Piven and Richard Cloward, *The New Class War: Reagan's Attack on the Welfare State and Its Consequences* (New York: Pantheon, 1982); Microelectronics Group, *Microelectronics: Capitalist Technology and the Working Class* (London: CSE, 1980); Karin Stallard, Barbara Ehrenreich, and Holly Sklar, *Poverty in the American Dream* (Boston: South End Press, 1983), including a useful organization and resource list.

21. Rae Lessor Blumberg, "A General Theory of Sex Stratification and Its Application to the Position of Women in Today's World Economy," paper delivered to Sociology Board, UCSC, February 1983. Also Blumberg, *Stratification: Socioeconomic and Sexual Inequality* (Boston: Brown, 1981). See also Sally Hacker, "Doing It the Hard Way: Ethnographic Studies in the Agribusiness and Engineering Classroom," California American Studies Association, Pomona, 1984; S. Hacker and Lisa Bovit, "Agriculture to Agribusiness: Technical Imperatives and Changing Roles," *Proceedings* of the Society for the History of Technology, Milwaukee, 1981; Lawrence Busch and William Lacy, *Science, Agriculture, and the Politics of Research* (Boulder, Colo.: Westview Press, 1983); Denis Wilfred, "Capital and Agriculture, a Review of Marxian Problematics," *Studies in Political Economy*, no. 7 (1982), pp. 127–154; Carolyn Sachs, *The Invisible Fanners: Women in Agricultural Production* (Totowa, N.J.: Rowman & Allanheld, 1983). Thanks to Elizabeth Bird, "Green Revolution Imperialism," I & II, ms. UCSC, 1984.

22. Cynthia Enloe, "Women Textile Workers in the Militarization of South-east Asia," in Nash and Fernandez-Kelly, *Women and Men*; Rosalind Petchesky, "Abortion, Anti-Feminism, and the Rise of the New Right," *Feminist Studies*, vol. 7, no. 2 (1981).

23. For a feminist version of this logic, see Sarah Blaffer Hrdy, *The Woman That Never Evolved* (Cambridge, Mass.: Harvard University Press, 1981). For an analysis of scientific women's story-telling practices, especially in relation to sociobiology, in evolutionary debates around child abuse and infanticide, see Donna Haraway, "The Contest for Primate Nature: Daughters of Man the Hunter in the Field, 1960–80," in Mark Kann, ed., *The Future of American Democracy* (Philadelphia: Temple University Press, 1983), pp. 175–208.

24. For the moment of transition of hunting with guns to hunting with cameras in the construction of popular meanings of nature for an American urban immigrant public, see Donna Haraway, "Teddy Bear Patriarchy," *Social Text*, forthcoming, 1985; Roderick

Nash, "The Exporting and Importing of Nature: Nature-Appreciation as a Commodity, 1850–1980," *Perspectives in American History*, vol. 3 (1979), pp. 517–560; Susan Sontag, *On Photography* (New York: Dell, 1977); and Douglas Preston, "Shooting in Paradise," *Natural History*, vol. 93, no. 12 (December 1984), pp. 14–19.

25. For crucial guidance for thinking about the political/cultural implications of the history of women doing science in the United States, see: Violet Haas and Carolyn Perucci, eds., *Women in Scientific and Engineering Professions* (Ann Arbor: University of Michigan Press, 1984); Sally Hacker, "The Culture of Engineering: Women, Workplace, and Machine," *Women's Studies International Quarterly*, vol. 4, no. 3 (1981), pp. 341–353; Evelyn Fox Keller, *A Feeling for the Organism* (San Francisco: Freeman, 1983); National Science Foundation, *Women and Minorities in Science and Engineering* (Washington, D.C.: NSF, 1982); Margaret Rossiter, *Women Scientists in America* (Baltimore: Johns Hopkins University Press, 1982).

26. John Markoff and Lenny Siegel, "Military Micros," paper presented at the UCSC Silicon Valley Research Project conference, 1983. High Technology Professionals for Peace and Computer Professionals for Social Responsibility are promising organizations.

27. Katie King, "The Pleasure of Repetition and the Limits of Identification in Feminist Science Fiction: Reimaginations of the Body after the Cyborg," California American Studies Association, Pomona, 1984. An abbreviated list of feminist science fiction underlying themes of this essay: Octavia Butler, *Wild Seed, Mind of My Mind, Kindred, Survivor;* Suzy McKee Charnas, *Motherliness;* Samuel Delany, *Tales of Nevèrÿon;* Anne McCaffery, *The Ship Who Sang, Dinosaur Planet;* Vonda McIntyre, *Super-luminal, Dreamsnake;* Joanna Russ, *Adventures of Alix, The Female Man;* James Tiptree, Jr., *Star Songs of an Old Primate, Up the Walls of the World;* John Varley, *Titan, Wizard, Demon.*

28. Mary Douglas, *Purity and Danger* (London: Routledge & Kegan Paul, 1966), *Natural Symbols* (London: Cresset Press, 1970).

29. French feminisms contribute to cyborg heteroglossia. Carolyn Burke, "Irigaray through the Looking Glass," *Feminist Studies*, vol. 7, no. 2 (Summer 1981), pp. 288–306; Luce Irigaray, *Ce sexe qui n'en est pas un* (Paris: Minuit, 1977); L. Irigaray, *Et l'une ne bouge pas sans l'autre* (Paris: Minuit, 1979); Elaine Marks and Isabelle de Courtivron, ed., *New French Feminisms* (Amherst: University of Massachusetts Press, 1980); *Signs*, vol. 7, no. 1 (Autumn, 1981), special issue on French feminism; Monique Wittig, *The Lesbian Body*, trans. David LeVay (New York: Avon, 1975; *Le corps lesbien*, 1973).

30. But all these poets are very complex, not least in treatment of themes of lying and erotic, decentered collective and personal identities. Susan Griffin, *Women and Nature: The Roaring Inside Her* (New York: Harper & Row, 1978); Audre Lorde, *Sister Outsider* (New York: Crossing Press, 1984); Adrienne Rich, *The Dream of a Common Language* (New York: Norton, 1978).

31. Jacques Derrida, *Of Grammatology*, trans. and introd. G. C. Spivak (Baltimore: Johns Hopkins University Press, 1976), esp. part II, "Nature, Culture, Writing"; Claude Lévi-Strauss, *Tristes Tropiques*, trans. John Russell (New York, 1961), esp. "The Writing Lesson."

32. Cherrie Moraga, *Loving in the War Years* (Boston: South End Press, 1983). The sharp relation of women of color to writing as theme and politics can be approached through: "The Black Woman and the Diaspora: Hidden Connections and Extended Acknowledgments," An International Literary Conference, Michigan State University, October 1985; Mari Evans, ed., *Black Women Writers: A Critical Evaluation* (Garden City, N.Y.: Doubleday/Anchor, 1984); Dexter Fisher, ed., *The Third Woman: Minority Women Writers of the United States* (Boston: Houghton Mifflin, 1980); several issues of *Frontiers*, esp. vol. 5 (1980), "Chicanas en el Ambiente Nacional" and vol. 7 (1983). "Feminisms in the Non-Western World"; Maxine Hong Kingston, *China Men* (New York: Knopf, 1977); Gerda Lerner, ed., *Black Women in White America: A Documentary History*

(New York: Vintage, 1973); Cherrie Moraga and Gloria Anzaldus, eds., *This Bridge Called My Back: Writings by Radical Women of Color* (Watertown, Mass.: Persephone, 1981); Robin Morgan, ed., *Sisterhood Is Global* (Garden City, N.Y.: Anchor/Doubleday, 1984). The writing of white women has had similar meanings: Sandra Gilbert and Susan Gubar, *The Madwoman in the Attic* (New Haven: Yale University Press, 1979); Joanna Russ, *How to Suppress Women's Writing* (Austin: University of Texas Press, 1983).

33. James Clifford argues persuasively for recognition of continuous cultural reinvention, the stubborn non-disappearance of those "marked" by Western imperializing practices; see "On Ethnographic Allegory: Essays," in Clifford and Marcus, 1985, and "On Ethnographic Authority," *Representations*, vol. 1, no. 2 (1983), pp. 118–146.

34. Page DuBois, *Centaurs and Amazons* (Ann Arbor: University of Michigan Press, 1982); Lorraine Daston and Katharine Park, "Hermaphrodites in Renaissance France," ms., n.d.; Katharine Park and Lorraine Daston, "Unnatural Conceptions: The Study of Monsters in 16th and 17th Century France and England," *Past and Present*, no. 92 (August 1981), pp. 20–54.

2

ECCE HOMO, AIN'T (AR'N'T) I A WOMAN, AND INAPPROPRIATE/D OTHERS: THE HUMAN IN A POST-HUMANIST LANDSCAPE

I want to focus on the discourses of suffering and dismemberment. I want to stay with the disarticulated bodies of history as figures of possible connection and accountability. Feminist theory proceeds by figuration at just those moments when its own historical narratives are in crisis. Historical narratives are in crisis now, across the political spectrum, around the world. These are the moments when something powerful— and dangerous—is happening. Figuration is about resetting the stage for possible pasts and futures. Figuration is the mode of theory when the more "normal" rhetorics of systematic critical analysis seem only to repeat and sustain our entrapment in the stories of the established disorders. Humanity is a modernist figure; and this humanity has a generic face, a universal shape. Humanity's face has been the face of man. Feminist humanity must have another shape, other gestures; but, I believe, we must have feminist figures of humanity. They cannot be man or woman; they cannot be the human as historical narrative has staged that generic universal. Feminist figures cannot, finally, have a name; they cannot be native. Feminist humanity must, somehow, both resist representation, resist literal figuration, and still erupt in powerful new tropes, new figures of speech, new turns of historical possibility. For this process, at the inflection point of crisis, where all the tropes turn again, we need ecstatic speakers. This essay tells a history of such a speaker who might figure the self-contradictory and necessary condition of a nongeneric humanity.

I want here to set aside the Enlightenment figures of coherent and masterful subjectivity, the bearers of rights, holders of property in the self, legitimate sons with access to language and the power to represent, subjects endowed with inner coherence and rational clarity, the masters of theory, founders of states, and fathers of families, bombs, and scientific theories—in short, Man as we have come to know and love him in the death-of-the-subject critiques. Instead, let us attend to another crucial strand of Western humanism thrown into crisis in the late twentieth century. My focus is the figure of a broken and suffering humanity, signifying—in ambiguity, contradiction, stolen symbolism, and unending chains of noninnocent translation—a possible hope. But also signifying an unending series of mimetic and counterfeit events implicated in the great genocides and holocausts of ancient and modern history. But, it is the very nonoriginality, mimesis, mockery, and brokenness that draw me to this figure and its mutants. This essay is the beginning of a project on figurations that have appeared in an array of internationalist, scientific, and feminist texts, which I wish to examine for their contrasting modernist, postmodernist, and amodernist ways of constructing "the human" after World War II. Here, I begin by reading Jesus and Sojourner Truth as Western trickster figures in a rich, dangerous, old, and constantly renewed tradition of Judeo-Christian humanism and end by asking how recent intercultural and multicultural feminist theory constructs possible postcolonial, nongeneric, and irredeemably specific figures of critical subjectivity, consciousness, and humanity—not in the sacred image of the same, but in the self-critical practice of "difference," of the I and we that is/are never identical to itself, and so has hope of connection to others.

The larger project that this essay initiates will stage an historical conversation among three groups of powerfully universalizing texts:

1.   two versions of United Nations discourses on human rights (the UNESCO statements on race in 1950 and 1951 and the documents and events of the UN Decade for Women from 1975–85);

2.   recent modernist physical-anthropological reconstructions of the powerful fiction of science, species man, and its science-fiction variant, the female man (pace Joanna Russ) (i.e., Man the Hunter of the 1950s and 1960s and Woman the Gatherer of the 1970s and 1980s); and

3.   the transnational, multi-billion-dollar, highly automated, postmodernist apparatus—a language technology, literally—for the

production of what will count as "the human" (i.e., the Human Genome Project, with all its stunning power to recuperate, out of the endless variations of code fragments, the singular, the sacred image of the same, the one true man, the standard—copyrighted, catalogued, and banked).

The whole tale might fit together at least as well as the plot of Enlightenment humanism ever did, but I hope it will fit differently, negatively, if you will. I suggest that the only route to a nongeneric humanity, for whom specificity—but emphatically not originality—is the key to connection, is through radical nominalism. We must take names and essences seriously enough to adopt such an ascetic stance about who we have been and might yet be. My stakes are high; I think "we"—that crucial material and rhetorical construction of politics and of history—need something called humanity. It is that kind of thing which Gayatri Spivak called "that which we cannot not want." We also know now, from our perspectives in the ripped-open belly of the monster called history, that we cannot name and possess this thing which we cannot not desire. Humanity, whole and part, is not autochthonous. Nobody is self-made, least of all man. That is the spiritual and political meaning of poststructuralism and postmodernism for me. "We," in these very particular discursive worlds, have no routes to connection and to noncosmic, nongeneric, nonoriginal wholeness than through the radical dismembering and displacing of our names and our bodies. So, how can humanity have a figure outside the narratives of humanism; what language would such a figure speak?

ECCE HOMO! THE SUFFERING SERVANT AS A FIGURE OF HUMANITY[1]

Isaiah 52.13–14:

Behold, my servant shall prosper, he shall be exalted and lifted up, and shall be very high. As many were astonished at him—his appearance was so marred, beyond human semblance, and his form beyond that of the sons of men—so shall he startle many nations.

Isaiah 53.2–4:

He had no form or comeliness that we should look at him, and no beauty that we should desire him. He was despised and rejected by men; a man of sorrows, and acquainted with grief, and as one from whom men hide their faces he was despised, and we esteemed him not. Surely he has borne our griefs and carried our sorrows; yet we esteemed him stricken, smitten by God, and afflicted. But he was wounded for our transgressions, he was bruised for our iniquities; upon him was the chastisement that made us whole, and with his stripes we are healed.

Isaiah 54.1:

For the children of the desolate one will be more than the children of her that is married, says the Lord. ("Is this a threat or a promise?" ask both women, looking tentatively at each other after a long separation.)

John 18.37–38:

Pilate said to him, "So, you are a king?" Jesus answered, "You say that I am a king. For this I was born, and for this I have come into the world, to bear witness to the truth. Everyone who is of the truth hears my voice." Pilate said to him, "What is truth?"

John 19.1–6:

Then Pilate took Jesus and scourged him. And the soldiers plaited a crown of thorns, and put it on his head, and arrayed him in a purple robe; they came up to him, saying, "Hail, King of the Jews!" and struck him with their hands. Pilate went out again, and said to them, "Behold I am bringing him out to you, that you may know I find no crime in him." So Jesus came out, wearing the crown of thorns and the purple robe. Pilate said to them, "Behold the man!" When the chief priests and officers saw him, they cried out, "Crucify him, crucify him!" Pilate said to them, "Take him yourselves and crucify him, for I find no crime in him."

John staged the trial before Pilate in terms of the suffering-servant passages from Isaiah. The events of the trial of Jesus in this nonsynoptic gospel probably are not historical, but theatrical in the strict sense: from the start, they *stage* salvation history, which then became the model for world history in the secular heresies of the centuries of European colonialism with its civilizing missions and genocidal discourses on common humanity. Pilate probably spoke publicly in Greek or Latin, those languages that became the standard of "universal" European scholarly humanism, and his words were translated by his officials into Aramaic, the language of the inhabitants of Palestine. Hebrew was already largely a ceremonial language, not even understood by most Jews in the synagogue. The earliest texts for John's gospel that we have are in Greek, the likely language of its composition (the Koiné, the common Greek spoken and understood throughout the Roman Empire in the early centuries of the Christian era). We don't have the first versions, if there ever were such things; we have endless, gap-filled, and overlaid transcriptions and translations that have grounded the vast apparatus of biblical textual and linguistic scholarship—that cornerstone of modern scholarly humanism, hermeneutics, and semiology and of the human sciences generally, most certainly including anthropology and ethnography. We are, indeed, peoples of the Book, engaged in a Derridean writing and reading practice from the first cries of prophecy and codifications of salvation history.

From the start we are in the midst of multiple translations and stagings of a figure of suffering humanity that was not contained within the cultures of the origin of the stories. The Christian narratives of the Son of Man circulated rapidly around the Mediterranean in the first century of the present era. The Jewish versions of the suffering servant inform some of the most powerful ethical cautions in Faustian transnational technoscience worlds. The presentation to the people of the Son of Man as a suffering servant, arrayed mockingly and mimetically in his true dress as a king and salvation figure, became a powerful image for Christian humanists. The suffering servant figure has been fundamental in twentieth-century liberation theology and Christian Marxism. The guises of the suffering servant never cease. Even in Isaiah, he is clothed in the ambiguities of prophecy. His most important counterfeit historically was Jesus himself, as John appropriated Isaiah into a theater of salvation history that would accuse the Jews of demanding the death of their king and savior in the root narrative of Christian anti-Semitism. The "Ecce homo!" was standardized in the Latin vulgate after many passages through the languages and transcriptions and codifications of the gospels. Jesus appears as a mime in many layers; crowned with thorns and in a purple cloak, he is in the mock disguise of a king before his wrongful execution as a criminal. As a criminal, he is counterfeit for a scapegoat, indeed, *the* scapegoat of salvation history. Already, as a carpenter he was in disguise.

This figure of the Incarnation can never be other than a trickster, a check on the arrogances of a reason that would uncover all disguises and force correct vision of a recalcitrant nature in her most secret places. The suffering servant is a check on man; the servant is the figure associated with the promise that the desolate woman will have more children than the wife, the figure that upsets the clarity of the metaphysics of light, which John the Evangelist too was so enamored of. A mother's son, without a father, yet the Son of Man claiming *the* Father, Jesus is a potential worm in the Oedipal psychoanalytics of representation; he threatens to spoil the story, despite or because of his odd sonship and odder kingship, because of his disguises and form-changing habits. Jesus makes of man a most promising mockery, but a mockery that cannot evade the terrible story of the broken body. The story has constantly to be preserved from heresy, to be kept forcibly in the patriarchal tradition of Christian civilization, to be kept from too much attention to the economies of mimicry and the calamities of suffering.

Jesus came to figure for Christians the union of humanity and divinity in a universal salvation narrative. But, the figure is complex and ambiguous from the start, enmeshed in translation, staging, miming, disguises, and evasions. "Ecce homo!" can, indeed must, be read ironically by "post-Christians" and other post-humanists as "Behold the man, the figure of humanity (Latin), the sign of the same (the Greek tones of homo-), indeed, the Sacred Image of the Same, but also the original mime, the actor of a history that mocks especially the recurrent tales that insist that 'man makes himself' in the deathly onanistic nightdream of coherent wholeness and correct vision."

BUT, "AIN'T I A WOMAN?"

> Well, children, whar dar is so much racket der must be something out o' kilter. I tink dat 'twixt de niggers of de Souf and de women at de Norf all a talkin 'bout rights, de white men will be in a fix pretty soon. But what's all dis here talkin' 'bout? Dat man ober dar say dat women needs to be helped into carriages, and lifted ober ditches, and to have de best places—and ain't I a woman? Look at me! Look at my arm! . . . I have plowed and planted and gathered into barns, and no man could head me—and ain't I a woman? I could work as much as any man (when I could get it), and bear de lash as well—and ain't I a woman? I have borne five children and I seen 'em mos all sold off into slavery, and when I cried with a mother's grief, none but Jesus hear—and ain't I a woman?[2]

Sojourner Truth is perhaps less far from Isaiash's spine-tingling prophecy than was Jesus. How might a modern John, or Johanna, stage her claim to be—as a black woman, mother, and former slave—the Son of Man, the fulfillment of the promise to unite the whole people under a common sign? What kind of sign is Sojourner Truth—forcibly transported, without a home, without a proper name, unincorporated in the discourses of (white) womanhood, raped by her owner, forcibly mated with another slave, robbed of her children, and doubted even in the anatomy of her body? A powerful speaker for feminism and abolitionism, Sojourner Truth's famous lines from her 1851 speech in Akron, Ohio, evoke the themes of the suffering servant in order to claim the status of humanity for the shockingly inappropriate/d figure[3] of New World black womanhood, the bearer of the promise of humanity for womanhood in general, and indeed, the bearer of the promise of humanity also for men. Called by a religious vision, the woman received her final names directly from her God when she left her home in New

York City in 1843 for the road to preach her own unique gospel. Born a slave around 1797 in Ulster County, New York, her Dutch master named her Isabella Baumfree. "When I left the house of bondage I left everything behind. I wa'n't goin' to keep nothin' of Egypt on me, an' so I went to the Lord an' asked him to give me a new name."[4] And Sojourner Truth emerged from her second birth a prophet and a scourge.

Sojourner Truth showed up repeatedly at women's suffrage and abolitionist meetings over the last half of the nineteenth century. She delivered her most famous speech at the women's rights convention in Ohio in 1851 in answer to white male antisuffrage provocateurs who threatened to disrupt the meeting. In another exchange, she took on the problem of the gender of Jesus—whose manhood had been used by a heckler, a clergyman, to argue against women's rights. Sojourner Truth noted succinctly that man had nothing to do with Jesus; he came from God and a woman. Pilate was not this vagrant preacher's unwilling and evasive judge; but another man authorized by the hegemonic powers of his civilization stood in for him. This free white man acted far more assertively than had the colonial bureaucrat of the Roman Empire, whose wife's dreams had troubled him about his queer prisoner.[5] Pilate's ready surrogate, an irate white male physician, spoke out in protest of her speaking, demanding that she prove she was a woman by showing her breasts to the *women* in the audience. Difference (understood as the divisive marks of authenticity) was reduced to anatomy;[6] but even more to the point, the doctor's demand articulated the racist/sexist logic that made the very flesh of the black person in the New World indecipherable, doubtful, out of place, confounding—ungrammatical.[7] Remember that Trinh Minh-ha, from a different diaspora over a hundred years later, wrote, "Perhaps, for those of us who have never known what life in a vernacular culture is/was and are unable to imagine what it can be/could have been, gender simply does not exist otherwise than grammatically in language."[8] Truth's speech was out of place, dubious doubly; she was female and black; no, that's wrong—she was a black female, a black woman, not a coherent substance with two or more attributes, but an oxymoronic singularity who stood for an entire excluded and dangerously promising humanity. The language of Sojourner Truth's body was as electrifying as the language of her speech. And both were enmeshed in cascading questions about origins, authenticity, and generality or universality. This Truth is a figure of nonoriginality, but s/he is not Derridean. S/he is Trinhian, or maybe Wittigian, and the difference matters.[9]

When I began to sketch the outlines of this essay, I looked for versions of the story of Sojourner Truth, and I found them written and rewritten in a long list of nineteenth-century and contemporary feminist texts.[10] Her famous speech, transcribed by a white abolitionist—*Ain't I a Woman?*—adorns posters in women's studies offices and women's centers across the United States. These lines seem to stand for something that unifies "women," but what exactly, especially in view of feminism's excavation of the terrible edifice of "woman" in Western patriarchal language and systems of representation—the one who can never be a subject, who is plot space, matrix, ground, screen for the act of man? Why does her *question* have more power for feminist theory 150 years later than any number of affirmative and declarative sentences? What is it about this figure, whose hard name signifies someone who could never be at home, for whom truth was displacement from home, that compels retelling and rehearing her story? What kind of history might Sojourner Truth inhabit?

For me, one answer to that question lies in Sojourner Truth's power to figure a collective humanity without constructing the cosmic closure of the unmarked category. Quite the opposite, her body, names, and speech—their forms, contents, and articulations—may be read to hold promise for a never-settled universal, a common language that makes compelling claims on each of us collectively and personally, precisely through their radical specificity, in other words, through the displacements and resistances to unmarked identity precisely as the means to claiming the status of "the human." The essential Truth would not settle down; that *was* her specificity. S/he was not everyman; s/he was inappropriate/d. This is a "postmodern" reading from some points of view, and it is surely not the only possible reading of her story. But, it is one that I hope to convince the reader is at the heart of the inter- and multi-cultural feminist theory in our time. In Teresa de Lauretis's terms, this reading is not so much postmodern or poststructuralist, as it is specifically enabled by feminist theory:

> That, I will argue, is precisely where the particular discursive and epistemological character of feminist theory resides: its being at once inside its own social and discursive determinations, and yet also outside and excessive to them. This recognition marks a further moment in feminist theory, its current stage of reconceptualization and elaborations of new terms; a reconceptualization of the subject as shifting and multiply organized across variable axes of difference; a rethinking of the relations between forms of oppression and modes of resistance and agency, and between practices of writing and modes of formal understanding—of doing theory; an emerging redefinition of marginality as location, of identity as disidentification.... I will use the term feminist theory, like the term consciousness or subject, in the singular as referring to a process of understanding that is premised on the historical specificity and the simultaneous, if often contradictory, presence of those differences in each of its instances and practices....[11]

Let us look at the mechanisms of Sojourner Truth's exclusions from the spaces of unmarked universality (i.e., exclusion from "the human") in modern white patriarchal discourse in order to see better how she seized her body and speech to turn "difference" into an organon for placing the painful realities and practices of de-construction, dis-identification, and dis-memberment in the service of a newly articulated humanity. Access to this humanity will be predicated on a subject-making discipline hinted at by Trinh:

> The difficulties appear perhaps less insurmountable only as I/i suc-
> ceed in making a distinction between difference reduced to identity-
> authenticity and difference understood also as critical difference from my-
> self.... Difference in such an insituable context is *that which undermines*
> *the very idea of identity,* deferring to infinity the layers whose totality forms
> "I." ... If feminism is set forth as a demystifying force, then it will have to
> question thoroughly the belief in its own identity.[12]

Hazel Carby clarified how in the New World, and specifically in the United States, black women were not constituted as "woman," as white women were.[13] Instead, black women were constituted simultaneously racially and sexually—as marked female (animal, sexualized, and without rights), but not as woman (human, potential wife, conduit for the name of the father)—in a specific institution, slavery, that excluded them from "culture" defined as the circulation of signs through the system of marriage. If kinship vested men with rights in women that they did not have in themselves, slavery abolished kinship for one group in a legal discourse that produced whole groups of people as alienable property.[14] MacKinnon defined woman as an imaginary figure, the object of another's desire, made real.[15] The "imaginary" figures made real in slave discourse were objects in another sense that made them different from either the Marxist figure of the alienated laborer or the "unmodified" feminist figure of the object of desire. Free women in U.S. white patriarchy were exchanged in a system that oppressed them, but white women *inherited* black women and men. As Hurtado noted, in the nineteenth century prominent white feminists were *married* to white men, while black feminists were *owned* by white men. In a racist patriarchy, white men's "need" for racially "pure" offspring positioned free and unfree women in incompatible, asymmetrical symbolic and social spaces.[16]

The female slave was marked with these differences in a most literal fashion—the flesh was turned inside out, "add[ing] a lexical dimension to the narratives of woman in culture and society."[17] These differences did not end with formal emancipation; they have had definitive consequences into the late twentieth century and will continue to do so until racism as a founding institution of the New World is ended. Spillers called these founding relations of captivity and literal mutilation "an American grammar" (68). Under conditions of the New World conquest, of slavery, and of their consequences up to the present, "the lexis of reproduction, desire, naming, mothering, fathering, etc. [are] all thrown into extreme crisis" (76). "Gendering, in its coeval reference to African-American women, *insinuates* an implicit and unresolved puzzle both within current feminist discourse *and* within those discursive communities that investigate the problematics of culture" (78).

Spillers foregrounded the point that free men and women inherited their *name* from the father, who in turn had rights in his minor children and wife that they did not have in themselves, but he did not own them in the full sense of alienable property. Unfree men and women inherited their *condition* from their mother, who in turn specifically did not control her children. They had no *name* in the sense theorized by Lévi-Strauss or Lacan. Slave mothers could not transmit a name; they could not be wives; they were outside the system of marriage exchange. Slaves were unpositioned, unfixed, in a system of names; they were, specifically, unlocated and so disposable. In these discursive frames, white women were not legally or symbolically *fully* human; slaves were not legally or symbolically human *at all*. "In this absence from a subject position, the captured sexualities provide a physical and biological expression of 'otherness' " (67). To give birth (unfreely) to the heirs of property is not the same thing as to give birth (unfreely) to property.[18]

This little difference is part of the reason that "reproductive rights" for women of color in the United States prominently hinge on comprehensive control of children—for example, their freedom from destruction through lynching, imprisonment, infant mortality, forced pregnancy, coercive sterilization, inadequate housing, racist education, drug addiction, drug wars, and military wars.[19] For American white women the concept of property in the self, the ownership of one's own body, in relation to reproductive freedom, has more readily focused on the field of events around conception, pregnancy, abortion, and birth because the system of white patriarchy turned on the control of legitimate children and the consequent constitution of white females as women. To have or not have children then becomes literally a subject-defining choice for such women. Black women specifically—and the women subjected to the conquest of the New World in general—faced a broader social field of reproductive unfreedom, in which their children did not inherit the status of human in the founding hegemonic discourses of U.S. society. The problem of the black mother in this context is not simply her own status as subject, but also the status of her children and her sexual partners, male and female. Small wonder that the image of uplifting the race and the refusal of the categorical separation of men and women—without flinching from an analysis of colored and white sexist oppression—have been prominent in New World black feminist discourse.[20]

The positionings of African-American women are not the same as those of other women of color; each condition of oppression requires specific analysis that both refuses the separations and insists on the non-identities of race, sex, sexuality, and class. These matters make starkly clear why an adequate feminist theory of gender must *simultaneously* be a theory of racial and sexual difference in specific historical conditions of production and reproduction. They also make clear why a theory and practice of sisterhood cannot be grounded in shared positionings in a gender system and the cross-cultural structural antagonism between coherent categories called women and men. Finally, they make clear why feminist theory produced by women of color has constructed alternative discourses of womanhood that disrupt the humanisms of many Western discursive traditions. "[I]t is our task to make a place for this different social subject. In so doing we are less interested in joining the ranks of gendered femaleness than gaining the *insurgent* ground as female social subject. Actually *claiming* the monstrosity of a female with the potential to 'name,' . . . 'Sapphire' might rewrite after all a radically different text of female empowerment."[21] And, perhaps, of empowerment of the problematic category of "humanity."

While contributing fundamentally to the breakup of any master subject location, the politics of "difference" emerging from this and other complex reconstructings of concepts of social subjectivity and their associated writing practices is deeply opposed to leveling relativisms. Non-feminist poststructuralist theory in the human sciences has tended to identify the breakup of "coherent" or masterful subjectivity as the "death of the subject." Like others in newly *unstably* subjugated positions, many feminists resist this formulation of the project and question its emergence at just the moment when raced/sexed/colonized speakers begin "for the first time," to claim, that is, with an "originary" authority, to represent themselves in institutionalized publishing practices and other kinds of self-constituting practice. Feminist deconstructions of the "subject" have been fundamental, and they are not nostalgic for masterful coherence. Instead, necessarily political accounts of constructed embodiments, like feminist theories of gendered racial subjectivities, have to take affirmative *and* critical account of emergent, differentiating, self-representing, contradictory social subjectivities, with their claims on action, knowledge, and belief. The point involves the commitment to transformative social change, the moment of hope embedded in feminist theories of gender and other emergent discourses about the breakup of masterful subjectivity and the emergence of inappropriate/d others.

"Alterity" and "difference" are precisely what "gender" is "grammatically" about, a fact that constitutes feminism as a politics defined by its fields of contestation and repeated refusals of master theories. "Gender" was developed as a category to explore what counts as a "woman," to problematize the previously taken for granted, to reconstitute what counts as "human." If feminist theories of gender followed from Simone de Beauvoir's thesis that one is not born a woman, with all the consequences of that insight, in the light of Marxism and psychoanalysis (and critiques of racist and colonial discourse), for understanding that any finally coherent subject is a fantasy, and that personal and collective identity is precariously and constantly socially reconstituted,[22] then the title of bell hooks' provocative 1981 book, echoing Sojourner Truth, *Ain't I a Woman*, bristles with irony, as the identity of "woman" is both claimed and deconstructed simultaneously. This is a woman worthy of Isaiah's prophecy, slightly amended:

> S/he was despised and rejected by men; a wo/man of sorrows, acquainted with grief, and as one from whom men hide their faces s/he was despised, and we esteemed him/her not.... As many were astonished at him/her—his/her appearance was so marred, beyond human semblance ... so shall s/he startle many nations.

This decidedly unwomanly Truth has a chance to refigure a nongeneric, nonoriginal humanity after the breakup of the discourses of Eurocentric humanism.

However, we cannot leave Sojourner Truth's story without looking more closely at the transcription of the famous *Ain't I a Woman* speech delivered in Akron in 1851. That written text represents Truth's speech in the white abolitionist's imagined idiolect of The Slave, the supposedly archetypical black plantation slave of the South. The transcription does not provide a southern Afro-American English that any linguist, much less actual speaker, would claim. But it *is* the falsely specific, imagined language that represented the "universal" language of slaves to the literate abolitionist public, and this is the language that has come down to us as Sojourner Truth's "authentic" words. This counterfeit language, undifferentiated into the many Englishes spoken in the New World, reminds us of a hostile notion of difference, one that sneaks the masterful unmarked categories in through the back door in the *guise* of the specific, which is made to be not disruptive or deconstructive, but typical. The undifferentiated black slave could figure for a humanist abolitionist discourse, and its descendants on the walls of women's studies offices, an ideal type, a victim (hero), a kind of plot space for the abolitionists' actions, a special human, not one that could bind up the whole people through her unremitting figuring of critical difference—that is, not an unruly agent preaching her own unique gospel of displacement as the ground of connection.

To reinforce the point, this particular former slave was not southern. She was born in New York and owned by a Dutchman. As a young girl, she was sold with some sheep to a Yankee farmer who beat her for not understanding English.[23] Sojourner Truth as an adult almost certainly spoke an Afro–Dutch English peculiar to a region that was once New Amsterdam. "She dictated her autobiography to a white friend and lived by selling it at lectures."[24] Other available transcriptions of her speeches are printed in "standard" late-twentieth-century American English; perhaps this language seems less racist, more "normal" to hearers who want to forget the diasporas that populated the New World, while making one of its figures into a "typical" hero. A modern transcription/invention of Sojourner Truth's speeches has put them into Afro-Dutch English; her famous question retroubles the car, "Ar'n't I a woman?"[25] The change in the shape of the words makes us rethink her story, the grammar of her body and life. The difference matters.

One nineteenth-century, friendly reporter decided he could not put Truth's words into writing at all: "She spoke but a few minutes. To report her words would have been impossible. As well attempt to report the seven apocalyptic thunders."[26] He went on, in fact, to transcribe/reconstruct her presentation, which included these often-quoted lines:

> When I was a slave away down there in New York [was New York *down* for Sojourner Truth?!], and there was some particularly bad work to be done, some colored woman was sure to be called upon to do it. And when I heard that man talking away there as he did almost a whole hour, I said to myself, here's one spot of work sure that's fit for colored folks to clean up after.[27]

Perhaps what most needs cleaning up here is an inability to hear Sojourner Truth's language, to face her specificity, to acknowledge her, but *not* as the voice of the seven apocalyptic thunders. Instead, perhaps we need to see her as the Afro–Dutch–English New World itinerant preacher whose disruptive and risk-taking practice led her "to leave the house of bondage," to leave the subject-making (and humanist) dynamics of master and slave, and seek new names in a dangerous world. This sojourner's truth offers an inherently unfinished but potent reply to Pilate's skeptical query—"What is truth?" She is one of Gloria Anzaldúa's *mestizas*,[28] speaking the unrecognized hyphenated languages, living in the borderlands of history and consciousness where crossings are never safe and names never original.

I promised to read Sojourner Truth, like Jesus, as a trickster figure, a shape changer, who might trouble our notions—all of them: classical, biblical, scientific, modernist, postmodernist, and feminist—of "the human," while making us remember why we cannot not want this problematic universal. Pilate's words went through cascades of transcriptions, inventions, and translations. The "Ecce homo!" was probably never spoken. But, no matter how they may have originated, these lines in a play about what counts as humanity, about humanity's possible stories, were from the beginning implicated in permanent translation and reinvention. The same thing is true of Sojourner Truth's affirmative question, "Ain't/Ar'n't I a (wo)man?" These were tricksters, forcing by their constant displacements, a reconstruction of founding stories, of any possible home. "We, lesbian, *mestiza*, inappropriate/d other are all terms for that excessive critical position which I have attempted to tease out and rearticulate from various texts of contemporary feminism: a position attained through practices of political and personal displacement across boundaries between sociosexual identities and communities, between bodies and discourses, by what I like to call the "eccentric subject."[29] Such excessive and mobile figures can never ground what used to be called "a fully human community." That community turned out to belong only to the masters. However, these eccentric subjects can call us to account for our imagined humanity, whose parts are always articulated through translation. History can have another shape, articulated through differences that matter.

NOTES

This paper was originally presented at the American Anthropological Association meetings, Washington, D.C., 19 November 1989. Its rhetorical shuttling between the genres of scholarly writing and religious speech is inspired by, and dedicated to, Cornel West. Thanks to grants from the Academic Senate of the University of California at Santa Cruz.

1. Thanks to Gary Lease for biblical guidance.
2. Quoted in bell hooks, *Ain't I a Woman: Black Women and Feminism* (Boston, Mass.: South End Press, 1981), p. 160.
3. I borrow Trinh's powerful sign, an impossible figure, the inappropriate/d other. Trinh T. Minh-ha, "She, the Inappropriate/d Other," *Discourse*, 8 (1986–87).
4. Gerda Lerner, in *Black Women in White America: A Documentary History*, edited by Gerda Lerner (New York: Random House, 1973), pp. 370–75.
5. Matthew 27. 19.
6. Trinh T. Minh-ha, *Woman, Native, Other: Writing, Postcoloniality, and Feminism* (Bloomington: Indiana University Press, 1989).
7. Hortense Spillers, "Mama's Baby, Papa's Maybe: An American Grammar Book," *Diacritics*, 17, 2 (1987), pp. 65–81.

8. Trinh T. Minh-ha, *Woman, Native, Other,* p. 114.
9. I am using "matter" in the way suggested by Judith Butler in her work in progress, *Bodies That Matter.* See also Monique Wittig, *The Lesbian Body,* translated by David LeVay (New York: Avon, 1975). The marked bodies and subjects theorized by Trinh, Butler, and Wittig evacuate precisely the heterosexist and racist idealism-materialism binary that has ruled in the generic Western philosophical tradition. The feminist theorists might claim a siblingship to Derrida here, but not a relation of derivation or identity.
10. A sample: bell hooks, *Ain't I a Woman;* Trinh T. Minh-ha, *Woman, Native, Other;* Angela Davis, *Women, Race, and Class* (New York: Random House, 1981); Gerda Lerner, *Black Women;* Paula Giddings, *When and Where I Enter: The Impact of Black Women on Race and Sex in America* (New York: Bantam Books, 1984); Bettina Aptheker, *Woman's Legacy: Essays on Race, Sex, and Class in American History* (Amherst: University of Massachusetts Press, 1982); Olive Gilbert, *Narrative of Sojourner Truth, a Northern Slave* (Battle Creek, Mich.: Review and Herald Office, 1884; reissued New York: Arno Press, 1968); Harriet Carter, "Sojourner Truth," *Chautauquan,* 7 (May 1889); Lillie B. Wyman, "Sojourner Truth," in *New England Magazine* (March 1901); Eleanor Flexner, *Century of Struggle: The Woman's Rights Movement in the United States* (Cambridge, Mass.: Harvard University Press, 1959); Edith Blicksilver, "Speech of Woman's Suffrage," in *The Ethnic American Woman* (Dubuque, Iowa: Kendall/Hunt, 1978). p. 335; Hertha Pauli, *Her Name Was Sojourner Truth* (New York: Appleton-Century-Crofts, 1962).
11. Teresa de Lauretis, "Eccentric Subjects," in *Feminist Studies,* 16 (Spring 1990), p. 116.
12. Trinh T. Minh-ha, *Woman, Native, Other,* pp. 89, 96.
13. Hazel V. Carby, *Reconstructing Womanhood: The Emergence of the Afro-American Woman Novelist* (New York: Oxford University Press, 1987).
14. Hortense Spillers, "Mama's Baby."
15. Catharine MacKinnon, "Feminism, Marxism, Method, and the State: An Agenda for Theory," *Signs,* 7, 3 (1982), pp. 515–44.
16. Aida Hurtado, "Relating to Privilege: Seduction and Rejection in the Subordination of White Women and Women of Color," *Signs,* 14, 4 (1989), pp. 833–55, 841.
17. Hortense Spillers, "Mama's Baby," pp. 67–68.
18. Hazel V. Carby, *Reconstructing Womanhood,* p. 53.
19. Aida Hurtado, "Relating to Privilege," p. 853.
20. Hazel V. Carby, *Reconstructing Womanhood,* pp. 6–7; bell hooks, *Ain't I a Woman;* bell hooks, *Feminist Theory: From Margin to Center* (Boston, Mass.: South End Press, 1984).
21. Hortense Spillers, "Mama's Baby," p. 80.
22. Rosalind Coward, *Patriarchal Precedents: Sexuality and Social Relations* (London: Routledge and Kegan Paul, 1983), p. 265.
23. Gerda Lerner, *Black Women,* p. 371.
24. Ibid., p. 372; Olive Gilbert, *Narrative of Sojourner.*
25. Edith Blicksilver, "Speech."
26. Quoted in Bettina Aptheker, *Woman's Legacy,* p. 34.
27. Ibid.
28. Gloria Anzaldúa, *Borderlands/La Frontera* (San Francisco: Spinsters, 1987).
29. Teresa de Lauretis, *Feminist Studies,* p. 145.

3

THE PROMISES OF MONSTERS:
A REGENERATIVE POLITICS FOR
INAPPROPRIATE/D OTHERS

If primates have a sense of humor, there is no reason why intellectuals may not share in it.

—(Plank, 1989)

I. A BIOPOLITICS OF ARTIFACTUAL REPRODUCTION

"The Promises of Monsters" will be a mapping exercise and travelogue through mindscapes and landscapes of what may count as nature in certain local/global struggles. These contests are situated in a strange, allochronic time—the time of myself and my readers in the last decade of the second Christian millenium—and in a foreign, allotopic place—the womb of a pregnant monster, here, where we are reading and writing. The purpose of this excursion is to write theory; i.e., to produce a patterned vision of how to move and what to fear in the topography of an impossible but all-too-real present, in order to find an absent, but perhaps possible, other present. I do not seek the address of some full presence; reluctantly, I know better. Like Christian in *Pilgrim's Progress*, however, I am committed to skirting the slough of despond and the parasite-infested swamps of nowhere to reach more salubrious environs.[1] The theory is meant to orient, to provide the roughest sketch for travel, by means of moving within and through a relentless artifactualism, which forbids any direct si(gh)tings of nature, to a science fictional, speculative factual, SF place called, simply, elsewhere. At least for those whom this essay addresses, "nature" outside artifactualism is not so much elsewhere as nowhere, a different matter altogether. Indeed, a reflexive artifactualism offers serious political and analytical hope. This essay's theory is modest. Not a systematic overview, it is a little siting device in a long line of such craft tools. Such sighting devices have been known to reposition worlds for their devotees—and for their opponents. Optical instruments are subject-shifters. Goddess knows, the subject is being changed relentlessly in the late twentieth century.

My diminutive theory's optical features are set to produce not effects of distance, but effects of connection, of embodiment, and of responsibility for an imagined elsewhere that we may yet learn to see and build here. I have high stakes in reclaiming vision from the technopornographers, those theorists of minds, bodies, and planets who insist effectively—i.e., in practice—that sight is the sense made to realize the fantasies of the phallocrats.[2] I think sight can be remade for the activists and advocates engaged in fitting political filters to see the world in the hues of red, green, and ultraviolet, i.e., from the perspectives of a still possible socialism, feminist and anti-racist environmentalism, and science for the people. I take as a self-evident premise that "science is culture."[3] Rooted in that premise, this essay is a contribution to the heterogeneous and very lively contemporary discourse of science studies *as* cultural studies. Of course, what science, culture, or nature—and their "studies"—might mean is far less self-evident.

Nature is for me, and I venture for many of us who are planetary fetuses gestating in the amniotic effluvia of terminal industrialism,[4] one of those impossible things characterized by Gayatri Spivak as that which we cannot not desire. Excruciatingly conscious of nature's discursive constitution as "other" in the histories of colonialism, racism, sexism, and class domination of many kinds, we nonetheless find in this problematic, ethno-specific, long-lived, and mobile concept something we cannot do without, but can never "have." We must find another relationship to nature besides reification and possession. Perhaps to give confidence in its essential reality, immense resources have been expended to stabilize and materialize nature, to police its/her boundaries. Such expenditures have had disappointing results. Efforts to travel into "nature" become tourist excursions that remind the voyager of the price of such displacements—one pays to see fun-house reflections of oneself. Efforts to preserve "nature" in parks remain fatally troubled by the ineradicable mark of the founding explusion of those who used to live there, not as innocents in a garden, but as people for whom the categories of nature and culture were not the salient ones. Expensive projects to collect "nature's" diversity and bank it seem to produce debased coin, impoverished seed, and dusty relics. As the banks hypertrophy, the nature that feeds the storehouses "disappears." The World Bank's record on environmental destruction is exemplary in this regard. Finally, the projects for representing and enforcing human "nature" are famous for their imperializing essences, most recently reincarnated in the Human Genome Project.

So, nature is not a physical place to which one can go, nor a treasure to fence in or bank, nor an essence to be saved or violated. Nature is not hidden and so does not need to be unveiled. Nature is not a text to be read in the codes of mathematics and biomedicine. It is not the "other" who offers origin, replenishment, and service. Neither mother, nurse, nor slave, nature is not matrix, resource, or tool for the reproduction of man.

Nature is, however, a *topos*, a place, in the sense of a rhetorician's place or topic for consideration of common themes; nature is, strictly, a commonplace. We turn to this topic to order our discourse, to compose our memory. As a topic in this sense, nature also reminds us that in seventeenth-century English the "topick gods" were the local gods, the gods specific to places and peoples. We need these spirits, rhetorically if we can't have them any other way. We need them in order to reinhabit, precisely, *common* places—locations that are widely shared, inescapably local, worldly, enspirited; i.e., topical. In this sense, nature is the place to rebuild public culture.[5] Nature is also a *trópos*, a trope. It is figure, construction, artifact, movement, displacement. Nature cannot pre-exist its construction. This construction is based on a particular kind of move—a *trópos* or "turn." Faithful to the Greek, as *trópos* nature is about turning. Troping, we turn to nature as if to the earth, to the primal stuff—geotropic, physiotropic. Topically, we travel toward the earth, a commonplace. In discoursing on nature, we turn from Plato and his heliotropic son's blinding star to see something else, another kind of figure. I do not turn from vision, but I do seek something other than enlightenment in these sightings of science studies as cultural studies. Nature is a topic of public discourse on which much turns, even the earth.

In this essay's journey toward elsewhere, I have promised to trope nature through a relentless artifactualism, but what does artifactualism mean here? First, it means that nature for us is *made*, as both fiction and fact. If organisms are natural objects, it is crucial to remember that organisms are not born; they are made in world-changing technoscientific practices by particular collective actors in particular times and places. In the belly of the local/global monster in which I am gestating, often called the postmodern world,[6] global technology appears to *denature* everything, to make everything a malleable matter of strategic decisions and mobile production and reproduction processes (Hayles, 1990). Technological decontextualization is ordinary experience for hundreds of millions if not billions of human beings, as well as other organisms. I suggest that this is not a *denaturing* so much as a *particular production* of nature. The preoccupation with productionism that has characterized so much parochial Western discourse and practice seems to have hypertrophied into something quite marvelous: the whole world is remade in the image of commodity production.[7]

How, in the face of this marvel, can I seriously insist that to see nature as artifactual is an *oppositional*, or better, a *differential* siting?[8] Is the insistence that nature *is* artifactual not more evidence of the extremity of the violation of a nature outside and other to the arrogant ravages of our technophilic civilization, which, after all, we were taught began with the heliotropisms of enlightenment projects to dominate nature with blinding light focused by optical technology?[9] Haven't ecofeminists and other multicultural and intercultural radicals begun to convince us that nature is precisely *not* to be seen in the guise of the Eurocentric productionism and anthropocentrism that have threatened to reproduce, literally, all the world in the deadly image of the Same?

I think the answer to this serious political and analytical question lies in two related turns: (1) unblinding ourselves from the sun-worshiping stories about the history of science and technology as paradigms of rationalism; and (2) refiguring the actors in the construction of the ethnospecific categories of nature *and* culture. The actors are not all "us." If the world exists for us as "nature," this designates a kind of relationship, an achievement among many actors, not all of them human, not all of them organic, not all of them technological.[10] In its scientific embodiments as well as in other forms, nature is made, but not entirely by humans; it is a co-construction among humans and non-humans. This is a very different vision from the postmodernist observation that all the world is denatured and reproduced in images or replicated in copies. That specific kind of violent and reductive artifactualism, in the form of a hyper-productionism actually practiced widely throughout the planet, becomes *contestable* in theory and other kinds of praxis, without recourse to a resurgent transcendental naturalism. Hyper-productionism refuses the witty agency of all the actors but One; that is a dangerous strategy— for everybody. But transcendental naturalism also refuses a world full of cacophonous agencies and settles for a mirror image sameness that only pretends to difference. The commonplace nature I seek, a public culture, has many houses with many inhabitants which/who can refigure the earth. Perhaps those other actors/actants, the ones who are not human, are our topick gods, organic and inorganic.[11]

It is this barely admissible recognition of the odd sorts of agents and actors which/whom we must admit to the narrative of collective life, including nature, that simultaneously, first, turns us decisively away from enlightenment-derived modern and postmodern premises about nature and culture, the social and technical, science and society and, second, saves us from the deadly point of view of productionism. Productionism and its corollary, humanism, come down to the story line that "man makes everything, including himself, out of the world that can only be resource and potency to his project and active agency."[12] This productionism is about man the tool-maker and -user, whose highest technical production is himself; i.e., the story line of phallogocentrism. He gains access to this wondrous technology with a subject-constituting, self-deferring, and self-splitting entry into language, light, and law. Blinded by the sun, in thrall to the father, reproduced in the sacred image of the same, his reward is that he is self-born, an autotelic copy. That is the mythos of enlightenment transcendence.

Let us return briefly to my remark above that organisms are not born, but they are made. Besides troping on Simone de Beauvoir's observation that one is not born a woman, what work is this statement doing in this essay's effort to articulate a relentless differential/oppositional artifactualism? I wrote that organisms are made as objects of knowledge in world-changing practices of scientific discourse by particular and always collective actors in specific times and places. Let us look more closely at this claim with the aid of the concept of the apparatus of bodily production.[13] Organisms are *biological* embodiments; as natural-technical entities, they are not pre-existing plants, animals, protistes, etc., with boundaries already established and awaiting the right kind of instrument to note them correctly. Organisms emerge from a discursive process. Biology is a discourse, not the living world itself. But humans are not the only actors in the construction of the entities of any scientific discourse; machines (delegates that can produce surprises) and other partners (not "pre- or extra-discursive objects," but partners) are active constructors of natural scientific objects. Like other scientific bodies, organisms are not *ideological* constructions. The whole point about discursive construction has been that it is *not* about ideology. Always radically historically specific, always lively, bodies have a different kind of specificity and effectivity; and so they invite a different kind of engagement and intervention.

Elsewhere, I have used the term "material-semiotic actor" to highlight the object of knowledge as an active part of the apparatus of bodily production, without *ever* implying immediate presence of such objects or, what is the same thing, their final or unique determination of what can count as objective knowledge of a biological body at a particular historical juncture. Like Katie King's objects called "poems," sites of literary production where language also is an actor, bodies as objects of knowledge are material-semiotic generative nodes. Their boundaries materialize in social interaction among humans and non-humans, including the machines and other instruments that mediate exchanges at crucial interfaces and that function as delegates for other actors' functions and purposes. "Objects" like bodies do not pre-exist as such. Similarly, "nature" cannot pre-exist as such, but neither is its existence ideological. Nature is a commonplace and a powerful discursive construction, effected in the interactions among material-semiotic actors, human and not. The siting/sighting of such entities is not about disengaged discovery, but about mutual and usually unequal structuring, about taking risks, about delegating competences.[14]

The various contending biological bodies emerge at the intersection of biological research, writing, and publishing; medical and other business practices; cultural productions of all kinds, including available metaphors and narratives; and technology, such as the visualization technologies that bring color-enhanced killer T cells and intimate photographs of the developing fetus into high-gloss art books, as well as scientific reports. But also invited into that node of intersection is the analogue to the lively languages that actively intertwine in the production of literary value: the coyote and protean embodiments of a world as witty agent and actor. Perhaps our hopes for accountability for techno-biopolitics in the belly of the monster turn on revisioning the world as coding trickster with whom we must learn to converse. So while the late twentieth-century immune system, for example, is a construct of an elaborate apparatus of bodily production, neither the immune system nor any other of biology's world-changing bodies—like a virus or an ecosystem—is a ghostly fantasy. Coyote is not a ghost, merely a protean trickster.

This sketch of the artifactuality of nature and the apparatus of bodily production helps us toward another important point: the corporeality of theory. Overwhelmingly, theory is bodily, and theory is literal. Theory is not about matters distant from the lived body; quite the opposite. Theory is *anything* but disembodied. The fanciest statements about radical decontextualization as the historical form of nature in late capitalism are tropes for the embodiment, the production, the literalization of experience in that specific mode. This is not a question of reflection or correspondences, but of technology, where the social and the technical implode into each other. Experience is a semiotic process—a semiosis (de Lauretis, 1984). Lives are built; so we had best become good craftspeople with the other worldly actants in the story. There is a great deal of rebuilding to do, beginning with a little more surveying with the aid of optical devices fitted with red, green, and ultraviolet filters.

Repeatedly, this essay turns on figures of pregnancy and gestation. Zoe Sofia (1984) taught me that every technology is a reproductive technology. She and I have meant that literally; ways of life are at stake in the culture of science. I would, however, like to displace the terminology of reproduction with that of generation. Very rarely does anything really get *reproduced;* what's going on is much more polymorphous than that. Certainly people don't reproduce, unless they get themselves cloned, which will always be very expensive and risky, not to mention boring. Even technoscience must be made into the paradigmatic model not of closure, but of that which is contestable and contested. That involves knowing how the world's agents and actants work; how they/we/it come into the world, and how they/we/it are reformed. Science becomes the myth not of what escapes agency and responsibility in a realm above the fray, but rather of accountability and responsibility for translations and solidarities linking the cacophonous visions and visionary voices that characterize the knowledges of the marked bodies of history. Actors, as well as actants, come in many and wonderful forms. And best of all, "reproduction"—or less inaccurately, the generation of novel forms— need not be imagined in the stodgy bipolar terms of hominids.[15]

If the stories of hyper-productionism and enlightenment have been about the reproduction of the sacred image of the same, of the one true copy, mediated by the luminous technologies of compulsory heterosexuality and masculinist self-birthing, then the differential artifactualism I am trying to envision might issue in something else. Artifactualism is askew of productionism; the rays from my optical device diffract rather than reflect. These diffracting rays compose *interference* patterns, not reflecting images. The "issue" from this generative technology, the result of a monstrous[16] pregnancy, might be kin to Vietnamese-American filmmaker and feminist theorist Trinh Minh-ha's (1986/7b; 1989) "inappropriate/d others."[17] Designating the networks of multicultural, ethnic, racial, national, and sexual actors emerging since World War II, Trinh's phrase referred to the historical positioning of those who cannot adopt the mask of either "self" or "other" offered by previously dominant, modern Western narratives of identity and politics. To be "inappropriate/d" does not mean "not to be in relation with"—i.e., to be in a special reservation, with the status of the authentic, the untouched, in the allochronic and allotopic condition of innocence. Rather to be an "inappropriate/d other" means to be in critical, deconstructive relationality, in a diffracting rather than reflecting (ratio)nality—as the means of making potent connection that exceeds domination. To be inappropriate/d is not to fit in the *taxon,* to be dislocated from the available maps specifying kinds of actors and kinds of narratives, not to be originally fixed by difference. To be inappropriate/d is to be neither modern nor postmodern, but to insist on the *a*modern. Trinh was looking for a way to figure "difference" as a "critical difference within," and not as special taxonomic marks grounding difference as apartheid. She was writing about people; I wonder if the same observations might apply to humans and to both organic and technological non-humans.

The term "inappropriate/d others" can provoke rethinking social relationality within artifactual nature—which is, arguably, global nature in the 1990s. Trinh Minh-ha's metaphors suggest another geometry and optics for considering the relations of difference among people and among humans, other organims, and machines than hierarchical domination, incorporation of parts into wholes, paternalistic and colonialist protection, symbiotic fusion, antagonistic opposition, or instrumental production from resource. Her metaphors also suggest the hard intellectual, cultural, and political work these new geometries will require. If Western patriarchal narratives have told that the physical body issued from the first birth, while man was the product of the heliotropic second birth, perhaps a differential, diffracted feminist allegory might have the "inappropriate/d others" emerge from a third birth into an SF world called elsewhere—a place composed from interference patterns. Diffraction does not produce "the same" displaced, as reflection and refraction do. Diffraction is a mapping of interference, not of replication, reflection, or reproduction. A diffraction pattern does not map where differences appear, but rather maps where the *effects* of difference appear. Tropically, for the promises of monsters, the first invites the illusion of essential, fixed position, while the second trains us to more subtle vision. Science fiction is generically concerned with the interpenetration of boundaries between problematic selves and unexpected others and with the exploration of possible worlds in a context structured by transnational technoscience. The emerging social subjects called "inappropriate/d others" inhabit such worlds. SF—science fiction, speculative futures, science fantasy, speculative fiction—is an especially apt sign under which to conduct an inquiry into the artifactual as a reproductive technology that might issue in something other than the sacred image of the same, something inappropriate, unfitting, and so, maybe, inappropriated.

Within the belly of the monster, even inappropriate/d others seem to be interpellated—called through interruption—into a particular location that I have learned to call a cyborg subject position.[18] Let me continue this travelogue and inquiry into artifactualism with an illustrated lecture on the nature of cyborgs as they appear in recent advertisements in *Science*, the journal of the American Association for the Advancement

Fig. 3.1.

of Science. These ad figures remind us of the corporeality, the mundane materiality, and literality of theory. These commercial cyborg figures tell us what may count as nature in technoscience worlds. Above all, they show us the implosion of the technical, textual, organic, mythic, and political in the gravity wells of science in action. These figures are our companion monsters in the *Pilgrim's Progress* of this essay's travelogue.

Consider Figure 3.1, "A Few Words about Reproduction from a Leader in the Field," the advertising slogan for Logic General Corporation's software duplication system. The immediate visual and verbal impact insists on the absurdity of separating the technical, organic, mythic, textual, and political threads in the semiotic fabric of the ad and of the world in which this ad makes sense. Under the unliving, orange-to-yellow rainbow colors of the earth-sun logo of Logic General, the biological white rabbit has its (her? yet, sex and gender are not so settled in this reproductive system) back to us. It has its paws on a keyboard, that inertial, old-fashioned residue of the typewriter that lets our computers feel natural to us, user-friendly, as it were.[19] But the keyboard is misleading; no letters are transferred by a mechanical key to a waiting solid surface. The computer-user interface works differently. Even if she doesn't understand the implications of her lying keyboard, the white rabbit is in her natural home; she is fully artifactual in the most literal sense. Like fruit flies, yeast, transgenic mice, and the humble nematode worm, *Caenorhabditis elegans*,[20] this rabbit's evolutionary story transpires in the lab; the lab is its proper niche, its true habitat. Both material system and symbol for the measure of fecundity, this kind of rabbit occurs in no other nature than the lab, that preeminent scene of replication practices.

With Logic General, plainly, we are not in a biological laboratory. The organic rabbit peers at its image, but the image is not her reflection, indeed, *especially* not her reflection. This is not Lacan's world of mirrors; primary identification and maturing metaphoric substitution will be produced with other techniques, other writing technologies.[21] The white rabbit will be translated, her potencies and competences relocated radically. The guts of the computer produce another kind of visual product than distorted, self-birthing reflections. The simulated bunny peers out at us face first. It is she who locks her/its gaze with us. She, also, has her paws on a grid, one just barely reminiscent of a typewriter, but even more reminiscent of an older icon of technoscience—the Cartesian coordinate system that locates the world in the imaginary spaces of rational modernity. In her natural habitat, the virtual rabbit is on a grid that insists on the world as a game played on a chess-like board. This rabbit insists that the truly rational actors will replicate themselves in a virtual world where the best players will not be Man, though he may linger like the horse-drawn carriage that gave its form to the railroad car or the typewriter that gave its illusory shape to the computer interface. The functional privileged signifier in this system will not be so easily mistaken for any primate male's urinary and copulative organ. Metaphoric substitution and other circulations in the very material symbolic domain will be more likely to be effected by a competent mouse. The iffy femaleness of both of the rabbits, of course, gives no confidence that the new players other to Man will be women. More likely, the rabbit that is interpellated into the world in this non-mirror stage, this diffracting moment of subject constitution, will be literate in a quite different grammar of gender. *Both* the rabbits here are cyborgs—compounds of the organic, technical, mythic, textual, and political—and they call us into a world in which we may not wish to take shape, but through whose "Miry Slough" we might have to travel to get elsewhere. Logic General is into a very particular kind of *écriture*. The reproductive stakes in this text are future life forms and ways of life for humans and unhumans. "Call toll free" for "a few words about reproduction from an acknowledged leader in the field."

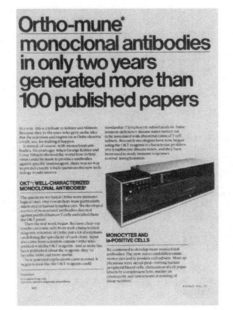

Fig. 3.2.

Ortho-mune*[TM]'s monoclonal antibodies expand our understanding of a cyborg subject's relation to the inscription technology that is the laboratory (Figure 3.2). In only two years, these fine monoclonals generated more than 100 published papers—higher than any rate of literary production by myself or any of my human colleagues in the human sciences. But this alarming rate of publication was achieved in 1982, and has surely been wholly surpassed by new generations of biotech mediators of literary replication. Never has theory been more literal, more bodily, more technically adept. Never has the collapse of the "modern" distinctions between the mythic, organic, technical, political, and textual into the gravity well, where the unlamented enlightenment transcendentals of Nature and Society also disappeared, been more evident.

LKB Electrophoresis Division has an evolutionary story to tell, a better, more complete one than has yet been told by physical anthropologists, paleontologists, or naturalists about the entities/actors/actants that structure niche space in an extra-laboratory world: "There are no missing links in MacroGene Workstation" (Figure 3.3). Full of promises, breaching the first of the ever-multiplying final frontiers, the prehistoric monster *Ichthyostega* crawls from the amniotic ocean into the future,

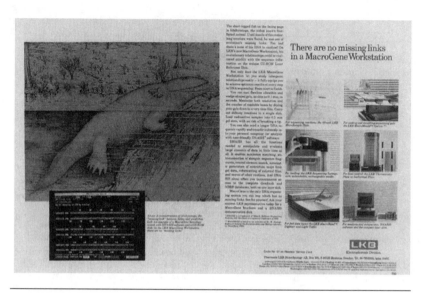

Fig. 3.3.

onto the dangerous but enticing dry land. Our no-longer-fish, not-yet-salamander will end up fully identified and separated, as man-in-space, finally disembodied, as did the hero of J. D. Bernal's fantasy in *The World, the Flesh, and the Devil*. But for now, occupying the zone between fishes and amphibians, *Ichthyostega* is firmly on the margins, those potent places where theory is best cultured. It behooves us, then, to join this heroic reconstructed beast with LKB, in order to trace out the transferences of competences—the metaphoric-material chain of substitutions—in this quite literal apparatus of bodily production. We are presented with a travel story, a *Pilgrim's Progress*, where there are no gaps, no "missing links." From the first non-original actor—the reconstructed *Ichthyostega*—to the final printout of the DNA homology search mediated by LKB's software and the many separating and writing machines pictured on the right side of the advertisement, the text promises to meet the fundamental desire of phallologocentrism for fullness and presence. From the crawling body in the Miry Sloughs of the narrative to the printed code, we are assured of full success—the compression of time into instantaneous and full access "to the complete GenBank . . . on one laser disk." Like Christian, we have conquered time and space, moving from entrapment in body to fulfillment in spirit, all in the everyday workspaces of the Electrophoresis Division, whose

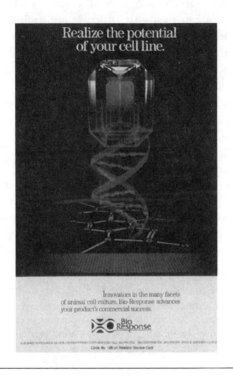

Fig. 3.4.

Hong Kong, Moscow, Antwerp, and Washington phone numbers are all provided. Electrophoresis: *pherein*—to bear or carry us relentlessly on.

Bio-Response, innovators in many facets of life's culture, interpellates the cyborg subject into the barely secularized, evangelical, Protestant Christianity that pervades American techno-culture: "Realize the potential of your cell line" (Figure 3.4). This ad addresses us directly. We are called into a salvation narrative, into history, into biotechnology, into our true natures: our cell line, ourselves, our successful product. We will testify to the efficacy of this culture system. Colored in the blues, purples, and ultraviolets of the sterilizing commercial rainbow—in which art, science, and business arch in lucrative grace—the virus-like crystalline shape mirrors the luminous crystals of New Age promises. Religion, science, and mysticism join easily in the facets of modern and postmodern commercial bio-response. The simultaneously promising and threatening crystal/virus unwinds its tail to reveal the language-like icon of the Central Dogma, the code structures of DNA that underlie all possible bodily response, all semiosis, all culture. Gem-like, the frozen, spiraling crystals of Bio-Response promise life itself. This is a jewel of great price—available from the Production Services office in Hayward, California. The imbrications of layered signifiers and signifieds forming cascading hierarchies of signs guide us through this mythic, organic, textual, technical, political icon.[22]

Finally, the advertisement from Vega Biotechnologies graphically shows us the final promise, "the link between science and tomorrow: Guaranteed. Pure." (Figure 3.5). The graph reiterates the ubiquitous grid system that is the signature and matrix, father and mother, of the modern world. The sharp peak is the climax of the search for certainty and utter clarity. But the diffracting apparatus of a monstrous artifactualism can perhaps interfere in this little family drama, reminding us that the modern world never existed and its fantastic guarantees are void. Both the organic and computer rabbits of Logic General might re-enter at this point to challenge all the passive voices of productionism. The oddly duplicated bunnies might resist their logical interpellation and instead hint at a neo-natalogy of inappropriate/d others, where the child will

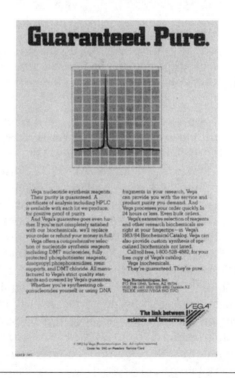

Fig. 3.5.

not be in the sacred image of the same. Shape-shifting, these interfering cyborgs might craft a diffracted logic of sameness and difference and utter a different word about reproduction, about the link between science and tomorrow, from collective actors in the field.

II. THE FOUR-SQUARE CYBORG: THROUGH ARTIFACTUALISM TO ELSEWHERE

It is time to travel, therefore, with a particular subset of shifted subjects, Cyborgs for Earthly Survival,[23] into the mindscapes and landscapes indicated at the beginning of this essay. To get through the artifactual to elsewhere, it would help to have a little travel machine that also functions as a map. Consequently, the rest of the "Promises of Monsters" will rely on an artificial device that generates meanings very noisily: A. J. Greimas's infamous semiotic square. The regions mapped by this clackety, structuralist meaning-making machine could never be mistaken for the transcendental realms of Nature or Society. Allied with Bruno Latour, I will put my structuralist engine to amodern purposes: this will not be a tale of the rational progress of science, in potential league with progressive politics, patiently unveiling a grounding nature, nor will it be a demonstration of the social construction of science and nature that locates all agency firmly on the side of humanity. Nor will the modern be superceded or infiltrated by the postmodern, because belief in something called the modern has itself been a mistake. Instead, the amodern refers to a view of the history of science as culture that insists on the absence of beginnings, enlightenments, and endings: the world has always been in the middle of things, in unruly and practical conversation, full of action and structured by a startling array of actants and of networking and unequal collectives. The much-criticized inability of structuralist devices to provide the narrative of diachronic history, of progress through time, will be my semiotic square's greatest virtue. The shape of my amodern history will have a different geometry, not of progress, but of permanent and multi-patterned interaction through which lives and worlds get built, human and unhuman. This Pilgrim's Progress is taking a monstrous turn.

I like my analytical technologies, which are unruly partners in discursive construction, delegates who have gotten into doing things on their own, to make a lot of noise, so that I don't forget all the circuits of competences, inherited conversations, and coalitions of human and unhuman actors that go into any semiotic excursions. The semiotic square, so subtle in the hands of a Fredric Jameson, will be rather more rigid and literal here (Greimas, 1966; Jameson, 1972). I only want it to keep

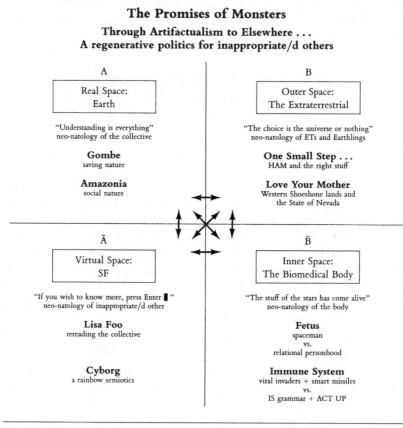

The Promises of Monsters
Through Artifactualism to Elsewhere . . .
A regenerative politics for inappropriate/d others

A	B
Real Space: **Earth**	**Outer Space:** **The Extraterrestrial**

"Understanding is everything"
neo-natology of the collective

"The choice is the universe or nothing"
neo-natology of ETs and Earthlings

Gombe
saving nature

One Small Step . . .
HAM and the right stuff

Amazonia
social nature

Love Your Mother
Western Shoshone lands and
the State of Nevada

Ā	B̄
Virtual Space: **SF**	**Inner Space:** **The Biomedical Body**

"If you wish to know more, press Enter ▌"
neo-natology of inappropriate/d other

"The stuff of the stars has come alive"
neo-natology of the body

Lisa Foo
rereading the collective

Fetus
spaceman
vs.
relational personhood

Cyborg
a rainbow semiotics

Immune System
viral invaders + smart missiles
vs.
IS grammar + ACT UP

Fig. 3.6.

four spaces in differential, relational separation, while I explore how certain local/global struggles for meanings and embodiments of nature are occurring within them. Almost a joke on "elementary structures of signification" ("Guaranteed. Pure."), the semiotic square in this essay nonetheless allows a contestable collective world to take shape for us out of structures of difference. The four regions through which we will move are A, Real Space or Earth; B, Outer Space or the Extraterrestrial; not-B, Inner Space or the Body; and finally, not-A, Virtual Space or the SF world oblique to the domains of the imaginary, the symbolic, and the real (Figure 3.6).

Somewhat unconventionally, we will move through the square clockwise to see what kinds of figures inhabit this exercise in science studies as cultural studies. In each of the first three quadrants of the square, I will begin with a popular image of nature and science that initially appears both compelling and friendly, but quickly becomes a sign of deep structures of domination. Then I will switch to a differential/oppositional image and practice that might promise something else. In the final quadrant, in virtual space at the end of the journey, we will meet a disturbing guide figure who promises information about psychic, historical, and bodily formations that issue, perhaps, from some other semiotic processes than the psychoanalytic in modern and postmodern guise. Directed by John Varley's (1986) story of that name, all we will have to do to follow this disquieting, amodern Beatrice will be to "Press Enter." Her job will be to instruct us in the neo-natology of inappropriate/d others. The goal of this journey is to show in each quadrant, and in the passage through the machine that generates them, metamorphoses and boundary shifts that give grounds for a scholarship and politics of hope in truly monstrous times. The pleasures promised here are not those libertarian masculinist fantasmics of the infinitely regressive practice of boundary violation and the accompanying *frisson* of brotherhood, but just maybe the pleasure of regeneration in less deadly, chiasmatic borderlands.[24] Without grounding origins and without history's illuminating and progressive tropisms, how might we map some semiotic possibilities for other topick gods and common places?

A. REAL SPACE: EARTH

In 1984, to mark nine years of underwriting the National Geographic Society's television specials, the Gulf Oil Corporation ran an advertisement entitled "Understanding Is Everything" (Figure 3.7). The ad referred to some of the most watched programs in the history of public television—the nature specials about Jane Goodall and the wild chimpanzees in Tanzania's Gombe National Park. Initially, the gently clasped hands of the ape and the young white woman seem to auger what the text proclaims—communication, trust, responsibility, and understanding across the gaps that have defined human existence in Nature and Society in "modern" Western narratives. Made ready by a scientific practice coded in terms of "years of patience," through a "spontaneous gesture of trust" *initiated by the animal,* Goodall metamorphoses in the ad copy from "Jane" to "Dr. Goodall." Here is a natural science, coded unmistakably feminine, to counter the instrumentalist excesses of a military-industrial-technoscience complex, where the code of science is stereotypically anthropocentric and masculine. The ad invites the viewer to forget Gulf's status as one of the Seven Sisters of big oil, ranking eighth among the Forbes 500 in 1980 (but acquired by Chevron by the

Fig. 3.7.

end of the decade's transnational capitalist restructuring). In response to the financial and political challenges mounted in the early 1970s by the Organization of Petroleum Exporting Countries (OPEC) and by ecological activism around the globe, by the late 1970s the scandal-ridden giant oil corporations had developed advertising strategies that presented themselves as the world's leading environmentalists—indeed, practically as the mothers of eco-feminism. There could be no better story than that of Jane Goodall and the chimpanzees for narrating the healing touch between nature and society, mediated by a science that produces full communication in a chain that leads innocently "from curiosity, to observation, to learning, to understanding."[25] Here is a story of blissful incorporation.

There is another repressed set of codes in the ad as well, that of race and imperialism, mediated by the dramas of gender and species, science and nature. In the National Geographic narrative, "Jane" entered the garden "alone" in 1960 to seek out "man's" closest relatives, to establish a knowing touch across gulfs of time. A natural family is at stake; the PBS specials document a kind of inter-species family therapy. Closing the distance between species through a patient discipline, where first the animals could only be known by their spoor and their calls, then by fleeting sightings, then finally by the animal's direct inviting touch, after which she could name them, "Jane" was admitted as "humanity's" delegate back into Eden. Society and nature had made peace; "modern science" and "nature" could co-exist. Jane/Dr. Goodall was represented almost as a new Adam, authorized to name not by God's creative hand, but by the animal's transformative touch. The people of Tanzania disappear in a story in which the actors are the anthropoid apes and a young British white woman engaged in a thoroughly modern sacred secular drama. The chimpanzees and Goodall are both enmeshed in stories of endangerment and salvation. In the post-World War II era the apes face biological extinction; the planet faces nuclear and ecological annihilation; and the West faces expulsion from its former colonial possessions. If only communication can be established, destruction can be averted. As Gulf Oil insists, "Our goal is to provoke curiosity about the world and the fragile complexity of its natural order; to satisfy that curiosity through observation and learning; to create an understanding of man's place in the ecological structure, and his responsibility to it—on the simple theory that no thinking person can share in the destruction of anything whose value he understands." Progress, rationality, and nature join in the great myth of modernity, which is so thoroughly threatened by a dozen looming apocalypses. A cross-species family romance promises to avert the threatened destruction.

Inaudible in the Gulf and National Geographic version, communication and understanding are to emerge in the communion between Jane/Dr. Goodall and the spontaneously trusting chimpanzee at just the historical moment when dozens of African nations are achieving their national independence, 15 in 1960 alone, the year Goodall set out for Gombe. Missing from the family romance are such beings as Tanzanians. African peoples seek to establish hegemony over the lands in which they live; to do that the stories of the natural presence of white colonists must be displaced, usually by extremely complex and dangerous nationalist stories. But in "Understanding Is Everything," the metonymic "spontaneous gesture of trust" from the animal hand to the white hand obliterates once again the invisible bodies of people of color who have never counted as able to represent humanity in Western iconography. The white hand will be the instrument for saving nature—and in the process be saved from a rupture with nature. Closing great gaps, the transcendentals of nature and society meet here in the metonymic figure of softly embracing hands from two worlds, whose innocent touch depends on the absence of the "other world," the "third world," where the drama actually transpires.

In the history of the life sciences, the great chain of being leading from "lower" to "higher" life forms has played a crucial part in the discursive construction of race as an object of knowledge and of racism as a living force. After World War II and the partial removal of explicit racism from evolutionary biology and physical anthropology, a good deal of racist and colonialist discourse remained projected onto the screen of "man's closest relatives," the anthropoid apes.[26] It is impossible to picture the entwined hands of a white woman and an African ape without evoking the history of racist iconography in biology and in European and American popular culture. The animal hand is metonymically the individual chimpanzee, all threatened species, the third world, peoples of color, Africa, the ecologically endangered earth—all firmly in the realm of Nature, all represented in the leathery hand folding around that of the white girl-woman under the Gulf sun logo shining on the Seven Sisters' commitment to science and nature. The spontaneous gesture of touch in the wilds of Tanzania authorizes a whole doctrine of representation. Jane, as Dr. Goodall, is empowered to speak for the chimpanzees. Science speaks for nature. Authorized by unforced touch, the dynamics of representation take over, ushering in the reign of freedom and communication. This is the structure of depoliticizing expert discourse, so critical to the mythic political structures of the "modern" world and to the mythic political despair of much "post-modernism," so undermined by fears about the breakdown of representation.[27] Unfortunately, representation, fraudulent or not, is a very resilient practice.

The clasping hands of the Gulf ad are semiotically similar to the elution peak in the Vega ad of Figure 3.5: "Guaranteed. Pure."; "Understanding Is Everything." There is no interruption in these stories of communication, progress, and salvation through science and technology. The story of Jane Goodall in Gombe, however, can be made to show its conditions of possibility; even in the footage of the National Geographic specials we see the young woman on a mountain top at night eating from a can of pork and beans, that sign of industrial civilization so crucial to the history of colonialism in Africa, as Orson Welles's voice-over speaks of the lonely quest for contact with nature! In one of Goodall's published accounts of the early days at Gombe, we learn that she and her mother, en route to the chimpanzee preserve, were stopped on the shores of Lake Tanganyika in the town of Kigoma, across from the no-longer-Belgian Congo, as *uhuru*, freedom, sounded across Africa. Goodall and her mother made 2000 spam sandwiches for fleeing Belgians before embarking for the "wilds of Tanzania" (Goodall, 1971, p. 27). It is also possible to reconstruct a history of Gombe as a research site in the 1970s. One of the points that stands out in this reconstruction is that people—research staff and their families,

African, European, and North American—considerably outnumbered chimpanzees during the years of most intense scientific work. Nature and Society met in one story; in another story, the structure of action and the actants take a different shape.

It is hard, however, to make the story of Jane Goodall and the wild chimpanzees shed its "modern" message about "saving nature," in both senses of nature as salvific and of the scientist speaking for and preserving nature in a drama of representation. Let us, therefore, leave this narrative for another colonized tropical spot in the Real/Earth quadrant in the semiotic square—Amazonia. Remembering that all colonized spots have, euphemistically stated, a special relation to nature, let us structure this story to tell something amodern about nature and society—and perhaps something more compatible with the survival of all the networked actants, human and unhuman. To tell this story we must disbelieve in both nature and society and resist their associated imperatives to represent, to reflect, to echo, to act as a ventriloquist for "the other." The main point is there will be no Adam—and no Jane—who gets to name all the beings in the garden. The reason is simple: there is no garden and never has been. No name and no touch is original. The question animating this diffracted narrative, this story based on little differences, is also simple: is there a consequential difference between a political semiotics of articulation and a political semiotics of representation?

The August 1990 issue of *Discover* magazine has a story entitled "Tech in the Jungle." A one and one-half page color photo of a Kayapó Indian, in indigenous dress and using a videocamera, dramatically accompanies the opening paragraphs. The caption tells us the man is "tap[ing] his tribesmen, who had gathered in the central Brazilian town of Altamira to protest plans for a hydroelectric dam on their territory" (Zimmer, 1990, 42–5). All the cues in the *Discover* article invite us to read this photo as the drama of the meeting of the "traditional" and the "modern," staged in this popular North American scientific publication for audiences who have a stake in maintaining belief in those categories. We have, however, as disbelieving members of those audiences, a different political, semiotic responsibility, one made easier by another publication, Susanna Hecht and Alexander Cockburn's *The Fate of the Forest* (1989; see also T. Turner, 1990) through which I propose to suggest articulations and solidarities with the *filming practice* of the Kayapó man, rather than to read the *photograph of him*, which will not be reproduced in this essay.[28]

In their book, which was deliberately packaged, published, and marketed in a format and in time for the 1989 December gift-giving season, a modest act of cultural politics not to be despised, Hecht and Cockburn have a central agenda. They insist on deconstructing the image of the tropical rain forest, especially Amazonia, as "Eden under glass." They do this in order to insist on locations of responsibility and empowerment in current conservation struggles, on the outcome of which the lives and ways of life for people and many other species depend. In particular, they support a politics not of "saving nature" but of "social nature," not of national parks and walled-off reserves, responding with a technical fix to whatever particular danger to survival seems most inescapable, but of a different organization of land and people, where the practice of justice restructures the concept of nature.

The authors tell a relentless story of a "social nature" over many hundreds of years, at every turn co-inhabited and co-constituted by humans, land, and other organisms. For example, the diversity and patterns of tree species in the forest cannot be explained without the deliberate, long-term practices of the Kayapó and other groups, whom Hecht and Cockburn describe, miraculously avoiding romanticizing, as "accomplished environmental scientists." Hecht and Cockburn avoid romanticizing because they do not invoke the category of the modern as the special zone of science. Thus, they do not have to navigate the shoals threatening comparisons of, according to taste, mere or wonderful "ethnoscience" with real or disgusting "modern science." The authors insist on visualizing the forest as the dynamic outcome of human as well as biological history. Only after the dense indigenous populations—numbering from six to twelve million in 1492—had been sickened, enslaved, killed, and otherwise displaced from along the rivers could Europeans represent Amazonia as "empty" of culture, as "nature," or, in later terms, as a purely "biological" entity.

But, of course, the Amazon was not and did not become "empty," although "nature" (like "man") is one of those discursive constructions that operates as a technology for making the world over into its image. First, there are indigenous people in the forest, many of whom have organized themselves in recent years into a regionally grounded, world-historical subject prepared for local/global interactions, or, in other terms, for building new and powerful collectives out of humans and unhumans, technological and organic. With all of the power to reconstitute the real implied in discursive construction, they have become a new discursive subject/object, the Indigenous Peoples of the Amazon, made up of national and tribal groups from Colombia, Ecuador, Brazil, and Peru, numbering about one million persons, who in turn articulate themselves with other organized groups of the indigenous peoples of the Americas. Also, in the forest are about 200,000 people of mixed ancestry, partly overlapping with the indigenous people. Making their living as petty extractors—of gold, nuts, rubber, and other forest products—they have a history of many generations in the Amazon. It is a complex history of dire exploitation. These people are also threatened by the latest schemes of world banks or national capitals from Brasília to Washington.[29] They have for many decades been in conflict with indigenous peoples over resources and ways of life. Their presence in the forest might be the fruit of the colonial fantasies of the *bandeirantes,* romantics, curators, politicians, or speculators; but their fate is entwined intimately with that of the other always historical inhabitants of this sharply contested world. It is from these desperately poor people, specifically the rubber tappers union, that Chico Mendes, the world-changing activist murdered on December 22, 1988, came.[30]

A crucial part of Mendes's vision for which he was killed was the union of the extractors and the indigenous peoples of the forest into, as Hecht and Cockburn argue, the "true defenders of the forest." Their position as defenders derives not from a concept of "nature under threat," but rather from a *relationship* with "the forest as the integument in their own elemental struggle to survive" (p. 196).[31] In other words, their authority derives *not* from the power to represent from a distance, *nor* from an ontological natural status, but from a constitutive social relationality in which the forest is an integral partner, part of natural/social embodiment. In their claims for authority over the fate of the forest, the resident peoples are articulating a social collective entity among humans, other organisms, and other kinds of non-human actors.

Indigenous people are resisting a long history of forced "tutelage," in order to confront the powerful representations of the national and international environmentalists, bankers, developers, and technocrats. The extractors, for example, the rubber tappers, are also independently articulating their collective viewpoint. Neither group is willing to see the Amazon "saved" by their exclusion and permanent subjection to historically dominating political and economic forces. As Hecht and Cockburn put it, "The rubber tappers have not risked their lives for extractive reserves so they could live on them as debt peons" (p. 202). "Any program for the Amazon begins with basic human rights: an end to debt bondage, violence, enslavement, and killings practiced by those who would seize the lands these forest people have occupied for generations. Forest people seek legal recognition of native lands and extractive reserves held under the principle of collective property, worked as individual holdings with individual returns" (p. 207).

At the second Brazilian national meeting of the Forest People's Alliance at Rio Branco in 1989, shortly after Mendes's murder raised the stakes and catapulted the issues into the international media, a program was formulated in tension with the latest Brazilian state policy called *Nossa Natureza*. Articulating quite a different notion of the first person plural relation to nature or natural surroundings, the basis of the program of the Forest People's Alliance is control by and for the peoples of the forest. The core matters are direct control of indigenous lands by native peoples; agrarian reform joined to an environmental program; economic and technical development; health posts; raised incomes; locally controlled marketing systems; an end to fiscal incentives for cattle ranchers, agribusiness, and unsustainable logging; an end to debt peonage; and police and legal protection. Hecht and Cockburn call this an "ecology of justice" that rejects a technicist solution, in whatever benign or malignant form, to environmental destruction. The Forest People's Alliance does not reject scientific or technical know-how, their own and others'; instead, they reject the "modern" political epistemology that bestows jurisdiction on the basis of technoscientific discourse. The fundamental point is that the Amazonian Biosphere is an irreducibly human/non-human collective entity.[32] There *will be* no nature without justice. Nature and justice, contested discursive objects embodied in the material world, will become extinct or survive together.

Theory here is exceedingly corporeal, and the body is a collective; it is an historical artifact constituted by human as well as organic and technological unhuman actors. Actors are entities which do things, have effects, build worlds in concatenation with other *unlike* actors.[33] Some actors, for example specific human ones, can try to reduce other actors to resources—to mere ground and matrix for their action; but such a move is contestable, not the necessary relation of "human nature" to the rest of the world. Other actors, human and unhuman, regularly resist reductionisms. The powers of domination do fail sometimes in their projects to pin other actors down; people can work to enhance the relevant failure rates. Social nature is the nexus I have called artifactual nature. The human "defenders of the forest" do not and have not lived in a garden; it is from a knot in the always historical and heterogeneous nexus of social nature that they articulate their claims. Or perhaps, it is within such a nexus that I and people like me narrate a possible politics of articulation rather than representation. It is our responsibility to learn whether such a fiction is one with which the Amazonians might wish to connect in the interests of an alliance to defend the rain forest and its human and non-human ways of life—because assuredly North Americans, Europeans, and the Japanese, among others, cannot watch from afar as if we were not actors, willing or not, in the life and death struggles in the Amazon.

In a review of *Fate of the Forest*, Joe Kane, author of another book on the tropical rain forest marketed in time for Christmas in 1989, the adventure trek *Running the Amazon* (1989),[34] raised this last issue in a way that will sharpen and clarify my stakes in arguing against a politics of representation generally, and in relation to questions of environmentalism and conservation specifically. In the context of worrying about ways that social nature or socialist ecology sounded too much like the multi-use policies in national forests in the United States, which have resulted in rapacious exploitation of the land and of other organisms, Kane asked a simple question: "[W]ho speaks for the jaguar?" Now, I care about the survival of the jaguar—and the chimpanzee, and the Hawaiian land snails, and the spotted owl, and a lot of other earthlings. I care a great deal; in fact, I think I and my social groups are particularly, but not uniquely, *responsible* if jaguars, and many other non-human, as well as human, ways of life should perish. But Kane's question seemed wrong on a fundamental level. Then I understood why. His question was precisely like that asked by some pro-life groups in the abortion debates: Who speaks for the fetus? What is wrong with both questions? And how does this matter relate to science studies as cultural studies?

Who speaks for the jaguar? Who speaks for the fetus? Both questions rely on a political semiotics of representation.[35] Permanently speechless, forever requiring the services of a ventriloquist, never forcing a recall vote, in each case the object or ground of representation is the realization of the representative's fondest dream. As Marx said in a somewhat different context, "They cannot represent themselves; they must be represented."[36] But for a political semiology of representation, nature and the unborn fetus are even better, epistemologically, than subjugated human adults. The effectiveness of such representation depends on distancing operations. The represented must be disengaged from surrounding and constituting discursive and non-discursive nexuses and relocated in the authorial domain of the representative. Indeed, the effect of this magical operation is to disempower precisely those—in our case, the pregnant woman and the peoples of the forest—who are "close" to the now-represented "natural" object. Both the jaguar and the fetus are carved out of one collective entity and relocated in another, where they are reconstituted as objects of a particular kind—as the ground of a representational practice that *forever* authorizes the ventriloquist. Tutelage will be eternal. The represented is reduced to the permanent status of the recipient of action, never to be a co-actor in an articulated practice among unlike, but joined, social partners.

Everything that used to surround and sustain the represented object, such as pregnant women and local people, simply disappears or re-enters the drama as an agonist. For example, the pregnant woman becomes *juridically* and *medically*, two very powerful discursive realms, the "maternal environment" (Hubbard, 1990). Pregnant women and local people are the *least* able to "speak for" objects like jaguars or fetuses because they get discursively reconstituted as beings with opposing "interests." Neither woman nor fetus, jaguar nor Kayapó Indian is an actor in the drama of representation. One set of entities becomes the represented, the other becomes the environment, often threatening, of the represented object. The *only* actor left is the spokesperson, the one who represents. The forest is no longer the integument in a co-constituted social nature; the woman is in no way a partner in an intricate and intimate dialectic of social relationality crucial to her own personhood, as well as to the possible personhood of her social—*but unlike*—internal co-actor.[37] In the liberal logic of representation, the fetus and the jaguar must be protected precisely from those closest to them, from their "surround." The power of life and death must be delegated to the epistemologically most disinterested ventriloquist, and it is crucial to remember that all of this *is* about the power of life and death.

Who, within the myth of modernity, is less biased by competing interests or polluted by excessive closeness than the expert, especially the scientist? Indeed, even better than the lawyer, judge, or national legislator, the scientist is the perfect representative of nature, that is, of the permanently and constitutively speechless objective world. Whether he be a male or a female, his passionless distance is his greatest virtue; this discursively constituted, structurally gendered distance legitimates his professional privilege, which in these cases, again, is the power to testify about the right to life and death. After Edward Said quoted Marx on representation in his epigraph to *Orientalism*, he quoted Benjamin Disraeli's *Tancred*, "The East is a career." The separate, objective world—non-social nature—is a career. Nature legitimates the scientist's career, as the Orient justifies the representational practices of the Orientalist, even as precisely "Nature" and the "Orient" are the *products* of the constitutive practice of scientists and orientalists.

These are the inversions that have been the object of so much attention in science studies. Bruno Latour sketches the double structure of representation through which scientists establish the objective status of their knowledge. First, operations shape and enroll new objects or allies through visual displays or other means called inscription devices. Second, scientists speak as if they were the mouthpiece for the speechless objects that they have just shaped and enrolled as allies in an agonistic field called science. Latour defines the actant as that which is represented; the objective world *appears* to be the actant solely by virtue of the operations of representation (Latour, 1987, pp. 70–74, 90). The authorship rests with the representor, even as he claims independent object status for the represented. In this doubled structure, the simultaneously semiotic and political ambiguity of representation is glaring. First, a chain of substitutions, operating through inscription devices, relocates power and action in "objects" divorced from polluting contextualizations and named by formal abstractions ("the fetus"). Then, the reader of inscriptions speaks for his docile constituencies, the objects. This is not a very lively world, and it does not finally offer much to jaguars, in whose interests the whole apparatus supposedly operates.

In this essay I have been arguing for another way of seeing actors and actants—and consequently another way of working to position scientists and science in important struggles in the world. I have stressed actants as collective entities doing things in a structured and structuring field of action; I have framed the issue in terms of articulation rather than representation. Human beings use names to point to themselves and other actors and easily mistake the names for the things. These same humans also think the traces of inscription devices are like names—pointers to things, such that the inscriptions and the things can be enrolled in dramas of substitution and inversion. But the things, in my view, do not pre-exist as ever-elusive, but fully pre-packaged, referents for the names. Other actors are more like tricksters than that. Boundaries take provisional, never-finished shape in articulatory practices. The potential for the unexpected from unstripped human and unhuman actants enrolled in articulations—i.e., the potential for generation—remains both to trouble and to empower technoscience. Western philosophers sometimes take account of the inadequacy of names by stressing the "negativity" inherent in all representations. This takes us back to Spivak's remark cited early in this paper about the important things that we cannot not desire, but can never possess—or represent, because representation depends on possession of a passive resource, namely, the silent object, the *stripped* actant. Perhaps we can, however, "articulate" with humans and unhumans in a social relationship, which for us is always language-mediated (among other semiotic, i.e., "meaningful," mediations). But, for our unlike partners, well, the action is "different," perhaps "negative" from our linguistic point of view, but crucial to the generativity of the collective. It is the empty space, the undecidability, the wiliness of other actors, the "negativity," that give me confidence in the *reality* and therefore ultimate *unrepresentability* of social nature and that make me suspect doctrines of representation and objectivity.

My crude characterization does not end up with an "objective world" or "nature," but it certainly does insist on the *world*. This world must always be articulated, from people's points of view, through "situated knowledges" (Haraway, 1988; 1991). These knowledges are friendly to science, but do not provide any grounds for history-escaping inversions and amnesia about how articulations get made, about their political semiotics, if you will. I think the world is precisely what gets lost in doctrines of representation and scientific objectivity. It is *because* I care about jaguars, among other actors, including the overlapping but non-identical groups called forest peoples and ecologists, that I reject Joe Kane's question. Some science studies scholars have been terrified to criticize their constructivist formulations because the only alternative seems to be some retrograde kind of "going back" to nature and to philosophical realism.[38] But above all people, these scholars should know that "nature" and "realism" are precisely the consequences of representational practices. Where we need to move is not "back" to nature, but *elsewhere*, through and within an artifactual social nature, which these very scholars have helped to make expressable in current Western scholarly practice. That knowledge-building practice might be articulated to other practices in "pro-life" ways that aren't about the fetus or the jaguar as nature fetishes and the expert as their ventriloquist.

Prepared by this long detour, we can return to the Kayapó man videotaping his tribesmen as they protest a new hydroelectric dam on their territory. The National Geographic Society, *Discover* magazine, and Gulf Oil—and much philosophy and social science—would have us see his practice as a double boundary crossing between the primitive and the modern. His representational practice, signified by his use of the latest technology, places him in the realm of the modern. He is, then, engaged in an entertaining contradiction—the preservation of an unmodern way of life with the aid of incongruous modern technology. But, from the perspective of a political semiotics of articulation, the man might well be forging a recent collective of humans and unhumans, in this case made up of the Kayapó, videocams, land, plants, animals, near and distant audiences, and other constituents; but no boundary violation is involved. The way of life is not unmodern (closer to nature); the camera is not modern or postmodern (in society). Those categories should no longer make sense. Where there is no nature and no society, there is no pleasure, no entertainment to be had in representing the violation of the boundary between them. Too bad for nature magazines, but a gain for inappropriate/d others.

The videotaping practice does not thereby become innocent or uninteresting; but its meanings have to be approached differently, in terms of the kinds of collective action taking place and the claims they make on others—such as ourselves, people who do not live in the Amazon. We *are all* in chiasmatic borderlands, liminal areas where new shapes, new kinds of action and responsibility, are gestating in the world. The man using that camera is forging a practical claim on us, morally and epistemologically, as well as on the other forest people to whom he will show the tape to consolidate defense of the forest. His practice invites further articulation—on terms shaped by the forest people. They will no longer be represented as Objects, not because they cross a line to represent themselves in "modern" terms as Subjects, but because they powerfully form articulated collectives.

In May of 1990, a week-long meeting took place in Iquitos, a formerly prosperous rubber boom-town in the Peruvian Amazon. COICA, the Coordinating Body for the Indigenous Peoples of the Amazon, had assembled forest people (from all the nations constituting Amazonia), environmental groups from around the world (Greenpeace, Friends of the Earth, the Rain Forest Action Network, etc.), and media organizations (*Time* magazine, CNN, NBC, etc.) in order "to find a common path on which we can work to preserve the Amazon forest" (Arena-De Rosa, 1990, pp. 1–2). Rain forest protection was formulated as a necessarily joint human rights-ecological issue. The fundamental demand by indigenous people was that they must be part of *all* international negotiations involving their territories. "Debt for nature" swaps were particular foci of controversy, especially where indigenous groups end up worse off than in previous agreements with their governments as a result of bargaining between banks, external conservation groups, and national states. The controversy generated a proposal: instead of a swap of debt-for-nature, forest people would support swaps of debt-for-indigenous-controlled territory, in which non-indigenous environmentalists would have a "redefined role in helping to develop the plan for conservation management of the particular region of the rain forest" (Arena-De Rosa, 1990). Indigenous environmentalists would also be recognized not for their quaint "ethnoscience," but for their *knowledge*.

Nothing in this structure of action rules out articulations by scientists or other North Americans who care about jaguars and other actors; but the patterns, flows, and intensities of power are most certainly changed. That is what articulation does; it is always a non-innocent, contestable practice; the partners are never set once and for all. There is no ventriloquism here. Articulation is work, and it may fail. All the people who care, cognitively, emotionally, and politically, must articulate their position in a field constrained by a new collective entity, made up of indigenous people and other human and unhuman actors. Commitment and engagement, not their invalidation, in an emerging collective are the conditions of joining knowledge-producing and world-building practices. This is situated knowledge in the New World; it builds on common places, and it takes unexpected turns. So far, such knowledge has not been sponsored by the major oil corporations, banks, and logging interests. That is precisely one of the reasons why there is so much work for North Americans, Europeans, and Japanese, among others, to do in articulation with those humans and non-humans who live in rain forests and in many other places in the semiotic space called earth.

B. OUTER SPACE: THE EXTRATERRESTRIAL

Since we have spent so much time on earth, a prophylactic exercise for residents of the alien "First World," we will rush through the remaining three quadrants of the semiotic square. We move from one topical commonplace to another, from earth to space, to see what turns our journeys to elsewhere might take.

An ecosystem is always of a particular type, for example, a temperate grassland or a tropical rain forest. In the iconography of late capitalism, Jane Goodall did not go to that kind of ecosystem. She went to the "wilds of Tanzania," a mythic "ecosystem" reminiscent of the original garden from which her kind had been expelled and to which she returned to commune with the wilderness's present inhabitants to learn how to survive. This wilderness was close in its dream quality to "space," but the wilderness of Africa was coded as dense, damp, bodily, full of sensuous creatures who touch intimately and intensely. In contrast, the extraterrestrial is coded to be fully general; it is about escape from the bounded globe into an anti-ecosystem called, simply, space. Space is not about "man's" origins on earth but about "his" future, the two key allochronic times of salvation history. Space and the tropics are both utopian topical figures in Western imaginations, and their opposed properties dialectically signify origins and ends for the creature whose mundane life is supposedly outside both: modern or postmodern man.

The first primates to approach that abstract place called "space" were monkeys and apes. A rhesus monkey survived an 83-mile-high flight in 1949. Jane Goodall arrived in "the wilds of Tanzania" in 1960 to encounter and name the famous Gombe Stream chimpanzees introduced to the National Geographic television audience in 1965. However, other chimpanzees were vying for the spotlight in the early 1960s. On January 31, 1961, as part of the United States man-in-space program, the chimpanzee HAM, trained for his task at Holloman Air Force Base, 20 minutes by car from Alamogordo, New Mexico, near the site of the first atom bomb explosion in July 1945, was shot into suborbital flight (Figure 3.8). HAM's name inevitably recalls Noah's youngest and only black son. But this chimpanzee's name was from a different kind of text. His name was an acronym for the scientific-military institution that launched him, *Holloman Aero-Medical*; and he rode an arc that

Fig. 3.8. HAM awaits release in his couch aboard the recovery vessel LSD *Donner* after his successful Mercury Project launch. Photograph by Henry Borroughs.

traced the birth path of modern science—the parabola, the conic section. HAM's parabolic path is rich with evocations of the history of Western science. The path of a projectile that does not escape gravity, the parabola is the shape considered so deeply by Galileo, at the first mythic moment of origins of modernity, when the unquantifiable sensuous and countable mathematical properties of bodies were separated from each other in scientific knowledge. It describes the path of ballistic weapons, and it is the trope for "man's" doomed projects in the writings of the existentialists in the 1950s. The parabola traces the path of Rocket Man at the end of World War II in Thomas Pynchon's *Gravity's Rainbow* (1973). An understudy for man, HAM went only to the boundary of space, in suborbital flight. On his return to earth, he was named. He had been known only as #65 before his successful flight. If, in the official birth-mocking language of the Cold War, the mission had to be "aborted," the authorities did not want the public worrying about the death of a famous and named, even if not quite human, astronaut. In fact, #65 did have a name among his handlers, Chop Chop Chang, recalling the stunning racism in which the other primates have been made to participate.[39] The space race's surrogate child was an "understudy for man in the conquest of space" (Eimerl and DeVore, 1965, p. 173). His hominid cousins would transcend that closed parabolic figure, first in the ellipse of orbital flight, then in the open trajectories of escape from earth's gravity.

HAM, his human cousins and simian colleagues, and their englobing and interfacing technology were implicated in a reconstitution of masculinity in Cold War and space race idioms. The movie *The Right Stuff* (1985) shows the first crop of human astronau(gh)ts struggling with their affronted pride when they realize their tasks were competently performed by their simian cousins. They and the chimps were caught in the same theater of the Cold War, where the masculinist, death-defying, and skill-requiring heroics of the old jet aircraft test pilots became obsolete, to be replaced by the media-hype routines of projects Mercury, Apollo, and their sequelae. After chimpanzee Enos completed a fully automated orbital flight on November 29, 1961, John Glenn, who would be the first human American astronaut to orbit earth, defensively "looked toward the future by affirming his belief in the superiority of astronauts over chimponauts." *Newsweek* announced Glenn's orbital flight of February 20, 1962, with the headline, "John Glenn: One Machine That Worked Without Flaw."[40] Soviet primates on both sides of the line of hominization raced their U.S. siblings into extraterrestrial orbit. The space ships, the recording and tracking technologies, animals, and human beings were joined as cyborgs in a theater of war, science, and popular culture.

Henry Burroughs's famous photograph of an interested and intelligent, actively participating HAM, watching the hands of a white, laboratory-coated, human man release him from his contour couch, illuminated the system of meanings that binds humans and apes together in the late twentieth century (Weaver, 1961). HAM is the perfect child, reborn in the cold matrix of space. *Time* described chimponaut Enos in his "fitted contour couch that looked like a cradle trimmed with electronics.[41] Enos and HAM were cyborg neonates, born of the interface of the dreams about a technicist automaton and masculinist autonomy. There could be no more iconic cyborg than a telemetrically implanted chimpanzee, understudy for man, launched from earth in the space program, while his conspecific in the jungle, "in a spontaneous gesture of trust," embraced the hand of a woman scientist named Jane in a Gulf Oil ad showing "man's place in the ecological structure." On one end of time and space, the chimpanzee in the wilderness modeled communication for the stressed, ecologically threatened and threatening, modern human. On the other end, the ET chimpanzee modeled social and technical cybernetic communication systems, which permit postmodern man to escape both the jungle and the city, in a thrust into the future made possible by the social-technical systems of the "information age" in a global context of threatened nuclear war. The closing image of a human fetus hurtling through space in Stanley Kubrick's *2001: A Space Odyssey* (1968) completed the voyage of discovery begun by the weapon-wielding apes at the film's gripping opening. It was the project(ile) of self-made, reborn man, in the process of being raptured out of history. The Cold War was simulated ultimate war; the media and advertising industries of nuclear culture produced in the bodies of animals—paradigmatic natives and aliens—the reassuring images appropriate to this state of pure war (Virilio and Lotringer, 1983).[42]

In the aftermath of the Cold War, we face not the end of nuclearism, but its dissemination. Even without our knowing his ultimate fate as an adult caged chimpanzee, the photograph of HAM rapidly ceases to entertain, much less to edify. Therefore, let us look to another cyborg image to figure possible emergencies of inappropriate/d others to challenge our rapturous mythic brothers, the postmodern spacemen.

At first sight, the T-shirt worn by anti-nuclear demonstrators at the Mother's and Others' Day Action in 1987 at the United States's Nevada nuclear test site seems in simple opposition to HAM in his electronic cradle (Figure 3.9). But a little unpacking shows the promising semiotic and political complexity of the image and of the action. When the T-shirt was sent to the printer, the name of the event was still the "Mother's Day Action," but not long after some planning participants objected. For many, Mother's Day was, at best, an ambivalent time for a women's action. The overdetermined gender coding of patriarchal nuclear culture all too easily makes women responsible for peace while men fiddle with their dangerous war toys without semiotic dissonance. With its commercialism and multi-leveled reinforcement of compulsory heterosexual reproduction, Mother's Day is also not everybody's favorite feminist holiday. For others, intent on reclaiming the holiday for other meanings, mothers, and by extension women in general, do have a special obligation to preserve children, and so the earth, from military destruction. For them, the earth is metaphorically mother and child, and in both figurations, a subject of nurturing and birthing. However, this was not an all-women's (much less all-mothers') action, although women organized and shaped it. From discussion, the designation "Mother's and Others' Day Action" emerged. But then, some thought that meant mothers and men. It took memory exercises in feminist analysis to rekindle shared consciousness that mother does not equal woman and vice versa. Part of the day's purpose was to recode Mother's Day to signify men's obligations to nurture

Fig. 3.9. Action T-shirt. Thanks to Noël Sturgeon for information and analysis.

the earth and all its children. In the spirit of this set of issues, at a time when Baby M and her many debatable—and unequally positioned—parents were in the news and the courts, the all-female affinity group which I joined took as its name the Surrogate Others. These surrogates were not understudies for man, but were gestating for another kind of emergence.

From the start, the event was conceived as an action that linked social justice and human rights, environmentalism, anti-militarism, and anti-nuclearism. On the T-shirt, there is, indeed, the perfect icon of the union of all issues under environmentalism's rubric: the "whole earth," the lovely, cloud-wrapped, blue, planet earth is simultaneously a kind of fetus floating in the amniotic cosmos and a mother to all its own inhabitants, germ of the future, matrix of the past and present. It is a perfect globe, joining the changeling matter of mortal bodies and the ideal eternal sphere of the philosophers. This snapshot resolves the dilemma of modernity, the separation of Subject and Object, Mind and Body. There is, however, a jarring note in all this, even for the most devout. That particular image of the earth, of Nature, could only exist if a camera on a satellite had taken the picture, which is, of course, precisely the case.

Who speaks for the earth? Firmly in the object world called nature, this bourgeois, family-affirming snapshot of mother earth is about as uplifting as a loving commercial Mother's Day card. And yet, it *is* beautiful, and it is ours; it must be brought into a different focus. The T-shirt is part of a complex collective entity, involving many circuits, delegations, and displacements of competencies. Only in the context of the space race in the first place, and the militarization and commodification of the whole earth, does it make sense to relocate that image as the special sign of an anti-nuclear, anti-militaristic, earth-focused politics. The relocation does not cancel its other resonances; it contests for their outcome.

I read Environmental Action's "whole earth" as a sign of an irreducible artifactual social nature, like the Gaia of SF writer John Varley and biologist Lynn Margulis. Relocated on this particular T-shirt, the satellite's eye view of planet earth provokes an ironic version of the question, who speaks for the earth (for the fetus, the mother, the jaguar, the object world of nature, all those who must be represented)? For many of us, the irony made it possible to participate—indeed, to participate as fully committed, if semiotically unruly, eco-feminists. Not everybody in the Mother's and Others' Day Action would agree; for many, the T-shirt image meant what it said, love your mother who is the earth. Nuclearism is misogyny. The field of readings in tension with each other is also part of the point. Eco-feminism and the non-violent direct action movement have been based on struggles over differences, not on identity. There is hardly a need for affinity groups and their endless process if sameness prevailed. Affinity is precisely *not* identity; the sacred image of the same is not gestating on this Mother's and Others' Day. Literally, enrolling the satellite's camera and the peace action in Nevada into a new collective, this Love Your Mother image is based on diffraction, on the processing of small but consequential differences. The processing of differences, semiotic action, is about ways of life.

The Surrogate Others planned a birthing ceremony in Nevada, and so they made a birth canal—a sixteen-foot long, three-foot diameter, floral polyester-covered worm with lovely dragon eyes. It was a pleasingly artifactual beast, ready for connection. The worm-dragon was laid under the barbed-wire boundary between the land on which the demonstrators could stand legally and the land on which they would be arrested as they emerged. Some of the Surrogate Others conceived of crawling through the worm to the forbidden side as an act of solidarity with the tunneling creatures of the desert, who had to share their subsurface niches with the test site's chambers. This surrogate birthing was definitely not about the obligatory heterosexual nuclear family compulsively reproducing itself in the womb of the state, with or without the underpaid services of the wombs of "surrogate mothers." Mother's and Others' Day was looking up.

It wasn't only the desert's non-human organisms with whom the activists were in solidarity as they emerged onto the proscribed territory. From the point of view of the demonstrators, they were quite legally on the test-site land. This was so not out of some "abstract" sense that the land was the people's and had been usurped by the war state, but for more "concrete" reasons: all the demonstrators had written permits to be on the land signed by the Western Shoshone National Council. The 1863 Treaty of Ruby Valley recognized the Western Shoshone title to ancestral territory, including the land illegally invaded by the U.S. government to build its nuclear facility. The treaty has never been modified or abrogated, and U.S. efforts to buy the land (at 15 cents per acre) in 1979 was refused by the only body authorized to decide, the Western Shoshone National Council. The county sheriff and his deputies, surrogates for the federal government, were, in "discursive" and "embodied" fact, trespassing. In 1986 the Western Shoshone began to issue permits to the anti-nuclear demonstrators as part of a coalition that joined anti-nuclearism and indigenous land rights. It is, of course, hard to make citizens' arrests of the police when they have you handcuffed and when the courts are on their side. But it is quite possible to join this ongoing struggle, which is very much "at home," and to articulate it with the defense of the Amazon. That articulation requires collectives of human and unhuman actors of many kinds.

There were many other kinds of "symbolic action" at the test site that day in 1987. The costumes of the sheriff's deputies and their nasty plastic handcuffs were also symbolic action—highly embodied symbolic action. The "symbolic action" of brief, safe arrest is also quite a different matter from the "semiotic" conditions under which most people in the U.S., especially people of color and the poor, are jailed. The difference is not the presence or absence of "symbolism," but the force of the respective collectives made up of humans and unhumans, of people, other organisms, technologies, institutions. I am not unduly impressed with the power of the drama of the Surrogate Others and the other affinity groups, nor, unfortunately, of the whole action. But I do take seriously the work to relocate, to diffract, embodied meanings as crucial work to be done in gestating a new world.[43] It is cultural politics, and it is technoscience politics. The task is to build more powerful collectives in dangerously unpromising times.

NOT-B. INNER SPACE: THE BIOMEDICAL BODY

The limitless reaches of outer space, joined to Cold War and post–Cold War nuclear technoscience, seem vastly distant from their negation, the enclosed and dark regions of the inside of the human body, domain of the apparatuses of biomedical visualization. But these two quadrants of our semiotic square are multiply tied together in technoscience's heterogeneous apparatuses of bodily production. As Sarah Franklin noted, "The two new investment frontiers, outer space and inner space, vie for the futures market." In this "futures market," two entities are especially interesting for this essay: the fetus and the immune system, both of which are embroiled in determinations of what may count as nature and as human, as separate natural object and as juridical subject. We have already looked briefly at some of the matrices of discourse about the fetus in the discussion of earth (who speaks for the fetus?) and outer space (the planet floating free as cosmic germ). Here, I will concentrate on contestations for what counts as a self and an actor in contemporary immune system discourse.

The equation of Outer Space and Inner Space, and of their conjoined discourses of extraterrestrialism, ultimate frontiers, and high technology war, is literal in the official history celebrating 100 years of the National Geographic Society (Bryan, 1987). The chapter that recounts the magazine's coverage of the Mercury, Gemini, Apollo, and Mariner voyages is called "Space" and introduced with the epigraph, "The Choice Is the Universe—or Nothing." The final chapter, full of stunning biomedical images, is titled "Inner Space" and introduced with the epigraph, "The Stuff of the Stars Has Come Alive."[44] The photography convinces the viewer of the fraternal relation of inner and outer space. But, curiously, in outer space, we see spacemen fitted into explorer craft or floating about as individuated cosmic fetuses, while in the supposed earthy space of our own interiors, we see non-humanoid strangers who are the means by which our bodies sustain our integrity and individuality, indeed our humanity in the face of a world of others. We seem invaded not just by the threatening "non-selves" that the immune system guards against, but more fundamentally by our own strange parts.

Lennart Nilsson's photographs, in the coffee table art book *The Body Victorious* (1987), as well as in many medical texts, are landmarks in the photography of the alien inhabitants of inner space[45] (Figure 3.10). The blasted scenes, sumptuous textures, evocative colors, and ET monsters of the immune landscape are simply *there*, inside *us*. A white extruding tendril of a pseudopodinous macrophage ensnares bacteria; the hillocks of chromosomes lie flattened on a blue-hued moonscape of some other planet; an infected cell buds myriads of deadly virus particles into the reaches of inner space where more cells will be victimized; the auto-immune disease-ravaged head of a femur glows against a sunset on a dead world; cancer cells are surrounded by the lethal mobil squads of killer T-cells that throw chemical poisons into the self's malignant traitor cells.

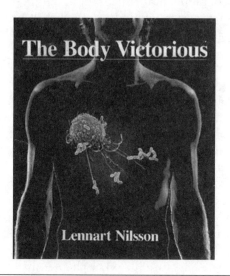

Fig. 3.10. Design for Lennart Nilsson book.

A diagram of the "Evolution of Recognition Systems" in a recent immunology textbook makes clear the intersection of the themes of literally "wonderful" diversity, escalating complexity, the self as a defended stronghold, and extraterrestrialism in inner space (Figure 3.11). Under a diagram culminating in the evolution of the mammals, represented without comment by a mouse and a *fully-suited spaceman*, is this explanation: "From the humble amoeba searching for food (top left) to the mammal with its sophisticated humoral and cellular immune mechanisms (bottom right), the process of 'self versus non-self recognition' shows a steady development, keeping pace with the increasing need of animals to maintain their integrity in a hostile environment. The decision at which point 'immunity' appeared is thus a purely semantic one" (Playfair, 1984, emphasis in the original). These are the "semantics" of defense and invasion. The perfection of the fully defended, "victorious" self is a chilling fantasy, linking phagocytotic amoeba and space-voyaging man cannibalizing the earth in an evolutionary teleology of post-apocalypse extraterrestrialism. When is a self enough of a self that its boundaries become central to institutionalized discourses in biomedicine, war, and business?

Images of the immune system as a battlefield abound in science sections of daily newspapers and in popular magazines, e.g., *Time* magazine's 1984 graphic for the AIDS virus's "invasion" of the cell-as-factory.

Evolution of recognition systems

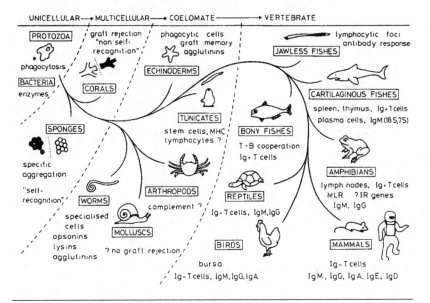

UNICELLULAR → MULTICELLULAR → COELOMATE → VERTEBRATE

PROTOZOA — graft rejection / phagocytic cells / lymphocytic foci antibody response

"non self- / graft memory /
recognition" / agglutinins

phagocytosis / JAWLESS FISHES

BACTERIA / ECHINODERMS

enzymes / CORALS

CARTILAGINOUS FISHES

spleen, thymus, Ig·Tcells
plasma cells, IgM(18 S,7S)

SPONGES / TUNICATES

specific / stem cells, MHC / BONY FISHES

aggregation / lymphocytes ? / T-B cooperation

"self- / Ig·T cells

recognition" / WORMS / ARTHROPODS / REPTILES / AMPHIBIANS

specialised / lymph nodes, Ig·Tcells
MLR   ?IR genes

cells / complement ? / Ig-Tcells, IgM,IgG / IgM, IgG

opsonins / MOLLUSCS

lysins / BIRDS / MAMMALS

agglutinins / ? no graft rejection

bursa / Ig-Tcells

Ig-Tcells, IgM,IgG,IgA / IgM, IgG, IgA, IgE, IgD

Fig. 3.11. From a 1984 immunology textbook.

The virus is a tank, and the viruses ready for export from the expropriated cells are lined up ready to continue their advance on the body as a productive force. The *National Geographic* explicitly punned on Star Wars in its graphic called "Cell Wars" (Jaret, 1986). The militarized, automated factory is a favorite convention among immune system technical illustrators and photographic processors. The specific historical markings of a Star Wars-maintained individuality are enabled by high-technology visualization technologies, which are also basic to conducting war, and commerce, such as computer-aided graphics, artificial intelligence software, and specialized scanning systems.

It is not just imagers of the immune system who learn from military cultures; military cultures draw symbiotically on immune system discourse, just as strategic planners draw directly from and contribute to video game practices and science fiction. For example, arguing for an elite special force within the parameters of "low-intensity conflict" doctrine, a U.S. army officer wrote: "The most appropriate example to describe how this system would work is the most complex biological model we know—the body's immune system. Within the body there exists a remarkably complex corps of internal bodyguards. In absolute numbers they are small—only about one percent of the body's cells. Yet they consist of reconnaissance specialists, killers, reconstitution specialists, and communicators that can seek out invaders, sound the alarm, reproduce rapidly, and swarm to the attack to repel the enemy. . . . In this regard, the June 1986 issue of *National Geographic* contains a detailed account of how the body's immune system functions" (Timmerman, 1987).

The circuits of competencies sustaining the body as a defended self—personally, culturally, and nationally—spiral through the fantasy entertainment industry, a branch of the apparatus of bodily production fundamental to crafting the important consensual hallucinations about "possible" worlds that go into building "real" ones. In Epcot Center of Walt Disney World, we may be interpellated as subjects in the new Met Life Pavilion, which is "devoted to dramatizing the intricacies of the human body." A special thrill ride, called "Body Wars," promises that we will "experience the wonders of life," such as encountering "the attack of the platelets."[46] This lively battle simulator is promoted as "family entertainment." The technology for this journey through the human body uses a motion-based simulator to produce three-dimensional images for a stationary observer. As in other forms of high-tech tourism, we can go everywhere, see everything and leave no trace. The apparatus has been adopted to teach medical anatomy at the University of Colorado Health Sciences Center. Finally, we should not forget that more Americans travel to the combined Disney worlds than voyage in most other mythrealizing machines, like Washington, D.C.[47] Met Life cautions those who journey on "Body Wars" that they may experience extreme vertigo from the simulated motion. Is that merely "symbolic action" too?

In the embodied semiotic zones of earth and outer space, we saw the diffraction patterns made possible by recomposed visualizing technologies, relocated circuits of competencies that promise to be more user-friendly for inappropriate/d others. So also, the inner spaces of the biomedical body are central zones of technoscientific contestation, i.e., of science as culture in the amodern frame of social nature. Extremely interesting new collectives of human and unhuman allies and actors are emerging from these processes. I will briefly sketch two zones where promising monsters are undergoing symbiogenesis in the nutrient media of technoscientific work: (1) theories of immune function based on laboratory research, and (2) new apparatuses of knowledge production being crafted by Persons with AIDS (PWAs) and their heterogeneous allies. Both sets of monsters generate distinctly diffracted views of the self, evident in beliefs and practices in relation to vulnerability and mortality.

Like non-violent direct action and environmentalism, immune system discourse is about the unequally distributed chances of life and death. Since sickness and mortality are at the heart of immunology, it is hardly surprising that conditions of battle prevail. Dying is not an easy matter crying out for "friendly" visualization. But battle is not the only way to figure the process of mortal living. Persons coping with the life-threatening consequences of infection with the HIV virus have insisted that they are *living* with AIDS, rather than accepting the status of *victims* (or prisoners of war?). Similarly, laboratory scientists also have built research programs based on non-militaristic, relational embodiments, rather than on the capabilities of the defended self of atomic individuals. They do this in order to construct IS articulations more effectively, not in order to be nice folks with pacifist metaphors.

Let me attempt to convey the flavor of the artifactual bodily object called the human immune system, culled from major textbooks and research reports published in the 1980s. These characterizations are part of working systems for interacting with the immune system in many areas of practice, including business decisions, clinical medicine, and lab experiments. With about 10 to the 12th cells, the IS has two orders of magnitude more cells than the nervous system. IS cells are regenerated throughout life from pluripotent stem cells. From embryonic life through adulthood, the immune system is sited in several morphologically dispersed tissues and organs, including the thymus, bone marrow, spleen, and lymph nodes; but a large fraction of its cells are in the blood and lymph circulatory systems and in body fluids and spaces. If ever there were a "distributed system," this is one! It is also a highly adaptable communication system with many interfaces.

There are two major cell lineages to the system: (1) The first is the *lymphocytes*, which include the several types of T cells (helper, suppressor, killer, and variations of all these) and the B cells (each type of which can produce only one sort of the vast array of potential circulating antibodies). T and B cells have particular specificities capable of recognizing almost any molecular array of the right size that can ever exist, no matter how clever industrial chemistry gets. This specificity is enabled by a baroque somatic mutation mechanism, clonal selection, and a polygenic receptor or marker system. (2) The second immune cell lineage is the *mononuclear phagocyte system*, including the multitalented macrophages, which, in addition to their other recognition skills and connections, also appear to share receptors and some hormonal peptide products with neural cells. Besides the cellular compartment, the immune system comprises a vast array of circulating acellular products, such as antibodies, lymphokines, and complement components. These molecules mediate communication among components of the immune system, but also between the immune system and the nervous and endocrine systems, thus linking the body's multiple control and coordination sites and functions. The genetics of the immune system cells, with their high rates of somatic mutation and gene product splicings and rearrangings to make finished surface receptors and antibodies, makes a mockery of the notion of a constant genome even within "one" body. The hierarchical body of old has given way to a network-body of amazing complexity and specificity. The immune system is everywhere and nowhere. Its specificities are indefinite if not infinite, and they arise randomly; yet these extraordinary variations are the critical means of maintaining bodily coherence.

In the early 1970s, winning a Nobel Prize for the work, Niels Jerne proposed a theory of immune system self-regulation, called the network theory, which deviates radically from notions of the body victorious and the defended self. "The network theory differs from other immunological thinking because it endows the immune system with the ability to regulate itself using only itself" (Golub, 1987; Jerne, 1985).[48] Jerne proposed that any antibody molecule must be able to act functionally as both antibody to some antigen *and* as antigen for the production of an antibody to itself, at another region of "itself." These sites have acquired a nomenclature sufficiently daunting to thwart popular understanding of the theory, but the basic conception is simple. The concatenation of internal recognitions and responses would go on indefinitely, in a series of interior mirrorings of sites on immunoglobulin molecules, such that the immune system would always be in a state of dynamic internal responding. It would never be passive, "at rest," awaiting an activating stimulus from a hostile outside. In a sense, there could be no *exterior* antigenic structure, no "invader," that the immune system had not already "seen" and mirrored internally. Replaced by subtle plays of partially mirrored readings and responses, self and other lose their rationalistic oppositional quality. A radical conception of *connection* emerges unexpectedly at the core of the defended self. Nothing in the model prevents therapeutic action, but the entities in the drama have different kinds of interfaces with the world. The therapeutic logics are unlikely to be etched into living flesh in patterns of DARPA's latest high-tech tanks and smart missiles.

Some of those logics are being worked out in and by the bodies of persons with AIDS and ARC. In their work to sustain life and alleviate pain in the context of mortal illness, PWAs engage in many processes of knowledge-building. These processes demand intricate code switching, language bridging, and alliances among worlds previously held apart. These "generative grammars" are matters of life and death. As one activist put it, "ACT UP's humor is no joke" (Crimp and Rolston, 1990, p. 20; see also Crimp, 1983). The AIDS Coalition to Unleash Power (ACT UP) is a collective built from many articulations among unlike kinds of actors—for example, activists, biomedical machines, government bureaucracies, gay and lesbian worlds, communities of color, scientific conferences, experimental organisms, mayors, international information and action networks, condoms and dental dams, computers, doctors, IV drug-users, pharmaceutical companies, publishers, virus components, counselors, innovative sexual practices, dancers, media technologies, buying clubs, graphic artists, scientists, lovers, lawyers, and more. The actors, however, are not all equal. ACT UP has an animating center—PWAs, who are to the damage wrought by AIDS and the work for restored health around the world as the indigenous peoples of the Amazon are to forest destruction and environmentalism. These are the actors with whom others must articulate. That structure of action is a fundamental consequence of learning to visualize the heterogeneous, artifactual body that is our "social nature," instead of narrowing our vision that "saving nature" and repelling alien invaders from an unspoiled organic eden called the autonomous self. Saving nature is, finally, a deadly project. It relies on perpetuating the structure of boundary violation and the falsely liberating *frisson* of transgression. What happened in the first Eden should have made that clear.

So, if the tree of knowledge cannot be forbidden, we had all better learn how to eat and feed each other with a little more savvy. That is the difficult process being engaged by PWAs, Project Inform, ACT UP, NIH, clinical practitioners, and many more actors trying to build responsible mechanisms for producing effective knowledge in the AIDS epidemic.[49] Unable to police the same boundaries separating insiders and outsiders, the world of biomedical research will never be the same again. The changes range across the epistemological, the commercial, the juridical, and the spiritual domains. For example, what is the status of knowledge produced through the new combinations of decision-making in experimental design that are challenging previous research conventions? What are the consequences of the *simultaneous* challenges to expert monopoly of knowledge *and* insistence on both the rapid improvement of the biomedical knowledge base and the equitable mass distribution of its fruits? How will the patently amodern hybrids of healing practices cohabit in the emerging social body? And, who will live and die as a result of these very non-innocent practices?

NOT-A. VIRTUAL SPACE: SF[50]

Articulation is not a simple matter. Language is the effect of articulation, and so are bodies. The articulata are jointed animals; they are not smooth like the perfect spherical animals of Plato's origin fantasy in the *Timaeus*.

The articulata are cobbled together. It is the condition of being articulate. I rely on the articulata to breathe life into the artifactual cosmos of monsters that this essay inhabits. Nature may be speechless, without language, in the human sense; but nature is highly articulate. Discourse is only one process of articulation. An articulated world has an undecidable number of modes and sites where connections can be made. The surfaces of this kind of world are not frictionless curved planes. Unlike things can be joined—and like things can be broken apart—and vice versa. Full of sensory hairs, evaginations, invaginations, and indentations, the surfaces which interest me are dissected by joints. Segmented invertebrates, the articulata are insectoid and worm-like, and they inform the inflamed imaginations of SF filmmakers and biologists. In obsolete English, to articulate meant to make terms of agreement. Perhaps we should live in such an "obsolete," amodern world again. To articulate is to signify. It is to put things together, scary things, risky things, contingent things. I want to live in an articulate world. We articulate; therefore, we are. Who "I" am is a very limited, in the endless perfection of (clear and distinct) Self-contemplation. Unfair as always, I think of it as the paradigmatic psychoanalytic question. "Who am I?" is about (always unrealizable) identity; always wobbling, it still pivots on the law of the father, the sacred image of the same. Since I am a moralist, the real question must have more virtue: who are "we"? That is an inherently more open question, one always ready for contingent, friction-generating articulations. It is a remonstrative question.

In optics, the virtual image is formed by the apparent, but not actual, convergence of rays. The virtual seems to be the counterfeit of the real; the virtual has effects by seeming, not being. Perhaps that is why "virtue" is still given in dictionaries to refer to women's chastity, which must always remain doubtful in patriarchal optical law. But then, "virtue" used to mean manly spirit and valor too, and God even named an order of angels the Virtues, though they were of only middling rank. Still, no matter how big the effects of the virtual are, they seem somehow to lack a proper ontology. Angels, manly valor, and women's chastity certainly constitute, at best, a virtual image from the point of view of late twentieth-century "postmoderns." For them, the virtual is precisely *not* the real; that's why "postmoderns" *like* "virtual reality." It seems transgressive. Yet, I can't forget that an obsolete meaning of "virtual" was having virtue, i.e., the inherent power to produce effects. "Virtu," after all, is excellence or merit, and it is still a common meaning of virtue to refer to having efficacy. The "virtue" of something is its "capacity." The virtue of (some) food is that it nourishes the body. Virtual space *seems* to be the negation of real space; the domains of SF *seem* the negation of earthly regions. But perhaps this negation is the real illusion.

"Cyberspace, absent its high-tech glitz, is the idea of virtual consensual community . . . A virtual community is first and foremost a community of belief."[51] For William Gibson (1986), cyberspace is "consensual hallucination experienced daily by billions. . . . Unthinkable complexity." Cyberspace seems to be the consensual hallucination of too much complexity, too much articulation. It is the virtual reality of paranoia, a well-populated region in the last quarter of the Second Christian Millenium. Paranoia is the belief in the unrelieved density of connection, requiring, if one is to survive, withdrawal and defense unto death. The defended self re-emerges at the heart of relationality. Paradoxically, paranoia is the condition of the impossibility of remaining articulate. In virtual space, the virtue of articulation—i.e., the power to produce connection—threatens to overwhelm and finally to engulf all possibility of effective action to change the world.

So, in our travels into virtual space, if we are to emerge from our encounter with the artifactual articulata into a livable elsewhere, we need a guide figure to navigate around the slough of despond. Lisa Foo, the principal character in a Hugo and Nebula award-winning short story by John Varley (1986), will be our unlikely Beatrice through the System. "If you wish to know more, press enter" (p. 286).[52]

With that fatal invitation, Varley's profoundly paranoid story begins and ends. The Tree of Knowledge is a Web, a vast system of computer connections generating, as an emergent property, a new and terrifyingly unhuman collective entity. The forbidden fruit is knowledge of the workings of this powerful Entity, whose deadly essence is extravagant connection. All of the human characters are named after computers, programs, practices, or concepts—Victor Apfel, Detective Osborne, and the hackers Lisa Foo and Charles Kluge. The story is a murder mystery. With a dubious suicide note, called up by responding to the command "press enter" on the screen of one of the dozens of personal computers in his house, which is also full of barrels of illicit drugs, Kluge has been found dead by his neighbor, Apfel. Apfel is a reclusive middle-aged epileptic, who had been a badly treated prisoner-of-war in Korea, leaving him with layers of psychological terror, including a fear and hatred of "orientals." When Los Angeles homicide Detective Osborne's men prove totally inept at deciphering the elaborate software running Kluge's machines, Lisa Foo, a young Vietnamese immigrant, now a U.S. citizen, is called in from Cal Tech; and she proceeds to play Sherlock Holmes to Osborne's Lestrade. The story is narrated from Apfel's point of view, but Foo is the tale's center and, I insist, its pivotal actor.

Insisting, I wish to exercise the license that is built into the anti-elitist reading conventions of SF popular cultures. SF conventions invite—or at least permit more readily than do the academically propagated, respectful consumption protocols for literature—rewriting as one reads. The books are cheap; they don't stay in print long; why not rewrite them as one goes? Most of the SF I like motivates me to engage actively with images, plots, figures, devices, linguistic moves, in short, with worlds, not so much to make them come out "right," as to make them move "differently." These worlds motivate me to test their virtue, to see if their articulations work—and what they work for. Because SF makes identification with a principal character, comfort within the patently constructed world, or a relaxed attitude toward language, especially risky reading strategies, the reader is likely to be more generous and more suspicious—*both* generous and suspicious, exactly the receptive posture I seek in political semiosis generally. It is a strategy closely aligned with the oppositional and differential consciousness theorized by Chela Sandoval and by other feminists insistent on navigating mined discursive waters.

Our first view of Lisa Foo is through Apfel's eyes; and for him, "[L]eaving out only the moustache, she was a dead ringer for a cartoon Tojo. She had the glasses, ears, and the teeth. But her teeth had braces, like piano keys wrapped in barbed wire. And she was five-eight or five-nine and couldn't have weighed more than a hundred and ten. I'd have said a hundred, but added five pounds for each of her breasts, so improbably large on her scrawny frame that all I could read of the message on her T-shirt was "POCK LIVE." It was only when she turned sideways that I saw the esses before and after" (pp. 241–42). Using such messages among the many other languages accessed by this intensely literate figure, Foo communicated constantly through her endless supply of T-shirts. Her breasts turned out to be silicone implants, and as Foo said, "I don't think I've ever been so happy with anything I ever bought. Not even the car [her Ferrari]" (p. 263). From Foo's childhood perspective, "the West . . . [is] the place where you buy tits" (p. 263).

When Foo and Apfel became lovers, in one of the most sensitively structured heterosexual, cross-racial relationships in print anywhere, we also learn that Foo's body was multiply composed by the history of Southeast Asia. Varley gave her a name that is an "orientalized" version of the computer term "fubar"—"fucked up beyond all recognition." Her Chinese grandmother had been raped in Hanoi by an occupying Japanese soldier in 1942. In Foo's mother's Vietnam, "Being Chinese was bad enough, but being half Chinese and half Japanese was worse . . . My father was half French and half Annamese. Another bad combination" (p. 275). Her mother was killed in the Tet offensive when Foo was ten. The girl became a street hustler and child prostitute in Saigon, where she was "protected" by a pedophilic white U.S. major. Refusing to leave Saigon with him, after Saigon "fell," Foo ended up in Pol Pot's Cambodia, where she barely survived the Khmer Rouge work camps. She escaped to Thailand, and "when I finally got the Americans to notice me, my Major was still looking for me" (p. 276). Dying of a cancer that might have been the result of his witnessing the atom bomb tests in Nevada early in his career, he sponsored her to the U.S. Her intelligence and hustling got her "tits by Goodyear" (p. 275), a Ferrari, and a Cal Tech education. Foo and Apfel struggle together within their respective legacies of multiple abuse, sexual and otherwise, and criss-crossing racisms. They are both multi-talented, but scarred, survivors. This story, its core figure and its narrator, will not let us dodge the scary issues of race/racism, gender/sexism, historical tragedy, and technoscience within the region of time we politely call "the late twentieth century." There is no safe place here; there are, however, many maps of possibility.

But, there is entirely *too much* connection in "Press Enter," and it is only the beginning. Foo is deeply in love with the power-knowledge systems to which her skills give her access. "This is money, Yank, she said, and her eyes glittered" (p. 267). As she traces the fascinating webs and security locks, which began in military computer projects but which have taken on a vastly unhuman life of their own, her love and her skills bring her too deep into the infinitely dense connections of the System, where she, like Kluge before her, is noticed. Too late, she tries to withdraw. Soon after, a clearly fake suicide note appears on her T-shirt on her ruined body. Investigation showed that she had rewired the microwave oven in Kluge's house to circumvent its security checks. She put her head in the oven, and she died shortly after in the hospital, her eyes and brain congealed and her breasts horribly melted. The promise of her name, "fu bar," was all-too-literally fulfilled—fucked up beyond all recognition. Apfel, who had been brought back into articulation with life in his love with Lisa Foo, retreated totally, stripping his house of all its wiring and any other means of connecting with the techno-webs of a world he now saw totally within the paranoid terms of infinite and alien connection. At the end, the defended self, alone, permanently hides from the alien Other.

It is possible to read "Press Enter" as a conventional heterosexual romance, bourgeois detective fiction, technophobic-technophilic fantasy, dragon-lady story, and, finally, white masculinist narrative whose condition of possibility is access to the body and mind of a woman, especially a "Third World" woman, who, here as elsewhere in misogynist and racist culture, is violently destroyed. Not just violently—superabundantly, without limit. I think such a reading does serious violence to the subtle tissues of the story's writing. Nonetheless, "Press Enter" induces in me, and in other women and men who have read the story with me, an irreconcilable pain and anger: Lisa Foo should not have been killed that way. It really is not alright. The text and the body lose all distinction. I fall out of the semiotic square and into the viciously circular thing-in-itself. More than anything else, that pornographic, gendered and colored death, that excessive destruction of her body, that total undoing of her being—that extravagant final connection—surpasses the limits of pleasure in the conventions of paranoid fiction and provokes the necessity of active rewriting as reading. I cannot read this story without rewriting it; that is one of the lessons of transnational, intercultural, feminist literacy. And the conclusion forces rewriting not just of itself, but of the whole human and unhuman collective that is Lisa Foo. The point of the differential/oppositional rewriting is not to make the story come out "right," whatever that would be. The point is to rearticulate the figure of Lisa Foo to unsettle the closed logics of a deadly racist misogyny. Articulation must remain open, its densities accessible to action and intervention. When the system of connections closes in on itself, when symbolic action becomes perfect, the world is frozen in a dance of death. The cosmos is finished, and it is One. Paranoia is the only possible posture; generous suspicion is foreclosed. To "press enter" is, in that world, a terrible mistake.

The whole argument of "The Promises of Monsters" has been that to "press enter" is not a fatal error, but an inescapable possibility for changing maps of the world, for building new collectives out of what is not quite a plethora of human and unhuman actors. My stakes in the textual figure of Lisa Foo, and of many of the actors in Varley's SF, are high. Built from multiple interfaces, Foo can be a guide through the terrains of virtual space, but only if the fine lines of tension in the articulated webs that constitute her being remain in play, open to the unexpected realization of an unlikely hope. It's not a "happy ending" we need, but a non-ending. That's why none of the narratives of masculinist, patriarchal apocalypses will do. The System is not closed; the sacred image of the same is not coming. The world is not full.

The final image of this excessive essay is *Cyborg*, a 1989 painting by Lynn Randolph, in which the boundaries of a fatally transgressive world, ruled by the Subject and the Object, give way to the borderlands, inhabited by human and unhuman collectives (Figure 3.12).[53] These borderlands suggest a rich topography of combinatorial possibility. That possibility is called the Earth, here, now, this elsewhere, where real, outer, inner, and virtual space implode. The painting maps the articulations among cosmos, animal, human, machine, and landscape in their recursive sidereal, bony, electronic, and geological skeletons. Their combinatorial logic is embodied; theory is corporeal; social nature is articulate. The stylized DIP switches of the integrated circuit board on the

Fig. 3.12. Lynn Randolph, "Cyborg" (1989). Photo by D. Caras.

human figure's chest are devices that set the defaults in a form intermediate between hardwiring and software control—not unlike the mediating structural-functional anatomy of the feline and hominid forelimbs,
especially the flexible, homologous hands and paws. The painting is replete with organs of touch and mediation, as well as with organs of vision.
Direct in their gaze at the viewer, the eyes of the woman and the cat center the whole composition. The spiraling skeleton of the Milky Way, our
galaxy, appears behind the cyborg figure in three different graphic displays made possible by high-technology visualizing apparatuses. In the
place of virtual space in my semiotic square, the fourth square is an imaging of the gravity well of a black hole. Notice the tic-tac-toe game, played
with the European male and female astrological signs (Venus won this
game); just to their right are some calculations that might appear in the
mathematics of chaos. Both sets of symbols are just below a calculation
found in the Einstein papers. The mathematics and games are like logical
skeletons. The keyboard is jointed to the skeleton of the planet Earth, on
which a pyramid rises in the left mid-foreground. The whole painting
has the quality of a meditation device. The large cat is like a spirit animal, a white tiger perhaps. The woman, a young Chinese student in the
United States, figures that which is human, the universal, the generic. The
"woman of color," a very particular, problematic, recent collective identity, resonates with local and global conversations.[54] In this painting, she

embodies the still oxymoronic simultaneous statuses of woman, "Third World" person, human, organism, communications technology, mathematician, writer, worker, engineer, scientist, spiritual guide, lover of the Earth. This is the kind of "symbolic action" transnational feminism have made legible. S/he is not finished.

We have come full circle in the noisy mechanism of the semiotic square, back to the beginning, where we met the commercial cyborg figures inhabiting technoscience worlds. Logic General's oddly recursive rabbits, forepaws on the keyboards that promise to mediate replication and communication, have given way to different circuits of competencies. If the cyborg has changed, so might the world. Randolph's cyborg is in conversation with Trinh Minh-ha's inappropriate/d other, the personal and collective being to whom history has forbidden the strategic illusion of self-identity. This cyborg does not have an Aristotelian structure; and there is no master-slave dialectic resolving the struggles of resource and product, passion and action. S/he is not utopian nor imaginary; s/he is virtual. Generated, along with other cyborgs, by the collapse into each other of the technical, organic, mythic, textual, and political, s/he is constituted by articulations of critical differences within and without each figure. The painting might be headed, "A few words about articulation from the actors in the field." Privileging the hues of red, green, and ultraviolet, I want to read Randolph's *Cyborg* within a rainbow political semiology, for wily transnational technoscience studies as cultural studies.

NOTES

1. "They drew near to a very Miry Slough. . . . The name of this Slow was Dispond" (John Bunyan, *Pilgrim's Progress,* 1678; quoted in the Oxford English Dictionary). The nonstandardization of spelling here should also mark, at the beginning of the "Promises of Monsters," the suggestiveness of words at the edge of the regulatory technologies of writing.

2. Sally Hacker, in a paper written just before her death ("The Eye of the Beholder: An Essay on Technology and Eroticism," manuscript, 1989), suggested the term "pornotechnics" to refer to the embodiment of perverse power relations in the artifactual body. Hacker insisted that the heart of pornotechnics is the military as an institution, with its deep roots and wide reach into science, technology, and erotics. "Technical exhilaration" is profoundly erotic; joining sex and power is the designer's touch. Technics and erotics are the cross hairs in the focusing device for scanning fields of skill and desire. See also Hacker (1989). Drawing from Hacker's arguments, I believe that control over technics is the enabling practice for class, gender, and race supremacy. Realigning the join of technics and erotics must be at the heart of anti-racist feminist practice. (cf. Haraway, 1989b; Cohn, 1987).

3. See the provocative publication that replaced *Radical Science Journal, Science as Culture,* Free Association Books, 26 Freegrove Rd., London N7 9RQ, England.

4.   This incubation of ourselves as planetary fetuses is not quite the same thing as pregnancy and reproductive politics in post-industrial, post-modern, or other posted locations, but the similarities will become more evident as this essay proceeds. The struggles over the outcomes are linked.

5.   Here I borrow from the wonderful project of the journal *Public Culture*, Bulletin of the Center for Transnational Cultural Studies, The University Museum, University of Pennsylvania, Philadelphia, PA 19104. In my opinion, this journal embodies the best impulses of cultural studies.

6.   I demure on the label "postmodern" because I am persuaded by Bruno Latour that within the historical domains where science has been constructed, the "modern" never existed, if by modern we mean the rational, enlightened mentality (the subject, mind, etc.) actually proceeding with an objective method toward adequate representations, in mathematical equations if possible, of the object—i.e., "natural"—world. Latour argues that Kant's *Critique,* which set off at extreme poles Things-in-Themselves from the Transcendental Ego, is what made us believe ourselves to be "modern," with escalating and dire consequences for the repertoire of explanatory possibilities of "nature" and "society" for Western scholars. The separation of the two transcendences, the object pole and the subject pole, structures " 'the political Constitution of Truth.' I call it 'modern,' defining modernity as the complete separation of the representation of things—science and technology—from the representation of humans—politics and justice." (Latour, 1993).

   Debilitating though such a picture of scientific activity should seem, it has guided research in the disciplines (history, philosophy, sociology, anthropology), science with a pedagogical and prophylactic vengeance, making culture seem other to science. Science alone could get the goods on nature by unveiling and policing her unruly embodiments. Thus, science studies, focused on the edifying object of "modern" scientific practice, have seemed immune from the polluting infections of cultural studies—but surely no more. To rebel against or to lose faith in rationalism and enlightenment, the infidel state of respectively modernists and postmodernists, is not the same thing as to show that rationalism was the emperor that had no clothes, that never was, and so there never was its other either. (There is a nearly inevitable terminological confusion here among modernity, the modern, and modernism. I use modernism to refer to a cultural movement that rebelled against the premises of modernity, while postmodernism refers less to rebellion than loss of faith, leaving nothing to rebel against.) Latour calls his position *a*modern and argues that scientific practice is and has been amodern, a sighting that makes the line between real scientific (West's) and ethnoscience and other cultural expressions (everything else) disappear. The difference reappears, but with a significantly different geometry—that of scales and volumes, i.e., the size differences among "collective" entities made of humans and non-humans—rather than in terms of a line between rational science and ethnoscience.

   This modest turn or tropic change does not remove the study of scientific practice from the agenda of cultural studies and political intervention, but places it decisively on the list. Best of all, the focus gets fixed clearly on inequality, right where it belongs in science studies. Further, the addition of science to cultural studies does not leave the notions of culture, society, and politics untouched, far from it. In particular, we cannot make a critique of science and its constructions of nature based on an ongoing belief in culture or society. In the form of social constructionism, that belief has grounded the major strategy of left, feminist, and anti-racist science radicals. To remain with that strategy, however, is to remain bedazzled by the ideology of enlightenment. It will not do to approach science as cultural or social construction, as if culture and society were transcendent categories, any more than nature or the object is. Outside the premises of enlightenment—i.e., of the modern—the binary pairs of culture and nature, science and society, the technical and the social all lose their co-constitutive, oppositional quality. Neither can explain the other. "But instead of providing the explanation, Nature and Society are now accounted for as the historical consequences of the movement of collective things. All the interesting realities are no longer captured by the two extremes but are to be found in the substitution, cross over, translations, through which actants shift their competences" (Latour, 1990, p. 170). When the pieties of belief in the modern are dismissed, both members of the binary pairs collapse into each other as into a black hole. But what happens to them in the black hole is, by definition, not visible from the shared terrain of modernity, modernism, or postmodernism. It will take a superluminal SF journey into elsewhere to find the interesting new vantage points. Where Latour and I fundamentally agree is that in that gravity well, into which Nature and Society as transcedentals disappeared, are to be found actors/actants of many and wonderful kinds. Their relationships constitute the artifactualism I am trying to sketch.

7. For quite another view of "production" and "reproduction" than that enshrined in so much Western political and economic (and feminist) theory, see Marilyn Strathern (1988, pp. 290–308).

8. Chela Sandoval develops the distinctions between oppositional and differential consciousness in her doctoral dissertation, "Dis-illusionment and the Poetry of the Future," University of California at Santa Cruz, 1984. See also Sandoval (1990).

9. My debt is extensive in these paragraphs to Luce Irigaray's wonderful critique of the allegory of the cave in *Spæculum de l'autre femme* (1974). Unfortunately, Irigaray, like almost all white Europeans and Americans after the mid-nineteenth-century consolidation of the myth that the "West" originated in a classical Greece unsullied by Semitic and African roots, transplants, colonizations, and loans, never questioned the "original" status of Plato's fathership of philosophy, enlightenment, and rationality. If Europe was colonized first by Africans, that historical narrative element would change the story of the birth of Western philosophy and science. Martin Bernal's extraordinarily important book, *Black Athena*, Vol. 1, *The Fabrication of Ancient Greece, 1785–1985* (1987), initiates a groundbreaking re-evaluation of the founding premises of the myth of the uniqueness and self-generation of Western culture, most certainly including those pinnacles of Man's self-birthing, science and philosophy. Bernal's is an account of the determinative role of racism and Romanticism in the fabrication of the story of Western rationality. Perhaps ironically, Martin Bernal is the son of J. D. Bernal, the major pre–World War II British biochemist and Marxist whose four-volume *Science in History* movingly argued the superior rationality of a science freed from the chains of capitalism. Science, freedom, and socialism were to be, finally, the legacy of the West. For all its warts, that surely would have been better than Reagan's and Thatcher's version! See Gary Wersky, *The Invisible College: The Collective Biography of British Socialist Scientists in the 1930s* (1978).

Famous in his own generation for his passionate heterosexual affairs, J. D. Bernal, in the image of enlightenment second birthing so wryly exposed by Irigaray, wrote his own vision of the future in *The Word, the Flesh, and the Devil* as a science-based speculation that had human beings evolving into disembodied intelligences. In her manuscript (May 1990) "Talking about Science in Three Colors: Bernal and Gender Politics in the Social Studies of Science," Hilary Rose discusses this fantasy and its importance for "science, politics, and silences." J. D. Bernal was also actively supportive of independent women scientists. Rosalind Franklin moved to his laboratory after her nucleic acid crystallographic work was stolen by the flamboyantly sexist and heroic James Watson on his way to the immortalizing, luminous fame of the *Double Helix* of the 1950s and 60s and its replicant of the 1980s and 90s, the Human Genome Project. The *story* of DNA has been an archetypal tale of blinding modern enlightenment and untrammeled, disembodied, autochthonous origins. See Ann Sayre (1975); Mary Jacobus (1982); Evelyn Fox Keller (1990).

10. For an argument that nature is a *social* actor, see Elizabeth Bird (1987).

11. Actants are not the same as actors. As Terence Hawkes (1977, p. 89) put it in his introduction to Greimas, actants operate at the level of function, not of character. Several characters in a narrative may make up a single actant. The structure of the narrative generates its actants. In considering what kind of entity "nature" might be, I am looking for a coyote and historical grammar of the world, where deep structure can be quite a surprise, indeed, a veritable trickster. Non-humans are not necessarily "actors" in the human sense, but they are part of the functional collective that makes up an actant. Action is not so much an ontological as a semiotic problem. This is perhaps as true for humans as non-humans, a way of looking at things that may provide exits from the methodological individualism inherent in concentrating constantly on who the agents and actors are in the sense of liberal theories of agency.

12. In this productionist story, women make babies, but this is a poor if necessary substitute for the real action in reproduction—the second birth through self-birthing, which requires the obstetrical technology of optics. One's relation to the phallus determines whether one gives birth to oneself, at quite a price, or serves, at an even greater price, as the conduit or passage for those who will enter the light of self-birthing. For a refreshing demonstration that women do not make babies everywhere, see Marilyn Strathern (1988), pp. 314–18.

13. I borrow here from Katie King's notion of the apparatus of literary production, in which the poem congeals at the intersection of business, art, and technology. See King (1990). See also Donna Haraway (1991), chaps. 8–10.

14. Latour has developed the concept of delegation to refer to the translations and exchanges between and among people doing science and their machines, which act as "delegates" in a wide array of ways. Marx considered machines to be "dead labor," but that notion, while still necessary for some crucial aspects of forced and reified delegation, is too unlively to get at the many ways that machines are part of *social* relations "through which actants shift competences" Latour (1990, p. 170). See also Bruno Latour (1994). Latour, however, as well as most of the established scholars in the social studies of science, ends up with too narrow a concept of the "collective," one built up out of only machines and scientists, who are considered in a very narrow time and space frame. But circulations of skills turn out to take some stranger turns. First, with the important exception of his writing and teaching in collaboration with the primatologist Shirley Strum, who has fought hard in her profession for recognition of primates as savvy social actors, Latour pays too little attention to the non-machine, *other* non-humans in the interactions. See Strum (1987).

    The "collective," of which "nature" in any form is one example from my point of view, is always an artifact, always social, not because of some transcendental Social that explains science or vice versa, but because of its heterogeneous actants/actors. Not only are not all of those actors/actants people; I agree there *is* a sociology of machines. But that is not enough; not all of the other actors/actants were *built* by people. The artifactual "collective" includes a witty actor that I have sometimes called coyote. The interfaces that constitute the "collective" must include those between humans and artifacts in the form of instruments and machines, a genuinely *social* landscape. But the interface between machines and *other* non-humans, as well as the interface between humans and *non-machine* non-humans, must also be counted in. Animals are fairly obvious actors, and their interfaces with people and machines are easier to admit and theorize. See Donna Haraway (1989a); Barbara Noske (1989). Paradoxically, from the perspective of the kind of artifactualism I am trying to sketch, animals lose their *object* status that has reduced them to things in so much Western philosophy and practice. They inhabit neither nature (as object) nor culture (as surrogate human), but instead inhabit a place called elsewhere. In Noske's terms (p. xi), they are other "worlds, whose otherworldliness must not be disenchanted and cut to our size but respected for what it is." Animals, however, do not exhaust the coyote world of non-machine non-humans. The domain of machine and non-machine non-humans (the unhuman, in my terminology) joins people in the building of the artifactual collective called nature. None of these actants can be considered as simply resource, ground, matrix, object, material, instrument, frozen labor; they are all more unsettling than that. Perhaps my suggestions here come down to re-inventing an old option within a non-Eurocentric Western tradition indebted to Egyptian Hermeticism that insists on the active quality of the world and on "animate" matter. See Martin Bernal (1987, pp. 121–60); Frances Yates (1964). Worldly and enspirited, coyote nature is a collective, cosmopolitan artifact crafted in stories with heterogeneous actants.

But there is a second way in which Latour and other major figures in science studies work with an impoverished "collective." Correctly working to resist a "social" explanation of "technical" practice by exploding the binary, these scholars have a tendency covertly to reintroduce the binary by worshipping only one term—the "technical." Especially, *any* consideration of matters like masculine supremacy or racism or imperialism or class structures are inadmissible because they are the old "social" ghosts that blocked real explanation of science in action. See Latour (1987). As Latour noted, Michael Lynch is the most radical proponent of the premise that there is no social explanation of a science but the technical content itself, which assuredly includes the interactions of people with each other in the lab and with their machines, but excludes a great deal that I would include in the "technical" content of science if one really doesn't want to evade a binary by worshipping one of its old poles. Lynch (1985); Latour (1990, p. 169n). I agree with Latour and Lynch that practice creates its own context, but they draw a suspicious line around what gets to count as "practice." They *never* ask how the *practices* of masculine supremacy, or many other systems of structured inequality, get *built* into and out of working machines. How and in what directions these transferences of "competences" work should be a focus of rapt attention. Systems of exploitation might be crucial parts of the "technical content" of science. But the SSS scholars tend to dismiss such questions with the assertion that they lead to the bad old days when science was asserted by radicals simply to "reflect" social relations. But in my view, such transferences of competences, or delegations, have nothing to do with reflections or harmonies of social organization and cosmologies, like "modern science." Their unexamined, consistent, and defensive prejudice seems part of Latour's (1990, pp. 164–69) stunning misreading of several moves in Sharon Traweek's *Beam Times and Life Times: The World of High Energy Physicists* (1988). See also Hilary Rose, "Science in Three Colours: Bernal and Gender Politics in the Social Studies of Science," unpublished manuscript, May 2, 1990.

The same blind spot, a retinal lesion from the old phallogocentric heliotropism that Latour *did* know how to avoid in other contexts, for example in his trenchant critique of the modern and postmodern, seems responsible for the abject failure of the social studies of science as an organized discourse to take account of the last twenty years of feminist inquiry. What counts as "technical" and what counts as "practice" should remain far from self-evident in science in action. For all of their extraordinary creativity, so far the mappings from most SSS scholars have stopped dead at the fearful seas where the worldly practices of inequality lap at the shores, infiltrate the estuaries, and set the parameters of reproduction of scientific practice, artifacts, and knowledge. If only it were a question of reflections between social relations and scientific constructions, how easy it would be to conduct "political" inquiry into science! Perhaps the tenacious prejudice of the SSS professionals is the punishment for the enlightenment transcendental, the social, that did inform the rationalism of earlier generations of radical science critique and is still all too common. May the topick gods save us from both the reified technical and the transcendental social!

15. See Lynn Margulis and Dorion Sagan (1986). This wonderful book does the cell biology and evolution for a host of inappropriate/d others. In its dedication, the text affirms "the combinations, sexual and parasexual, that bring us out of ourselves and make us more than we are alone" (p. v). That should be what science studies as cultural studies do, by showing how to visualize the curious collectives of humans and unhumans that make up naturalsocial (one word) life. To stress the point that all the actors in these generative, dispersed, and layered collectives do not have human form and function—and should not be anthropomorphized—recall that the Gaia hypothesis with which Margulis is associated is about the tissue of the planet as a living entity, whose metabolism and genetic exchange are effected through webs of prokaryotes. Gaia is a society; Gaia is nature; Gaia did not read the *Critique*. Neither, probably, did John Varley. See his Gaea hypothesis in the SF book, *Titan* (1979). Titan is an alien that is a world.

16. Remember that *monsters* have the same root as *to demonstrate*; monsters signify.

17. Trinh T. Minh-ha, ed., 1986/7b, *She, the Inappropriate/d Other*. See also her *Woman, Native, Other: Writing Postcoloniality and Feminism* (1989).

18. Interpellate: I play on Althusser's account of the call which constitutes the production of the subject in ideology. Althusser is, of course, playing on Lacan, not to mention on God's interruption that calls Man, his servant, into being. Do we have a vocation to be cyborgs? Interpellate: *Interpellatus*, past participle for "interrupted in speaking"—effecting transformations like Saul into Paul. Interpellation is a special kind of interruption, to say the least. Its key meaning concerns a procedure in a parliament for asking a speaker who is a member of the government to provide an explanation of an act or policy, usually leading to a vote of confidence. The following ads interrupt us. They insist on an explanation in a confidence game; they force recognition of how transfers of competences are made. A cyborg subject position results from and leads to interruption, diffraction, reinvention. It is dangerous and replete with the promises of monsters.

19. In *King Solomon's Ring*, Konrad Lorenz pointed out how the railroad car kept the appearance of the horse drawn carriage, despite the different functional requirements and possibilities of the new technology. He meant to illustrate that biological evolution is similarity conservative, almost nostalgic for the old, familiar forms, which are reworked to new purposes. Gaia was the first serious bricoleuse.

20. For a view of the manufacture of particular organisms as flexible model systems for a universe of research practice, see Barbara R. Jasny and Daniel Koshland, Jr., eds., *Biological Systems* (1990). As the advertising for the book states, "The information presented will be especially useful to graduate students and to all researchers interested in learning the limitations and assets of biological systems currently in use," *Science* 248 (1990), p. 1024. Like all forms of protoplasm collected in the extra-laboratory world and brought into a technoscientific niche, the organic rabbit (not to mention the simulated one) and its tissues have a probable future of a particular sort—as a commodity. Who should "own" such evolutionary products? If seed protoplasm is collected in peasants' fields in Peru and then used to breed valuable commercial seed in a "first world" lab, does a peasant cooperative or the Peruvian state have a claim on the profits? A related problem about proprietary interest in "nature" besets the biotechnology industry's development of cell lines and other products derived from removed human tissue, e.g., as a result of cancer surgery. The California Supreme Court recently reassured the biotechnology industry that a patient, whose cancerous spleen was the source of a product, Colony Stimulating Factor, that led to a patent that brought its scientist-developer stock in a company worth about $3 million, did not have a right to a share of the bonanza. Property in the self, that lynchpin of liberal existence, does not seem to be the same thing as proprietary rights in one's body or its products—like fetuses

or other cell lines in which the courts take a regulatory interest. See Marcia Barinaga (1990, p. 239).

21. Here and throughout this essay, I play on Katie King's play on Jacques Derrida's *Of Grammatology* (1976). See King (1990), and also King's manuscript "Feminism and Writing Technologies" (available online from King's website at the University of Maryland).

22. Roland Barthes, *Mythologies* (1972) is my guide here and elsewhere.

23. Peace-activist and scholar in science studies, Elizabeth Bird came up with the slogan and put it on a political button in 1986 in Santa Cruz, California.

24. I am indebted to another guide figure throughout this essay, Gloria Anzaldúa, *Borderlands, La Frontera: The New Mestiza* (1987) and to at least two other travelers in embodied virtual spaces, Ramona Fernandez, "Trickster Literacy: Multiculturalism and the (Re) Invention of Learning," Qualifying Essay, History of Consciousness, University of California at Santa Cruz, 1990, and Allucquére R. Stone, "Following Virtual Communities," unpublished essay, History of Consciousness, University of California at Santa Cruz. The ramifying "virtual consensual community" (Sandy Stone's term in another context) of feminist theory that incubates at UCSC densely infiltrates my writing.

25. For an extended reading of *National Geographic*'s Jane Goodall stories, *always to be held in tension with other versions of Goodall and the chimpanzees at Gombe,* see Haraway, "Apes in Eden, Apes in Space," in *Primate Visions* (1989, pp. 133–95). Nothing in my analysis should be taken as grounds to oppose primate conservation or to make claims about the other Jane Goodalls; those are complex matters that deserve their own careful, materially specific consideration. My point is about the semiotic and political frames within which survival work might be approached by geo-politically differentiated actors.

26. My files are replete with recent images of cross-species ape-human family romance that fail to paper over the underlying racist iconography. The most viciously racist image was shown to me by Paula Treichler: an ad directed to physicians by the HMO Premed in Minneapolis, from the *American Medical News*, August 7, 1987. A white-coated white man, stethoscope around his neck, is putting a wedding ring on the hand of an ugly, very black, gorilla-suited female dressed in a white wedding gown. White clothing does not mean the same thing for the different races, species, and genders! The ad proclaims, "If you've made an unholy HMO alliance, perhaps we can help." The white male physician (man) tied to the black female patient (animal) in the inner cities by HMO marketing practices in relation to Medicaid policies must be freed. There is no woman in this ad; there is a hidden threat disguised as an ape female, dressed as the vampirish bride of scientific medicine (a single white tooth gleams menacingly against the black lips of the ugly bride)—another illustration, if we needed one, that black women do not have the *discursive* status of woman/human in white culture. "All across the country, physicians who once had visions of a beautiful marriage to an HMO have discovered the honeymoon is over. Instead of quality care and a fiscally sound patient-base, they end up accepting reduced fees and increased risks." The codes are transparent. Scientific medicine has been tricked into a union with vampirish poor black female patients. Which risks are borne by whom goes unexamined. The clasped hands in this ad carry a different surface message from the Gulf ad's, but their enabling semiotic structures share too much.

27. At the oral presentation of this paper at the conference on "Cultural Studies Now and in the Future," Gloria Watklins/bell hooks pointed out the painful current U.S. discourse on African-American men as "an endangered species." Built into that awful metaphor is a relentless history of animalization and political infantilization. Like other "endangered species," such people cannot speak for themselves, but must be

spoken for. They must be represented. Who speaks for the African-American man as "an endangered species"? Note also how the metaphor applied to black *men* justifies anti-feminist and misogynist rhetoric about and policy toward black women. They actually become one of the forces, if not the chief threat, endangering African-American men.

28. Committing only a neo-imperialist venial sin in a footnote, I yield to voyeuristic temptation just a little: in *Discover* the videocam and the "native" have a relation symmetrical to that of Goodall's and the chimpanzee's hands. Each photo represents a touch across time and space, and across politics and history, to tell a story of salvation, of saving man and nature. In this version of cyborg narrative, the touch that joins portable high technology and "primitive" human parallels the touch that joins animal and "civilized" human.

29. It is, however, important to note that the present man in charge of environmental affairs in the Amazon in the Brazilian government has taken strong, progressive stands on conservation, human rights, destruction of indigenous peoples, and the links of ecology and justice. Further, current proposals and policies, like the government's plan called *Nossa Natureza* and some international aid and conservation organizations' activities and ecologists' understandings, have much to recommend them. In addition, unless arrogance exceeds all bounds, *I* can hardly claim to adjudicate these complex matters. The point of my argument is not that whatever comes from Brasilia or Washington is bad and whatever from the forest residents is good—a patently untrue position. Nor is it my point that nobody who doesn't come from a family that has lived in the forest for generations has any place in the "collectives, human and unhuman," crucial to the survival of lives and ways of life in Amazonia and elsewhere. Rather, the point is about the self-constitution of the indigenous peoples as *principal* actors and agents, with whom others must interact—in coalition and in conflict—not the reverse.

30. For the story of Mendes's life work and his murder by opponents of an extractive reserve off limits to logging, see Andrew Revkin (1990).

31. Further references are parenthetical in the text.

32. Similar issues confront Amazonians in countries other than Brazil. For example, there are national parks in Colombia from which native peoples are banned from their historical territory, but to which loggers and oil companies have access under park multi-use policy. This should sound very familiar to North Americans, as well.

33. Revising and displacing his statements, I am again in conversation with Bruno Latour here, who has insisted on the social status of both human and non-human actors. "We use actor to mean anything that is made by some other actor the source of an action. It is in no way limited to humans. It does not imply will, voice, self-consciousness or desire." Latour makes the crucial point that "figuring" (in words or in other matter) non-human actors as if they were like people is a semiotic operation; non-figural characterizations are quite possible. The likeness or unlikeness of actors is an interesting problem opened up by placing them firmly in the shared domain of social interaction. Bruno Latour (1994).

34. Kane's review appeared in the *Voice Literary Supplement,* February 1990, and Hecht and Cockburn replied under the title "Getting Historical," *Voice Literary Supplement,* March 1990, p. 26.

35. My discussion of the politics of representation of the fetus depends on twenty years of feminist discourse about the location of responsibility in pregnancy and about reproductive freedom and constraint generally. For particularly crucial arguments for this essay, see Jennifer Terry (1989); Valerie Hartouni (1991); and Rosalind Pollock Petchesky (1987).

36. *The Eighteenth Brumaire of Louis Bonaparte.* Quoted in Edward Said (1978, p. xiii), as his opening epigraph.

37.   Marilyn Strathern describes Melanesian notions of a child as the "finished repository of the actions of multiple others," and not, as among Westerners, a resource to be constructed into a fully human being through socialization by others. Marilyn Strathern, "Between Things: A Melanesianist's Comment on Deconstructive Feminism," unpublished manuscript. Western feminists have been struggling to articulate a phenomenology of pregnancy that rejects the dominant cultural framework of productionism/reproductionism, with its logic of passive resource and active technologist. In these efforts the woman-fetus nexus is refigured as a knot of relationality within a wider web, where liberal individuals are not the actors, but where complex collectives, including non-liberal social persons (singular and plural), are. Similar refigurings appear in ecofeminist discourse.

38.   See the fall 1990 newsletter of the Society for the Social Study of Science, *Technoscience* 3, no. 3, pp. 20, 22, for language about "going back to nature." A session of the 4S October meetings is titled "Back to Nature." Malcolm Ashmore's abstract, "With a Reflexive Sociology of Actants, There Is No Going Back," offers "fully comprehensive insurance against going back," instead of other competitors' less good "ways of not going back to Nature (or Society or Self)." All of this occurs in the context of a crisis of confidence among many 4S scholars that their very fruitful research programs of the last 10 years are running into dead ends. They are. I will refrain from commenting on the blatant misogyny in the Western scholar's textualized terror of "going back" to a phantastic nature (figured by science critics as "objective" nature. Literary academicians figure the same terrible dangers slightly differently; for both groups such a nature is definitively pre-social, monstrously not-human, and a threat to their careers). Mother nature always waits, in these adolescent boys' narratives, to smother the newly individuated hero. He forgets this weird mother is his creation; the forgetting, or the inversion, is basic to ideologies of scientific objectivity and of nature as "eden under glass." It also plays a yet-to-be-examined role in some of the best (most reflexive) science studies. A theoretical gender analysis is indispensable to the reflexive task.

39.   *Time*, February 10, 1961, p. 58. The caption under HAM's photograph read "from Chop Chop Chang to No. 65 to a pioneering role." For HAM's flight and the Holloman chimps' training see Weaver (1961) and *Life Magazine*, February 10, 1961. *Life* headlined, "From Jungles to the Lab: The Astrochimps." All were captured from Africa; that means many other chimps died in the "harvest" of babies. The astrochimps were chosen over other chimps for, among other things, "high IQ." Good scientists all.

40.   *Time*, December 8, 1961, p. 50; *Newsweek*, March 5, 1962, p. 19.

41.   *Time*, December 8, 1961, p. 50.

42.   See also Chris Gray, "Postmodern War," Qualifying Exam, History of Consciousness, UCSC, 1988.

43.   For indispensable theoretical and participant-observation writings on eco-feminism, social movements, and non-violent direct action, see Barbara Epstein (1991).

44.   For a fuller discussion of the immune system, see Haraway, "The Biopolitics of Postmodern Bodies," in *Simians, Cyborgs, and Women* (1991).

45.   Recall that Nilsson shot the famous and discourse-changing photographs of fetuses (really abortuses) as glowing back-lit universes floating free of the "maternal environment." Nilsson (1977).

46.   Advertising copy for the Met Life Pavilion. The exhibit is sponsored by the Metropolitan Life and Affiliated Companies. In the campground resort at Florida's Walt Disney World, we may also view the "endangered species island," in order to learn the conventions for "speaking for the jaguar" in an eden under glass.

47.   Ramona Fernandez, "Trickster Literacy," Qualifying Exam, History of Consciousness, UCSC, 1990, wrote extensively on Walt Disney World and the multiple cultural literacies required and taught on-site for successfully traveling there. Her essay described the visualizing technology and medical school collaboration in its development and use.

See the *Journal of the American Medical Association* 260, no. 18 (November 18, 1988), pp. 2776–83.

48. Building an unexpected collective, Jerne (1985) drew directly from Noam Chomsky's theories of structural linguistics. The "textualized" semiotic body is not news by the late twentieth century, but what kind of textuality is put into play still matters!

49. See, for example, the recent merger of Project Inform with the Community Research Alliance to speed the community-based testing of promising drugs—and the NIH's efforts to deal with these developments: *PI Perspective,* May 1990. Note also the differences between President Bush's Secretary of Health and Human Services, Lewis Sullivan, and Director of the National Institute of Allergy and Infectious Diseases, Anthony Fauci, on dealing with activists and PWAs. After ACT UP demonstrations against his and Bush's policies during the secretary's speech at the AIDS conference in San Francisco in June 1990, Sullivan said he would have no more to do with ACT UP and instructed government officials to limit their contacts. (Bush had been invited to address the international San Francisco conference, but his schedule did not permit it. He was in North Carolina raising money for the ultra-reactionary senator Jesse Helms at the time of the conference.) In July 1990, at the ninth meeting of the AIDS Clinical Trials Group (ACTG), at which patient activists participated for the first time, Fauci said that he would work to include the AIDS constituency at every level of the NIAID process of clinical trials. He urged scientists to develop the skills to discuss freely in those contexts ("Fauci," 1990). Why is constructing this kind of scientific articulation "softer"? I leave the answer to readers' imaginations informed by decades of feminist theory.

50. This quadrant of the semiotic square is dedicated to A. E. Van Vogt's *Players of Null-A* (1974), for their non-Aristotelian adventures. An earlier version of "The Promises of Monsters" had the imagination, not SF, in virtual space. I am indebted to a questioner who insisted that the imagination was a nineteenth-century faculty that is in political and epistemological opposition to the arguments I am trying to formulate. As I am trying vainly to skirt psychoanalysis, I must also skirt the slough of the romantic imagination.

51. Allucquére R. Stone, "Following Virtual Communities," unpublished manuscript, History of Consciousness, UCSC, 1990.

52. Thanks to Barbara Ige, graduate student in the Literature Board, UCSC, for conversations about our stakes in the figure of Lisa Foo.

53. Oil on canvas, 36" by 28", photo by D. Caras. In conversation with the 1985 essay "A Manifesto for Cyborgs" (in Haraway, 1991), Randolph painted her *Cyborg* while at the Bunting Institute and exhibited it there in a spring 1990 solo exhibition, titled "A Return to Alien Roots." The show incorporated, from many sources, "traditional religious imagery with a postmodern secularized context." Randolph paints "images that empower women, magnify dreams, and cross racial, class, gender, and age barriers" (exhibition brochure). Living and painting in Texas, Randolph was an organizer of the Houston Area Artists' Call Against U.S. Intervention in Central America. The human model for *Cyborg* was Grace Li, from Beijing, who was at the Bunting Institute in the fateful year of 1989.

54. I borrow this use of "conversation" and the notion of transnational feminist literacy from Katie King's concept of women and writing technologies. See note 21.

REFERENCES

Anzaldúa, Gloria. 1987. *Borderlands, La Frontera: The New Mestiza.* San Francisco: Spinsters/Aunt Lute.

Arena-DeRosa, James. 1990. "Indigenous Leaders Host U.S. Environmentalists in the Amazon." *Oxfam America News,* Summer/Fall:1–2.

Barinaga, Marcia. 1990. "A Muted Victory for the Biotech Industry." *Science* 249 (20 July).

Barthes, Roland. 1972. *Mythologies*. Trans. A. Lavers. London: Cape.

Bernal, Martin. 1987. *Black Athena, vol. 1, The Fabrication of Ancient Greece, 1785–1985*. London: Free Association Books.

Bird, Elizabeth. 1987. "The Social Construction of Nature: Theoretical Approaches to the History of Environmental Problems." *Environmental Review* 11(4): 255–64.

Bryan, C. 1987. *The National Geographic Society: 100 Years of Adventure and Discovery*. New York: Abrams.

Cohn, C. 1987. "Sex and Death in the Rational World of Defense Intellectuals." *Signs* 12(4): 687–718.

Crimp, Douglas. 1983. "On the Museum's Ruins," in H. Foster, ed. *The Anti-Aesthetic*. Port Townsend, WA: Bay Press, pp. 43–56.

Crimp, Douglas, and A. Rolston. 1990. *AIDSDEMOGRAPHICS*. Seattle: Bay Press.

de Lauretis, Teresa. 1984. *Alice Doesn't*. Bloomington: Indiana University Press.

Eimerl, Sarel, and Irven DeVore. 1965. *The Primates*. New York: Time, Inc.

Epstein, Barbara. 1991. *Political Protest and Cultural Revolution*. Berkeley: University of California Press.

"Fauci Gets Softer on Activists." 1990. *Science* 249:244.

Franklin, Sarah. 1988. "Life Story: The Gene as Fetish Object on TV," *Science as Culture* 3:92–100.

Gibson, William. 1986. *Neuromancer*. New York: Ace.

Golub, Edward. 1987. *Immunology: A Synthesis*. Sunderland, MA: Sinauer.

Goodall, Jane. 1971. *In the Shadow of Man*. Boston: Houghton Mifflin.

Greimas, A. J. 1966. *Sémantique Structurale*. Paris: Larousse.

Hacker, Sally. 1989. *Pleasure, Power, and Technology: Some Tales of Gender, Engineering, and the Cooperative Workplace*. Boston: Unwin Hyman.

Haraway, Donna. 1988. "Situated Knowledge." *Feminist Studies* 14(3): 575–99.

———. 1989a. *Pimate Visions: Gender, Race, and Nature in the World of Modern Science*. New York: Routledge.

———. 1989b. "Technics, Erotics, Vision, Touch: Fantasies of the Designer Body." Talk presented at the meetings of the Society for the History of Technology, 13 October.

———. 1991. *Simians, Cyborgs, and Women: The Reinvention of Nature*. New York: Routledge.

Hartouni, Valerie. 1991. "Containing Women: Reproductive Discourse in the 1980s." In *Technoculture*, edited by C. Penley and A. Ross. Minneapolis: University of Minnesota Press.

Hawkes, Terence. 1977. *Structuralism and Semiotics*. Berkeley: University of California Press.

Hayles, N. Katherine. 1990. *Chaos Bound: Orderly Disorder in Contemporary Literature and Science*. Ithaca: Cornell University Press, 265–95.

Hecht, Susanna, and Alexander Cockburn. 1989. *The Fate of the Forest: Developers, Destroyers, and Defenders of the Amazon*. New York: Verso.

Hubbard, Ruth. 1990. "Technology and Childbearing." In *The Politics of Women's Biology*. New Brunswick, NJ: Rutgers University Press.

Irigaray, Luce. 1974. *Spéculum de l'autre femme*. Paris: Minuit. 1985. *Speculum of the Other Woman*. Trans. G. Gill. Ithaca: Cornell University Press.

Jacobus, Mary. 1982. "Is There a Woman in This Text?" *New Literary History* 14:117–41.

Jameson, Fredric. 1972. *The Prison-House of Language*. Princeton: Princeton University Press.

Jaret, Peter. 1986. "Our Immune System: The Wars Within." *National Geographic* 169(6): 701–35.

Jasny, Barbara, and Daniel Koshland, Jr., eds. 1990. *Biological Systems*. Washington, D.C.: AAAS Books.

Jerne, N. 1985. "The Generative Grammar of the Immune System." *Science* 229:1057–59.

Kane, Joe. 1989. *Running the Amazon*. New York: Knopf.

Keller, Evelyn Fox. 1990. "From Secrets of Life to Secrets of Death." In *Body/Politics: Women and the Discourse of Science,* edited by M. Jacobus, E. Fox Keller, and S. Shuttlworth. New York: Routledge, 177–91.

King, Katie. 1990. "A Feminist Apparatus of Literary Production." *Text* 5: 91–103.

———. 1994. *Theory in Its Feminist Travels: Conversations in U.S. Women's Movements.* Bloomington: Indiana University Press.

Latour, Bruno. 1987. *Science in Action: How to Follow Scientists and Engineers Through Society.* Cambridge, MA: Harvard University Press.

———. 1990. "Postmodern? No, Simply Amodern! Steps towards an Anthropology of Science." *Studies in the History and Philosophy of Science* 21(1): 145–71.

———. 1992. "One More Turn after the Social Turn. Easing Science Studies into the Non-Modern World." In *The Social Dimensions of Science,* edited by E. Mullin. Notre Dame: Notre Dame University Press.

———. 1992. "Where Are the Missing Masses? The Sociology of a Few Mundane Artifacts." In *Shaping Technology / Building Society. Studies in Sociotechnical Change,* edited by W. Bijker and J. Law. Cambridge MA: The MIT Press, 225–58.

———. 1993. *We Have Never Been Modern,* transl. Catherine Porter. Cambridge: Harvard University Press.

Lynch, Michael. 1985. *Art and Artifact in Laboratory Science: A Study of Shop Work and Shop Talk in a Research Laboratory.* London: Routledge.

Margulis, Lynn, and Dorian Sagan. 1986. *Origins of Sex: Three Billion Years of Genetic Recombination.* New Haven: Yale University Press.

Nilsson, Lennart. 1977. *A Child Is Born.* New York: Dell.

———. 1987. *The Body Victorious: The Illustrated Story of Our Immune System and Other Defenses of the Huan Body.* New York: Delacourt.

Noske, Barbara. 1989. *Humans and Other Animals: Beyond the Boundaries of Anthropology.* London: Pluto Press.

Petchesky, Rosalind Pollock. 1987. "Fetal Images: The Power of Visual Culture in the Politics of Reproduction." *Feminist Studies* 13(2): 263–92.

Playfair, J. 1984. *Immunology at a Glance,* 3rd ed. Oxford: Blackwell.

Revkin, Andrew. 1990. *The Burning Season.* New York: Houghton Mifflin.

Said, Edward. 1978. *Orientalism.* New York: Random House. 1985. Harmondsworth: Penguin.

Sandoval, Chela. 1990. "Feminism and Racism." In *Making Face, Making Soul: Hciendo Caras,* edited by Gloria Anzaldúa. San Francisco: Aunt Lute, 55–71.

Sayre, Ann. 1975. *Rosalind Franklin and DNA.* New York: Norton.

Sofia, Zoe. 1984. "Exterminating Fetuses: Abortion, Disarmament, and the Sexo-semiotics of Extraterrestrialism." *Diacritics* 14(2): 47–59.

Strathern, Marilyn. 1988. *The Gender of the Gift.* Berkeley: University of California Press.

Strum, S. 1987. *Almost Home: A Journey into the World of Baboons.* New York: Random House.

Terry, Jennifer 1989. "The Body Invaded: Medical Surveillance of Women as Reproducers." *Socialist Review* 19(3): 13–43.

Timmerman, Col. Frederick. 1987. "Future Warriors." *Military Review,* September: 44–55.

Traweek, Sharon. 1988. *Beam Times and Life Times: The World of High Energy Physicists.* Cambridge, MA: Harvard University Press.

Trinh T. Minh-ha, 1986/7. *She, The Inappropriated Other. Discourse* 8.

———. 1989. *Woman, Native, Other: Writing Post-coloniality and Feminism.* Bloomington: Indiana University Press.

Turner, T. 1990. "Visual Media, Cultural Politics, and Anthropological Practice: Some Implications of Recent Uses of Film and Video among the Kaiapo of Brazil." *Commission on Visual Anthropology Review,* Spring.

Van Vogt, A. E. 1974 [1948]. *Players of Null-A.* New York: Berkeley Books.

Varley, John. 1979. *Titan.* New York: Berkeley Books.

———. 1986. "Press Enter." In *Blue Champagne*. New York: Berkeley Books.

Virilio, Paul, Sylvere Lotringer. 1983. *Pure War*. New Yorke: Semiotext(e).

Weaver, Kenneth. 1961. "Countdown for Space." *National Geographic* 119(5): 702–34.

Wersky, Gary. 1978. *The Invisible College: The Collective Biography of British Socialist Scientists in the 1930s*. London: Allen Lane.

Yates, Frances. 1964. *Giordano Bruno and the Hermetic Tradition*. London: Routledge.

Zimmer, Carl. 1990. "Tech in the Jungle." *Discover*, August: 42.

4

OTHERWORLDLY CONVERSATIONS; TERRAN TOPICS; LOCAL TERMS

Therefore the Lord God sent him forth from the garden of Eden, to till the
ground from whence he was taken. So he drove out the man; and he placed
at the east of the garden of Eden Cherubims, and a flaming sword which
turned every way, to keep the way of the tree of life.

—Genesis 3:23–24

Nothing is ultimately contextual; all is constitutive, which is another way of
saying that all relationships are dialectical.

—Robert Young, *Darwin's Metaphor*

Animals are not lesser humans; they are other worlds.

—Barbara Noske, *Humans and Other Animals*

Although, of course, I longed in the normal way for exploration, I found my
first world oddly disconcerting. . . . It is only in circumstances like these that
we realise how much we ourselves are constructed bilaterally on either–or
principles. Fish rather than echinoderms . . . It was quite a problem to get
through to those radial entities.

—Naomi Mitchison, *Memoirs of a Spacewoman*

Nature is for me, and I venture for many of us who are planetary foe-
tuses gestating in the amniotic effluvia of terminal industrialism and
militarism, one of those impossible things characterized during a talk in
1989 in California by Gayatri Spivak as that which we cannot not desire.
Excruciatingly conscious of nature's constitution as Other in the histories
of colonialism, racism, sexism and class domination of many kinds, many
people who have been both ground to powder and formed in European
and Euro-American crucibles none the less find in this problematic,

ethno-specific, long-lived and globally mobile concept something we cannot do without, but can never "have." We must find another relationship to nature besides reification, possession, appropriation and nostalgia. No longer able to sustain the fictions of being either subjects or objects, all the partners in the potent conversations that constitute nature must find a new ground for making meanings together.[1]

Perhaps to give confidence in its essential reality, immense resources have been expended to stabilize and materialize nature, to police its/her boundaries. From one reading of Genesis 3:23–24, it looks like God established the first nature park in the neolithic First World, now become the oil-rich Third World, complete with an armed guard to keep out the agriculturalists. From the beginning such efforts have had disappointing results. Efforts to travel into "nature" become tourist excursions that remind the voyager of the price of such displacements—one pays to see fun-house reflections of oneself. Efforts to preserve "nature" in parks remain fatally troubled by the ineradicable mark of the founding expulsion of those who used to live there, not as innocents in a garden, but as people for whom the categories of nature and culture were not the salient ones.

Expensive projects to collect "nature's" diversity and bank it seem to produce debased coin, impoverished seed and dusty relics. As the banks hypertrophy, the nature that feeds the storehouses disappears. The World Bank's record on environmental destruction is exemplary in this regard. Finally, the projects for representing and enforcing human "nature" are famous for their imperializing essences, most recently replicated in the Human Genome Project. It seems appropriate that a core computer project for storing the record of human unity and diversity, GenBank, the U.S. depository for DNA sequence data, is located at the national laboratories at Los Alamos, New Mexico, site of the Manhattan Project and a major U.S. weapons laboratory since the Second World War.

So, nature is not just a physical place to which one can go, nor a treasure to fence in or bank, nor an essence to be saved or violated. Nature is not hidden and so does not need to be unveiled. Nature is not a text to be read in the codes of mathematics and biomedicine. It is not the Other who offers origin, replenishment and service. Neither mother, nurse, lover, nor slave, nature is not matrix, resource, mirror, nor tool for the reproduction of that odd, ethnocentric, phallogocentric, putatively universal being called Man. Nor for his euphemistically named surrogate, the "human."

Nature is, however, a *topos*, a place, in the sense of a rhetorician's place or topic for consideration of common themes; nature is, strictly, a commonplace. We turn to this topic to order our discourse, to compose our memory. As a topic in this sense, nature also reminds us that in seventeenth-century English the "topick gods" were the local gods, the gods specific to places and peoples. We need these spirits rhetorically if we can't have them any other way. We need them in order to reinhabit, precisely, *common* places—locations that are widely shared, inescapably local, worldly, enspirited; that is, topical. In this sense, nature is the place in which to rebuild public culture.[2]

Nature is also a *trópos,* a trope. It is figure, construction, artefact, movement, displacement. Nature cannot pre-exist its construction, its articulation in heterogeneous social encounters, where all of the actors are not human and all of the humans are not "us," however defined. Worlds are built from such articulations. Fruitful encounters depend on a particular kind of move—a *trópos* or "turn." Faithful to the Greek, as *trópos* nature is about turning. Troping, we turn to nature as if to the earth, to the tree of life—geotropic, physiotropic. We turn in the hope that the park police, the cherubims, are on strike against God and that both swords and ploughshares might be beaten into other tools, other metaphors for possible conversations about inhabitable terran other-worlds. Topically, we travel toward the earth, a commonplace. Nature is a topic of public discourse on which much turns, even the earth.

THREE STORIES

Less grandly, I turn to a little piece of this work of worldbuilding—telling stories. When I grow up, or, as we used to say, after the revolution, I know what I want to do. I want to have charge of the animal stories in the *Reader's Digest,* reaching twenty or so million people monthly in over a dozen languages. I want to write the stories about morally astute dogs, endangered people, instructive beetles, marvellous microbes and co-habitable houses of difference. With my friends, I want to write natural history at the end of the second Christian millennium to see if some other stories are possible, ones not premised on the divide between nature and culture, armed cherubims, and heroic quests for secrets of life and secrets of death.[3]

Following Ursula LeGuin, and inspired by some of the chapters in the evolutionary tales of woman-the-gatherer, I want to engage in a carrier-bag practice of storytelling, in which the stories do not reveal secrets acquired by heroes pursuing luminous objects across and through the plot matrix of the world. Bag-lady storytelling would instead proceed by putting unexpected partners and irreducible details into a frayed, porous carrier bag. Encouraging halting conversations, the encounter transmutes and reconstitutes all the partners and all the details. The stories do not have beginnings or ends; they have continuations, interruptions and reformulations—just the kind of survivable stories we could use these days. And, perhaps, my beginning with the transmogrification of LeGuin's Carrier-Bag Theory of Fiction[4] to the bag-lady practice of storytelling can remind us that the lurking dilemma in all of these tales is comprehensive homelessness, the lack of a common place, and the devastation of public culture.

In the United States, storytelling about nature, whatever problematic kind of category that is, remains an important practice for forging and expressing basic meanings. The profusion of nature television specials is a kind of collective video-Bridgewater Treatise, producing secularized natural theology within late capitalism. A recent visit to the San Diego Zoo confirmed my conviction that people reaffirm many of their beliefs about each other and about what kind of planet the earth can be by telling each other what they think they are seeing as they watch the animals. So, I would like to begin this meditation on three books, by Robert Young, Barbara Noske and Naomi Mitchison, with a few stories that reveal some of the investments I bring to reading their work.

A few years ago I was visiting my highschool friend, who lived with her husband and three sons, aged sixteen, fourteen and eleven, near Milwaukee, Wisconsin. Periodically throughout the weekend, the two older boys teased each other mercilessly about a highschool dance that was coming up; each boy tried to get under his brother's skin by queer-baiting him relentlessly. In this middle-class, white, American community, their patent nervousness about dating girls was enacted in "playful" insults about each other's not-yet-fully-consolidated gender allegiances and identities. In confused, but numbingly common moves, they accused each other of being simultaneously a girl and a queer. From my point of view, they were performing a required lesson in the compulsory heterosexuality of my culture and theirs.

I found the whole scene personally deeply painful for many reasons, not least the profoundly poor manners and disrespect the parents allowed, knowing the gay, lesbian and bisexual makeup of my life, family and community. My world is sustained by queer confederacies. Lacking courage and feeling disoriented, late in the weekend I told my friend what I thought was happening. Shocked, she said the boys were much too young to be taught anything about homosexuality and homophobia, and in any case what they were doing was just natural. Despite the fact that I was the godmother of the older boy, I culpably shut up, leaving his moral education to the proven sensibilities of his milieu.

Later that day, knowing my interest in another kind of nature and hoping to heal our dis-ease with each other by a culturally appropriate, therapeutic trip "outside civilization," my friend and her husband took me to a beautiful small lake in the wooded countryside. With high spirits, if little zoological erudition, we began talking about some ducks across the lake. We could see very little, and we knew less. In instant solidarity, my friend and her husband narrated that the four ducks in view were in two reproductive, heterosexual pairs. It quickly sounded like they had a modest mortgage on the wetlands around that section of the lake and were about to send their ducklings to a good school to consolidate their reproductive investment. I demurred, mumbling something about the complexity and specificity of animal behaviour and society. Meanwhile, I, of course, held that the ducks were into queer communities. I knew better; I knew they were *ducks,* even though I was embarrassed not to know their species. I knew ducks deserved our recognition of their *non-human* cultures, subjectivities, histories and material lives. They had enough problems with all the heavy metals and organic solvents in those lakes without having to take sides in our ideological struggles too. Forced to live in our ethno-specific constructions of nature, the birds could ill afford the luxury of getting embroiled in what counts as natural for the nearby community.

None the less, furious at each other, both my friends and I were sure we were right in our self-interested and increasingly assertive stories about the ducks. After all, we could *see* what they were doing; they were right across the lake; we had positive knowledge about them. They were objects performing on our stage, called nature. They had been appropriated into our shamefully displaced struggle, which belonged where earlier in the day we were too "chicken" to put it—directly over the homophobia, compulsory heterosexuality and commitments to normalizing particular kinds of families in *our* lives. We avoided building needed, contested, situated knowledges among ourselves by—once again, in ways historically sanctioned in middle-class, Anglo cultures—objectifying nature.

More sophisticated scientific accounts of animal behaviour published in the best technical journals and popularized in the most expensive public television series patently do much the same thing. But not always; sometimes, rarely and preciously, we—those of us gestating in techno-scientific media—manage to tell some very non-innocent stories about, and even with, the animals, rather than about our "natural" selves. Meanwhile, I'm still sure I was *more* right than my friends about those ducks, whoever they were. And, while queer-bashing remains a popular sport, I still feel the pain and know my complicity in those particular boys' natural development.

A second story: once upon a time, early in graduate school in biology in the mid-1960s, I was tremendously moved, intellectually and emotionally, by an ordinary lecture on the enzymes of the electron transport system (ETS). These biological catalysts are involved in energy-processing in cells complicated enough to have elaborate, internal, membrane-bound organelles (little organs) to partition and enlarge their activities. Using new techniques, the process was being studied experimentally *in vitro* in structural-functional complexes of membrane sub-units prepared from the cellular organelles, called mitochondria. The membrane sub-units were dis-assembled and re-assembled to be analysed both by electron microscopy and biochemistry. The result was a stunning narrative and visual imagery of structural-functional complexity of the type that has always made biology, including molecular biology, a beautiful science for me. The apparatus of production of these written and oral accounts and visual artefacts was rigorously analytical and biotechnical. There was no way around elaborate machine mediations, complete with all their encasements of dead labour, intentional and unintentional delegations, unexpected agencies, and past and present, pain-fraught, socio-technical histories.

After the lecture, on a walk around town, I felt a surging high. Trees, weeds, dogs, invisible gut parasites, people—we all seemed bound together in the ultra-structural tissues of our being. Far from feeling alienated by the reductionistic techniques of cell biology, I realized to my partial embarrassment, but mainly pleasure, that I was responding *erotically* to the connections made possible by the knowledge-producing practices, and their constitutive narratives, of techno-science. So, who is surprised: when were love and knowledge not co-constitutive? I refused, then and now, to dismiss the specific pleasure experienced on that walk as epistemological sado-masochism, rooted in alienation and objectifying scientific reductionism or in ignorant denial of the terrible histories of domination built into what we politely call "modern science." I was *not* experiencing a moment of romantic postmodern rapture in the techno-sublime. Machine, organism and human embodiment all were articulated—brought into a *particular* co-constitutive relationship—in complex ways that forced me to recognize a historically specific, conjoined discipline of love, power and knowledge. Through its enabling constraints, that is, through lab practice in cell biology, this discipline was making possible—unequally—particular kinds of subjectivity and systematic artefactual embodiments, for which people in my worlds had to be responsible.

This knowing love could not be innocent; it did not originate in a garden. But neither did it originate in *expulsion* from a garden. Not about secrets—of life or death—this knowing love took shape in quite particular, historical—social intercourse, or "conversation," among machines, people, other organisms and parts of organisms. All those feminists like me still in the closet—that is, those who have not come out to acknowledge the viscous, physical, erotic pleasure we experienced from dis-harmonious conversations about abstract ideas, auto repair and possible worlds that took place in local consciousness-raising groups in the early 1970s—might have a thrill of self-recognition in thinking about the electron transport system. Our desires are very heterogeneous, indeed, as are our embodiments. We may not be ducks; but, as natural-technical terran constructs, we are certainly ETs.

And a final story: when Alexander Berkman and Sojourner Truth, my and my lover's half-labrador mutts, were just over a year old, we all went to obedience training together in a small town in northern California. Although we had discoursed on dog training from library books for a year, none of us had ever before been to obedience training, that amazing institution which domesticates people and their canine companions to agree to cohabit particular stories important to civic peace. It was late in our lives together to seek institutionalized obedience training. One of us already showed signs of criminality, or at least bore the marks of a shared incoherent relation to authority, of the kind that could result in mayhem and legally mandated death sentences for dogs and nasty fines for people. That is, one of us seemed intent on murdering conspecifics (other dogs) in any and all circumstances, and the rest of us were handling the situation badly.

In some important situations in the 1980s in California, we four didn't seem to speak the same language, either within or across species boundaries. We needed help. So, with a motley assortment of other cross-specific pairs of mammals, of types that had shared biological and social histories for a couple of tens of thousands of years, we entered a commercial pedagogical relationship with the dog, Goody-goody, and her human, Perfection, who seemed to have mastered the political problem of paying consequential attention to each other. They seemed to have a story to tell us.

In her discussion of the language games of training, Vicki Hearne invoked Wittgenstein's injunction that "to imagine a language is to imagine a form of life."[5] A professional trainer and an incurable intellectual, Hearne was looking for a philosophically responsible language for talking about the stories inhabited by trainers and companion animals like dogs and horses. She was convinced that the training relationship is a moral one that requires the personhood of all the partners. But, although Hearne did not affirm this point, the moral relationship cannot rely on a shared *anthropomorphic* personhood. Only some of the partners are people, and the form of life the conversants construct is neither purely canine nor purely human. Further, personhood is only one local, albeit historically broadly important, way of being a subject. And, like most moral relationships, this one cannot rely on ignorance of radical hetero-geneity in the commitment to equality-as-sameness.

Certainly, however, in the training relationship animals and people are constructing a historically specific form of life, and therefore a language. They are engaged in making some effective meanings rather than others. Hearne's moral universe had such premises as: dogs have the right to the consequences of their actions, and, biting (by the dog) is a response to incoherent authority (the human's). She envisioned certain civil rights, like those enjoyed by seeingeye dogs and their people, for other dogs and people who had achieved superb off-lead control.

I quibble about discussing this matter in terms of people's control of the dogs, not out of a fetishized fear of control, and of naming who exercises it over whom, but out of a sense that my available languages for discussing control and its directional arrows mis-shapes the forms of attention and response achieved by serious dogs and trainers. By *mis-shapes,* I do not mean *mis-represents,* but, more seriously, I mean that the language of unidirectional "human control over dog" instrumentally is part of producing an incoherent and even dangerous relationship that is not conducive to civil peace within or across species. A convinced sceptic about the ideologies of representation anyway, I am not interested in worrying too much about the accurate portrayal of training relationships. But I am very much concerned about the instrumentality of languages, since they are forms of life.

Sojourner, Alexander—the canine reincarnation of the lover of Emma Goldman and the anarchist who shot Frick in 1892 after the Homestead strike—Rusten, and I were serious about training, but we were very unskilled. We should have met Vicki Hearne, but at that point we had her only in *The New Yorker.* We needed more on-the-ground skill.

Instead, we blundered into an appalling conversation that makes those heterosexually construed ducks look untouched by human tongues. As long as you didn't listen to the English that Perfection used to explain to the other humans what was happening, but attended only to the other semiotic processes, like gesture, touch and unadorned verbal command, Goody-goody and Perfection had a pretty good story for lots of ordinary events in inter-species life. But, like lots of sheltered folk, they weren't good with anarchists and criminals; they relied, or at least Perfection did, on escalating force and languages of stripped-down subjugation. The result was stunning escalation of the potential for violence in our dog. The conversation was going quite wrong. We later met some trainer humans and social-worker dogs who taught us how to work on reliable obedience in challenging circumstances—such as the mere existence of other dogs in the world. But, our first encounter with obedience training posed in stark terms the fact that forms of inter-species domestic life can go very wrong.

My growing suspicions that our incoherence was only increasing in this particular attempt at training reached their apogee near graduation time, when Goody-goody and Perfection demonstrated how a human could examine any spot on a dog's body if necessary. This exercise could be crucial in emergencies, where pain and injury to the dog could put both human and animal at risk. The class was very attentive. While Perfection was touching Goody-goody in every imaginable place, opening and closing orifices, and generally showing how few boundaries were necessary when trust and good authority existed, the conversants seemed to me to be involved in a complex intercourse of gesture, touch, eye movement, tone of voice and many other modalities. But, while grasping a paw and holding it up for our view, what Perfection was saying to us went something like this: "See this paw? It may look like Goody-goody's paw, but it's really my paw. I own this paw, and I can do anything I want with it. If you are to be able to do what you see us doing, you too must accept that form of appropriation of your dog's body."

It was my opinion that day, and still is, that if Perfection had really acted on her explicit words, Goody-goody and she would have achieved nothing. Their other conversation belied the discourse she provided the students. If my lover and I had been better at attending to that other conversation, we might have been able to get further on our needed communication with Alexander and Sojourner on difficult subjects. We were actually very good at the physical examination language game. In our harder task, the one involving our dog's tendency to attack other dogs, with his sister aiding and abetting, we were deterred by mis-shaping words that Perfection did not follow in her relations with Goody-goody, but did impose in both physical and verbal relationships with at least some other dogs and people. Maybe she just had that trouble with criminals, anarchists and socialists. That population is, it must be said, quite large. Unpromisingly, my household went to obedience-training graduation wearing "question authority" buttons. We had not yet built a coherent conversation, inter- or intra-specific, for the crucial subject of authority.

So, my opening stories have been about three forms of life and three conversations involving historically located people and other organisms or parts of organisms, as well as technological artefacts. All of these are stories about demarcation and continuity among actors, human and not, organic and not. The stories of the "wild ducks in nature," of the "reductive" methodologies and the ETS in cell biology, and of the problem of "discourse" between people and companion species collectively raise the problems that will concern us for the rest of this article, as we turn to Young, Noske and Mitchison. Is there a common context for discussion of what counts as nature in techno-science? What kind of topic is the "human" place in "nature" by the late twentieth century in worlds shaped by techno-science? How might inhabitable narratives about science and nature be told, without denying the ravages of the dedication of techno-science to militarized and systematically unjust relations of knowledge and power, while refusing to replicate the apocalyptic stories of Good and Evil played out on the stages of Nature and Science?

DIGESTING DISCOURSES

In *Darwin's Metaphor: Nature's Place in Victorian Culture,* a richly textured, scholarly and politically passionate book that collects a series of still vitally important essays written across the 1970s on the nineteenth-century British debates on "man's place in nature,"[6] Robert Young depicts the structure and consequences of the broad, common cultural context within which the intellectuals' struggle over the demarcation between man, God and nature took place. The twentieth-century phrase "science *and* society" would not have made much sense to the participants in the earlier debates, for whom the parts were not two preconstituted, oppositional entities, *science* and *society* held apart by a deceiving conjunction. The phrase should not make sense in the 1990s either, but for different reasons from those that pertained in Darwin's world. My debased goal of writing the animal stories for the *Reader's Digest* should be seen in the context of a very different social scene from the nineteenth-century contestation for shared meanings and inhabitable stories. In those halcyon days, I might have aspired instead to have written for the *Edinburgh Review*.

Young's fundamental insistence is that probing deeper into a scientific debate leads inexorably to the wider issues of a culture. If we inquire insistently enough—if we take doing cultural studies seriously—"we may learn something about the nature of science itself, and thereby illuminate the way societies set agendas in their broad culture, including science, as part of the pursuit of social priorities and values" (p. 122). For Young, to understand the nineteenth-century debates which he is exploring, and more generally, to engage any important issues in the history of science as culture, to sequester the scientific debate from the social, political, theological and economic ones is to falsify all the parts and "to mystify oppression in the form of science" (p. 192).

Young's reference in the 1970s for these arguments was the debate over "internalist versus externalist" approaches to the history of science that exercised scholars throughout the decade. Can science be understood to have "insides" and "outsides" that justify separating off "contents" of scientific "discovery" from "contexts" of "construction"? All of Young's essays are rigorous, principled objections to the dichotomy as scholarly obfuscation and political mystification. In 1992, I would still be hard put to recommend a more richly argued invitation to and enactment of politically engaged, holistic, scholarly work than Young's 1973 essay, "The historiographic and ideological contexts of the nineteenth-century debate on man's place in nature." I found in the 1970s, and still find, that Robert Young's cogency about the need to confront the content of the sciences with a non-reductionist, social-historical analysis and to avoid easy answers to the relations of science and ideology is indispensable to all my projects as a critical intellectual.

That cogency is certainly indispensable to understanding science *as* culture, rather than science *and* culture. As he put it much earlier than those now cited in science studies for injunctions that sound similar, but lack Young's crucial political edge, "Nothing is ultimately contextual; all is constitutive, which is another way of saying that all relationships are dialectical" (p. 241). Following a Marxist tradition, especially in the work of Georg Lukács, "whose analysis of reification provides the tools for looking more closely at the ways in which science has been used for the purpose of reconciling people to the status quo," Young builds his book around the premiss that "nature is a social category" (p. 242). I will come back to this indispensable and highly problematic assertion.

Young's general view of the nineteenth-century debate is that its questions were not allocated to specialist disciplines, but were deeply embedded in a shared (although class-differentiated) cultural context, for which the relation of God and His creation, that is, theism and the fate of natural theology, was the organizing centre. For example, the fine structure of Darwin's scientific discourse on "selection" shows that theological and philosophical issues were constitutive, not contextual. The common intellectual context of the debate about "man's place in nature" from the early 1800s to the 1880s took shape in a widely read periodical literature, in which theological, geological, biological, literary and political questions were complexly knotted together. The debate about "man's place in nature" was not an integrated whole, but a rich web. The threads were sustained by a material apparatus of production of a common culture among the intelligentsia, including potent reading, writing and publishing practices.

Through the 1870s and 1880s, the breakup of the common intellectual context through specializations and disciplinizations familiar to an observer from the late twentieth century was reflected by—and partly *effected* by—a very different structure of writing and publishing practices. As Young put it, "The common intellectual context came to pieces in the 1870s and 1880s, and this fragmentation was reflected in the development of specialist societies and periodicals, increasing professionalization, and the growth of general periodicals of a markedly lower intellectual standard" (p. 128). Thus was born my desire to write for *National Geographic, Omni,* and, to return to the deepest shame and hope for an academic used to audiences of a few hundred souls, the *Reader's Digest,* or even *The National Enquirer.*

But if there has been a disruption in one context—and its constitutive literary, social and material technologies—Young notes a stream of continuity from the early nineteenth to the late twentieth century that is fed by the very specializations in practice and the literary debasements which he describes: current biotechnology, perhaps especially genetic engineering and the profusion of genome projects to appropriate an organism's DNA sequences in a particular historical form—one amenable to property and commodity relations—must be seen as, in significant part, the "harvest of Darwinism" that has reshaped biological culture at its roots. "The current context for reflecting on these matters is a period in which biotechnology is harvesting and commercializing the long-term fruits of Darwinism and making commodities out of the least elements of living nature—amino acids and genes" (p. 247). "Darwinism provides the unifying thread and themes from Malthus to the commodification of the smallest elements of living nature in genetic engineering. With this set of interrelations go the social forms of technocracy, information processing, and the disciplines that are recasting how we think of humanity in terms of cybernetics, information theory, systems theory, and 'communication and control' " (p. xiii).

So, our "common context" is not theism—the relations of a creator God to His product—but constructivism and productionism. Constructivism and productionism are the consequences of the material relocation of the narratives and practices of creation and their ensuing legal relations onto "man" and "nature" in (how else can I say it?) white capitalist heterosexist patriarchy (hereafter WCHP, an acronym whose beauty fits its referent). The nineteenth-century debate about the demarcation between God's creative action and nature's laws, and so between man and nature, mind and body, was resolved by a commitment to the principle of the uniformity of nature and scientific naturalism. In the context of the founding law of the Father, nature's *capacities* and nature's *laws* were identical. Narratively, this identification entails the escalating dominations built into stories of the endless transgressions of forbidden boundaries—the erotic *frisson* of man's projects of transcendence, prominently including techno-science. "Science did not replace God: God became identified with the laws of nature" (p. 240).

God did not interfere in His creation, not even in those previously reliable reservoirs for His action called biological design and mental function. But the deep European monotheist, patriarchal, cultural commitment to relating to the world as made, designed, and structured by the prohibitions of law remained. A recent element in the stories, *progress* was inserted into the body of nature and deeply tied to a particular kind of conception of the uniformity of nature as a *product*. By the late twentieth century, very few cracks, indeed, are allowed to show in the solid cultural complex of WCHP constructionism.

In March 1988, Charles Cantor, then head of the U.S. Department of Energy's Human Genome Center at the Lawrence Berkeley Laboratory, made these matters clear in his talk at the National Institute of Medicine entitled "The Human Genome Project: problems and prospects." In the context of explaining the different material modes of existence of various kinds of genetic maps (genetic linkage maps, physical maps stored in Yeast Artificial Chromosome (YAC) libraries or "cosmid libraries," and database sequence information existing only in computers and their printouts), Cantor noted why having the physical maps mattered so much: "You own the genome at this point." I wanted Cantor to explore further the socio-technical relations of physical libraries to sequence data; this exploration would show us something about the late twentieth-century "common context" for demarcation debates between "nature," "man" and, if not God, at least the supreme engineer. The "realization" of the value of the genome requires its full materialization in a particular historical form. Instrumentalism and full constructivism are not disembodied concepts. To make and store the genome is to appropriate it as a specific kind of entity. This is historically specific human self-production and self-possession or ownership.

The long tradition of methodological individualism and liberty based in property in the self comes to a particular kind of fruition in this discourse. To patent something, one must hold the key to making it. That is what bestows the juridical right of private appropriation of the product of no longer simply *given*, but rather fully technically *replicated*, "nature." In the Human Genome Project, generic "man's place in nature" really does become the universal "human place in nature" in a particular form: species existence as fully specified process and product. The body is matrix, superfluous, or obstructive; the programme is the prize.

In this mutated, but still masculinist, heroic narrative, the relation between sex and gender is one of the many worlds that is transformed in
this concrete socio-technical project underway in Europe, Japan and the
United States.

Like toys in other games, Genes "R" Us, and "we" (who?) are our self-
possessed products in this apotheosis of technological humanism. There
is only one Actor, and we are It. Nature mutates into its binary opposite, culture, and vice versa, in such a way as to displace the entire
nature/culture (and sex/gender) dialectic with a new discursive field,
where the actors who count are their own instrumental objectifications.
Context is content with a vengeance. Nature is the programme; we replicated it; we own it; we are it. Nature and culture implode into each other
and disappear into the resulting black hole. Man makes himself, indeed,
in a cosmic onanism. The nineteenth-century transfer of God's creative
role to natural processes, within a multiply stratified industrial culture
committed to relentless constructivism and productionism, bears fruit
in a comprehensive biotechnological harvest, in which control of the
genome is control of the game of life—legally, mythically and technically.
The stakes are very unequal chances for life and death on the planet. I
honestly don't think Darwin would have been very happy about all this.

Let us return to Young's affirmation of Lukács's proposition that nature is a social category. In the face of the implosion described above,
that formulation seems inadequate in a basic way. In the Marxist radical
science movement of the 1970s, Young formulated the problem in these
terms: "In the nineteenth century, the boundaries between humanity
and nature were in dispute. On the whole, nature won, which means
that reification won. It is still winning, but some radicals are trying to
push back the boundaries of reifying scientism as far as they can, and a
critical study of the development of the models which underlie reifying
rationalizations may be of service to them as they begin to place science
in history—the history of people and events" (p. 246). I would rather
say not that "nature" won, but that the man/nature game is the problem.
But this is a quibble within my analysis so far; Young and I are united in
identifying crucial parts of the structure of reification.

To oppose reification, Young appealed to a Marxist modification of
the premise, "Man (i.e., human praxis) is the *measure* of all things"
(p. 241). But, deeply influenced by the practices of an anti-imperialist
environmentalism that joins justice and ecology, and of a multi-cultural
feminism that insists on a different imagination of relationality, both
social movements that took deep root after Young wrote this essay, I
think that human praxis formulated in this way is precisely part of the
problem.

In 1973, Young sought a theory of mediations between nature and man. But nature remained either a product of human praxis (nature's state as transformed by the history of people and events), or it was a pre-social category not yet in relation to the transforming relation of human labour. What nature could not be in these formulations of Marxist humanisms is a social partner, a social agent with a history, a conversant in a discourse where all of the actors are not "us." A theory of "mediations" is not enough. If "human praxis is the measure of all things," then the conversation and its forms of life spell trouble for the planet. And, less consequentially for others but dear to my heart, I'll never get to have a coherent conversation with my anarchist mongrel dog, Alexander Berkman. In Lukács's and Young's story in the 1970s, nature could only be matrix or product, while man had to be the sole agent, exactly the masculinist structure of the human story, including the versions that narrate both the planting and the harvest of Darwinism.

We are in troubled waters, but not ones utterly unnavigated by European craft, not to mention other traditions. But, animism has a bad name in the language games I need to enter as a critical intellectual in techno-science worlds, and besides, animism is patently a kind of human representational practice. Still, efforts to figure the world in lively terms pervade hermeticism in early modern Europe, and some important radical and feminist work has tried to reclaim that tradition. There is not really much help for us in that history, I fear. However, I think we must engage in forms of life with non-humans—both machines and organisms—on livelier terms than those provided by harvesting Darwinism or Marxism. Refiguring conversations with those who are not "us" must be part of that project. We have got to strike up a coherent conversation where humans are not the measure of all things, and where no one claims unmediated access to anyone else. Humans, at least, need a different *kind* of theory of mediations.

OTHERWORLDLY CONVERSATIONS

It is that project that enlivens Barbara Noske's book, *Humans and Other Animals: Beyond the Boundaries of Anthropology*. Noske thoroughly warps the organizing field of humanist stories about nature and culture. Her situation as a radical Western intellectual in the late 1980s, where animal rights movements, environmentalism, feminism and antinuclearism restructure the intellectual and moral heritage of the left, stands in historical contrast to Young's a decade earlier. Noske's discussion of Darwinism is much poorer scholarship compared to Young's fine-grained analysis, but she has her finger on a key political—epistemological—moral problem that I don't think the Young of those essays could broach. If he had broached the trouble in our relationship with other organisms in the way Noske does, he certainly could not have resolved the issue as she does.

Noske is consumed by the scandal of the particular kind of object status of animals enforced in the Western histories and cultures she discusses. In Marxist formulations, reification refers to the re-presentation to human labourers of the product of their labour—that is, of the means through which they make themselves historically—in a particular, hostile form. In capitalist relations of production, the human activity embodied in the product of labour is frozen, appropriated, and made to reappear as It, the commodity form that dominates and distorts social life. In that frame, reification is not a problem for domestic animals, but, for example, for tenant farmers, who objectify their labour in the products of animal husbandry and then have the fruit of that labour appropriated by another, who represents it to the worker in a commodity form. But, more fundamentally, the farmer is represented to *himself* in the commodity form. The paradigmatic reification within a Marxist analysis is of the worker *himself,* whose own life-making activity, his labour power, is taken from him and represented in a coercive commodity form. He becomes It.

Noske is after another sense of objectification. For her, the Marxist analysis cannot talk about the animals at all. In that frame, animals have no history; they are matrix or raw material for human self-reformation, which can go awry, for example, in capitalist relations of production. Animals are not part of the *social* relationship at all; they never have any status but that of not-human; not subject, therefore object.

But, the kind of "not subject, not human, therefore object" that animals are made to be is also not like the status occupied by women within patriarchal logics and histories. Feminist analysis that either affirms or resists women's identification with animals as nature and as object has not really gotten the point about animals either, from Noske's provocative point of view. In an important stream of Anglo-feminist theory, woman as such does not suffer reification in the way the Marxist describes the process for the worker.[7] In masculinist sexual orders, woman is not a subject separated from the product of her life-shaping activity; her problem is much worse. She *is* a projection of another's desire, who then haunts man as his always elusive, seductive, unreliable Other. Woman as such is a kind of illusionist's projection, while mere women bear the violent erasures of that history-making move. There is nothing of her own for her to reappropriate; she is an object in the sense of being another's project.

The kind of objectification of animals that Noske is trying to understand is also not like the history of racial objectification in the West, although the status of slavery in the New World came dizzyingly close to imposing on the enslaved the same kind of animal object status borne by beasts, and by nature in general within colonizing logics. In African-American slavery, for example, slaves were fully alienable property. Slave women were not like white women—the conveyers of property through legitimate marriage.[8] Both slave men and women were the property itself. Slave women and men suffered both sexual and racial objectifications in a way that transformed both, but still the situation was not like that of non-human animals. Slave liberation depended on making the human subjecthood of the slaves an effective historical achievement. In that history-remaking process, what counts as human, that is, the story of "man," gets radically recast.

But no matter how recast, this human family drama is not the process of re-establishing the terms of relationality that concerns animals. The last thing they "need" is human subject status, in whatever cultural-historical form. That is the problem with much animal rights discourse. The best animals could get out of that approach is the "right" to be permanently represented, as lesser humans, in human discourse, such as the law—animals would get the right to be permanently "orientalized." As Marx put it in another context and for other beings, "They cannot represent themselves; they must be represented." Lots of well-intentioned, but finally imperialist ecological discourse takes that shape. Its tones resonate with the pro-life/anti-abortion question, "Who speaks for the foetus?" The answer is, anybody but the pregnant woman, especially if that anybody is a legal, medical or scientific expert. Or a father. Facing the harvest of Darwinism, we do not need an endless discourse on who speaks for animals, or for nature in general. We have had enough of the language games of fatherhood. We need other terms of conversation with animals, a much less respectable undertaking. The point is not new representations, but new *practices*, other forms of life rejoining humans and not-humans.

So, in the human-animal relationship gone awry, the analogy to other objectifications, so often invoked in radical discourse, breaks down systematically. That is the beauty of Noske's argument. There is *specific* work to be done if we are to strike up a coherent form of life, a conversation, with other animals. "It may all boil down to a form of anthropocentric colonizing, where everything and everyone is still being measured by a human and Western yardstick. In the context of our law systems, animals are bound to appear as human underlings. However, animals are not lesser humans; they are other worlds, whose otherworldliness must not be disenchanted and cut to our size, but must be respected for what it is" (p. xi, my punctuation). Great, but how? And how especially if there is no *outside* of language games?

Trying to get a grip on this matter, Noske achieves four things that I value highly. First, she starts out by formulating the historicity of all the partners in the stories. Animals have been active in their relations to humans, not just the reverse. Domestication, a major focus of Noske's discussion, is paradigmatic for her argument. Although an unequal relationship, domestication is a two-way matter. Domestication refers to the situation in which people actively force changes in the seasonal subsistence cycles of animals to make them coincide with particular human needs. Emphasizing the active aspect and the changing and specific ecologies of both species, the definition Noske uses insists on a historically dynamic continuum of human-animal relations of domestication. From this point of view, capture, taming, and reproductive isolation are relatively recent developments.

Second, in her analysis of contemporary factory-animal domestication, Noske formulates a very useful concept, the "animal-industrial complex."[9] Animals are forced to "specialize" in one "skill" in a way that would chill the harshest human labour-process deskillers. "The animal's life-time has truly been converted into "working-time": into round-the-clock production" (p. 17). The design of animals as laboratory research models is one of the most extreme examples of domestication. Not only has the animal been totally incorporated into human technology; it has become a fully designed instance of human technology.

Noske doesn't discuss genetic engineering, but her argument would readily accommodate those intensifications of reshaping the animals (and humans) to productionist purposes. As indeed, from the point of view of dominant narratives about the human genome initiative, humans themselves are the reading and writing technologies of their genes. Nature *is* a technology, and that is a very particular sort of embodied social category. "We" (who?) have become an instance of "our" (whose?) technology. The "Book of Life" (the genome in the title image used by the NOVA television programme "Decoding the Book of Life," 1988) is the law of life, and the law is paradigmatically a technical affair. Noske agrees with the Dutch philosopher Ton Lemaire that this full objectification of "nature" could only be complete with the full "autonomization" of the human subject. Autonomy and automaton are more than aural puns. Fully objectified, we are at last finished subjects—or finished *as* subjects. The world of "autonomous" subjects is the world of objects, and this world works by the law of the annihilation of defended selves imploding with their deadly projections.

Here, Noske, Young and I are very much in the same conversation. The notion of the "animal-industrial complex" makes it easy to discuss some of the crucial issues. The consequences of these forms of relating rest on humans and animals, but differently. At the very least, it must be admitted that "animal exploitation cannot be tolerated without damaging the principle of inter-subjectivity" (p. 38). Here we are getting to the heart of the matter. What is inter-subjectivity between radically different kinds of subjects? The word *subject* is cumbersome, but so are all the alternatives, such as agent, partner or person. How do we designate radical otherness at the heart of ethical relating? That problem is more than a human one; as we will see, it is intrinsic to the story of life on earth.

Noske's third achievement is, then, to state unequivocally that a coherent conversation between people and animals depends on our recognition of their "otherworldly" subject status. In a discussion of various concepts of culture in anthropology and biology, Noske notes that both traditions can only see animal behaviour as the outcome of mechanisms. They cannot take account of animals socially constructing their worlds, much less constructing ours. Biology, in particular, does not have the methodological equipment to recognize "things socially and culturally created and which in turn shape the creators" (p. 86).

In her final chapter, "Meeting the Other: towards an anthropology of animals," Noske describes the history of Western writing about "wolf children," very young children believed to be somehow lost from human communities, raised by other social animals, and then found by people. She is interested in how to hear the stories of and about animal-adopted children. So she asks if, instead of asking if people can "de-animalize" the children by restoring, or teaching for the first time, fully human language, we can instead ask what kind of social thing happened when a human child acquired a specific *non-human* socialization? She imagines that the children did not become "human," but they did become social beings. Even in stories of less extreme situations, such as the tales of white, middle-class, professional homes that contain young apes and human children, the children experience animal acculturation, as well as the reverse. For Noske, these situations suggest not so much "human-animal communication" as "animal-human communication." None of the partners is the same afterwards.

Noske's fourth achievement for me was her use of Sandra Harding's *The Science Question in Feminism* to shift the focus to "the animal question in feminism" (pp. 102–16). Noske insists that some feminists' positive identification with animals, including their embracing our own femaleness, and other feminists' resistance to such supposed biological essentialism are both wrong-headed as long as the terms of the troubled relationship of "women and nature" are seen within the inherited, eth-nocentric subject/object frame that generates the problem of biological reductionism. Noske argues for a feminist position *vis à vis* animals that posits continuity, connection and conversation, but without the frame that leads inexorably to "essentialism". "Essentialism" depends on re-ductive identification, rather than ethical relation, with other worlds, including with ourselves. It is the paradox of continuity *and* alien rela-tionality that sustains the tension in Noske's book and in her approach to feminism. Once the world of subjects and objects is put into ques-tion, that paradox concerns the congeries, or curious confederacy, that is the self, as well as selves' relations with others. A promising form of life, conversation defies the autonomization of the self, as well as the objectification of the other.

TRAVEL TALK

Science fiction offers a useful writing practice within which to take up Noske's arguments. Re-published in an explicitly feminist context by The Women's Press in London in 1985, *Memoirs of a Spacewoman* was the first SF novel written by Naomi Mitchison. The story of space explo-ration, told from the point of view of a woman xenobiologist and com-munications expert named Mary, was first copyrighted in 1962, when the author was 63 years old and in the midst of a rich career as a na-tional and international political activist and writer. Her references in the 1960s were to a different generation of women's consideration of science and politics from that represented by her later publishers and readers. Daughter of the important British physiologist, J.S. Haldane, and sis-ter of one of the architects of the modern evolutionary synthesis, J.B.S. Haldane, Mitchison could hardly have avoided her large concerns with forms of life. She came, in short, from the social world that produced the Darwins and the Huxleys, those familial arbiters of authoritative terran and otherworldly conversations. Sexual experimentation; political rad-icalism; unimpeded scientific literacy; literary self-confidence; a grand view of the universe from a rich, imperialist, intellectual culture—these were Mitchison's birthright. She wrote that legacy into her spacewoman's memoirs.

Foregrounding the problem of imperialism, which was the silent, if deeply constitutive, axis in Victorian debates on "man's place in nature," Mitchison set her xenobiologist a most interesting task: to make con-tact with "otherworlds," adhering to only one serious restriction in the deployment of her psychological, linguistic, physical and technologi-cal skills—non-interference. Knowledge would not come from scientific detachment, but from scientific connection. Exploring her garden of delights, space-woman Mary had to obey only one little restriction. "Contacts" could take any number of forms—linguistic, sexual, emotional, cognitive, mathematical, aesthetic, mechanical or, in principle, just about anything else. The novel's erotic fusions, odd couplings and curious progeny structure both its humour and its serious side. Communication, naturally, is inherently about desire; but there's the rub. How could conversation occur, in any form, if the rule of non-interference were to be strictly interpreted? The question of power cannot be evaded, least of all in "communication." This was the moral problem for Mary's world: "Humans were beginning to run out of serious moral problems about the time that space exploration really got going" (p. 16). But no more.

The rule of non-interference wasn't strictly interpreted, of course; so the story could continue. The delicate shades of interference turned out to be what really mattered narratively. "The difficulty seems to be that in the nursery world we take ourselves for granted as stable personalities, as completely secure. Impossible that we should ever deviate, that interference should ever be a temptation" (p. 19). Every explorer found out otherwise rather quickly. So, the imperative of non-interference constituted the law, the symbolic matrix within which subjects could be called into position for "conversation." To obey the founder's law is always impossible; that is the point of the tragicomic process of becoming a social subject webbed with others. Not to eat of the tree of life in Mitchison's book is not to know the necessary, impossible situation of the communicator's task. Communication, even with ourselves, *is* xenobiology: otherworldly conversation, terran topics, local terms, situated knowledges. "It all works out in the end. But the impact of other worlds on this apparently immovable stability comes as a surprise. Nobody enjoys their first personality changes" (p. 19). Neither, presumably, do those with whom contact is made.

In Althusser's sense, in *Memoirs of a Spacewoman* subjects are interpellated, or hailed, into being in a world where the law is not the policeman's "Hey, you!" or the father's "Thou shalt not know," but a deceptively gentler moralist's command, "Be fruitful and multiply; join in conversation, but know that you are not the only subjects. In knowing each other, your worlds will never be the same." Interference is static, noise, interruption in communication; and yet, interference, making contact, is the implicit condition of leaving the nursery world. "Although, of course, I longed in the normal human way for exploration, I found my first world oddly disconcerting... It is only in circumstances like these that we realise how much we ourselves are constructed bi-laterally on either-or principles. Fish rather than echinoderms... It was quite a problem to get through to those radial entities" (pp. 19, 20, 23). The subject-making action—and the moral universe—really begins once those bilateral and radial entities establish touch. And that's only the beginning: "I think about my children, but I think less about my four dear normals than I think about Viola. And I think about Ariel. And the other" (p. 16).

THREE BILLION YEARS

But, what if we went back to another beginning, to the early days of living organisms on earth a few billion years ago? That seems a good place to end this meditation on natural conversation as heterogeneous intercourse. Might those yuppie Wisconsin ducks have a legitimate queer birthright after all, and might there be a respectable material foundation to my sexual pleasure in mitochondrial respiratory enzymes? Using Lynn Margulis and Dorion Sagan's *Origins of Sex: Three Billion Years of Genetic Recombination*[10] as my guide, I will tell a very different concluding story from Cantor's version of the human genome project or corporate biotechnology's harvest of Darwinism.

As elsewhere, biology in my narrative is also a rich field of metaphors for ethno-specific cultural and political questions. My bag-lady version of Margulis and Sagan's authoritative account of the promiscuous origins of cells that have organelles[11] is about metaphor-work. Doing such work is part of my vocation to prepare for my job at the *Reader's Digest* after the revolution. I think this kind of metaphor-work could tell us something interesting about the metaphor-tools "we" (who?) might need for a usable theory of the subject at the end of the second Christian millennium.

Consider, then, the text given us by the existence, in the hindgut of a modern Australian termite, of the creature named *Mixotricha paradoxa*, a mixed-up, paradoxical, microscopic bit of "hair" (*trichos*) (Figure 4.1). This little filamentous creature makes a mockery of the notion of the bounded, defended, singular self out to protect its genetic investments. The problem our text presents is simple: what constitutes *M. paradoxa*? Where does the protist[12] stop and somebody else start in that insect's teeming hindgut? And what does this paradoxical individuality tell us about beginnings? Finally, how might such forms of life help us imagine a usable language?

M. paradoxa is a nucleated microbe with several distinct internal and external prokaryotic symbionts, including two kinds of motile spirochetes, which live in various degrees of structural and functional integration. All the associated creatures live in a kind of obligatory confederacy.

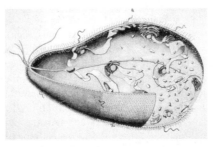

Fig. 4.1. *Mixotricha paradoxa.* Drawing by Christie Lyons; photo courtesy of Lynn Margulis.

From Margulis and Sagan's "symbiogenetic" point of view, this kind of confederacy is fundamental to life's history. Such associations probably arose repeatedly. The ties often involved genetic exchanges, or recombinations, that in turn had a history dating back to the earliest bacteria that had to survive the gene-damaging environment of ultraviolet light before there was an oxygen atmosphere to shield them.

> That genetic recombination began as a part of an enormous health delivery system to ancient DNA molecules is quite evident. Once healthy recombinants were produced, they retained the ability to recombine genes from different sources. As long as selection acted on the recombinants, selection pressure would retain the mechanism of re-combination as well. (Margulis and Sagan, p. 60)

I like the idea of gene exchange as a kind of prophylaxis against sun burn. It puts the heliotropic West into perspective.

Protists like *M. paradoxa* seem to show in mid-stream the ubiquitous, life-changing association events that brought motile, oxygen-using or photosynthetic bacteria into other cells, perhaps originally on an opportunistic hunt for a nutritious meal or a secure medium for their metabolic transactions. But, some predators settled down inside their prey and struck up quite a conversation. Mitochondria, those oxygen-using organelles with the interesting respiratory enzymes integrated into membrane structures, probably joined what are now modern cells in this way.

> With the elapse of time, the internal enemies of the prey evolved into microbial guests, and, finally, supportive adopted relatives. Because of a wealth of molecular biological and biochemical evidence supporting these models, the mitochondria of today are best seen as descendents of cells that evolved within other cells. (Margulis and Sagan, p. 71)

The story of heterogeneous associations at various levels of integration repeated itself many times at many scales.

> Clones of eukaryotic cells in the form of animals, plants, fungi, and pro-toctists seem to share a symbiotic history... From an evolutionary point of view, the first eukaryotes were loose confederacies of bacteria that, with continuing integration, became recognizable as protists, unicellular eukaryotic cells... The earliest protists were likely to have been most like bacterial communities... At first each autopoietic [self-maintaining] community member replicated its DNA, divided, and remained in contact with other members in a fairly informal manner. *Informal* here refers to the number of partners in these confederacies: they varied. (Margulis and Sagan, p. 72)

Indeed, they varied. So, speaking as a multicellular, eukaryotic, bilaterally symmetrical confederacy, a fish, in short, I want to learn to strike up interesting intercourse with possible subjects about livable worlds. In nineteenth-century bourgeois U.S. law, such sexually suspect doings were called criminal conversation. Mitchison's spacewoman understood: "Although, of course, I longed in the normal human way for exploration, I found my first world oddly disconcerting... "

NOTES

This essay is a meditation on three works: *Darwin's Metaphor: Nature's Place in Victorian Culture,* by Robert M. Young, Cambridge: Cambridge University Press, 1985; *Humans and Other Animals: Beyond the Boundaries of Anthropology,* by Barbara Noske, London: Pluto, 1989; *Memoirs of a Spacewoman,* by Naomi Mitchison, London: The Women's Press, 1976 [1962].

1. See Katie King (1994), *Theory in Its Feminist Travels: Conversations in U.S. Women's Movements.* Bloomington: Indiana University Press.
2. Here I borrow from the wonderful project of the journal *Public Culture,* Bulletin of the Center for Transnational Cultural Studies, at the University of Pennsylvania. In my opinion, this journal embodies the best impulses of cultural studies. It is available from The University Museum, University of Pennsylvania, Philadelphia, PA 19104, USA.
3. See Evelyn Fox Keller (1990), "From secrets of life to secrets of death," in Mary Jacobus, E. F. Keller and Sally Shuttleworth, eds., *Body/Politics: Women and the Discourses of Science.* New York: Routledge, pp. 177–91.
4. Ursula LeGuin (1989), "The Carrier-Bag Theory of Fiction," in D. du Pont, ed., *Women of Vision.* New York: St Martin's Press, pp. 1–11.
5. Vicki Hearne (1986), *Adam's Task: Calling Animals by Name.* New York: Knopf, p. 4.
6. In his Preface to the 1985 reprint of his essays, Young gives his justification for not dealing with the language of the pseudo-universal "man" in his revisions:

   > I cannot resolve the question of gender in these essays: "man's place in nature" was the rhetoric of the period, and "he" had characteristic resonances which it would be anachronistic to expunge, and this set the style. (p. xvii)

   I disagree not with Young's decision to keep "man" and "he," but with the absence of sustained discussion of precisely what *difference* the "characteristic resonances" and "style" made to nineteenth-century discourse and to Young's discourse. Feminist demands are not to expunge offensive material, but to require precise analysis of how the unmarked categories work—and how we continue to inherit the trouble. That analysis could not proceed if the problem were made harder to see by covering up "man" with a euphemistic and anachronistic "human." Some of Young's most important discussions, for example, in the essay on "Malthus and the evolutionists," originally published in 1969, before recent feminist theory could have made a difference, by the mid-1980s merited at least a footnote on how feminist analyses require the restructuring of historical understanding of the debates about natural theology, human perfection, and evolution. Minimally, Malthus's argument against Godwin's version of future human perfection through the complete transcendence of need, especially sex, and Malthus's doctrine on the private ownership of women and children in the institution of marriage were intrinsic to the establishment of a constitutively self-invisible masculinist discourse in natural theology.

Similarly, in "Natural theology, Victorian periodicals, and the fragmentation of a common context," I waited for some discussion of how the processes of specialization and publication fundamentally restructured the gender fabric of the practice of evolutionary biology. The 1985 postscript to that essay might have been a place to say something about how feminist theory makes one rethink the issues of "common context" and "fragmentation." I also think Young should have revised some of his notes, especially for "The historiographic and ideological contexts of the nineteenth-century debate on man's place in nature," themselves a real treasure for which I remain in his debt politically and professionally, to take better account of feminist theory in the field. It is because his notes are otherwise so exhaustive that I am critical of the very thin attention to feminist reformulations of science studies debates (see note 174.2 on p. 273; here would have been an opportunity).

Robert Young's notes helped train me in the history of science; that's why I am disappointed in this aspect of his revisions for the 1985 book. The unexamined *commitment* to masculinism in the chief texts of the history of science to which Young reacted at Cambridge remained present in too much of the radical science movement and its literatures. The same *commitment* to masculinism is evident in the canonized texts of the current social studies of science orthodoxy, e.g., the important books by Steve Shapin and Simon Schaffer, *The Leviathan and the Air-Pump*, Princeton, NJ: Princeton University Press, 1985, and by Bruno Latour, *Science in Action*, Cambridge, MA: Harvard University Press, 1987. This trouble must not be allowed to persist in the movement to address science as culture, in which Young is a creative leader.

7. I am indebted here to Catharine MacKinnon (1982) "Feminism, Marxism, method, and the state: an agenda for theory," *Signs* 7(3): 515–44.

8. Here I rely heavily on Hazel Carby (1987), *Reconstructing Womanhood*, New York: Oxford University Press, and Hortense Spillers (1987) "Mama's baby, Papa's maybe: an American grammar book," *Diacritics* 17(2): 65–81.

9. To my mind, on this subject as elsewhere in her interesting and rich book, Noske makes sweeping generalizations and does not ask carefully enough how her claims should be limited or modified. Her discussion of the history of "objectifying" Western science is particularly stereotypical in this regard. Other discussions, like those about the history of primate behavioural studies, are much better. But these issues are quibbles in relation to the fundamental and synthetic project of her book, which remains unique in green (environmentalist), red (socialist), purple (feminist) and ultraviolet (scientific) literatures. Noske's book is firmly located in critical conversation with social movements and with natural and social sciences on the tricky problem of anthropocentrism.

10. Margulis, L. and D. Sagan (1986), *Origins of Sex: Three Billion Years of Genetic Recombination*. New Haven, CT: Yale University Press.
11. Such cells are called eukaryotes; they have a membrane-bound nucleus and other differentiated internal structures. Prokaryotes, or bacteria, do not have a nucleus to house their genetic material, but keep their DNA naked in the cell.
12. A protist is a single-celled, eukaryotic micro-organism, such as the familiar amoeba. Plants, animals and fungi descended from such beginnings.

5

TEDDY BEAR PATRIARCHY: TAXIDERMY IN THE GARDEN OF EDEN, NEW YORK CITY, 1908–1936

Nature teaches law and order and respect for property. If these people cannot go to the country, then the Museum must bring nature to the city.[1]

I started my thoughts on the legend of Romulus and Remus who had been suckled by a wolf and founded Rome, but in the jungle I had my little Lord Greystoke suckled by an ape.[2]

EXPERIENCE

In the heart of New York City stands Central Park—the urban garden designed by Frederick Law Olmsted to heal the overwrought or decadent city dweller with a prophylactic dose of nature. Across from the park the Theodore Roosevelt Memorial presides as the central building of the American Museum of Natural History, a monumental reproduction of the Garden of Eden.[3] In the Garden, Western "man" may begin again the first journey, the first birth from within the sanctuary of nature. Founded just after the Civil War and dedicated to popular education and scientific research, the American Museum of Natural History is the place to undertake this genesis, this regeneration. Passing through the Museum's Roosevelt Memorial atrium into the African Hall, opened in 1936, the ordinary citizen enters a privileged space and time: the Age of Mammals in the heart of Africa, scene origins.[4] A hope is implicit in every architectural detail; in immediate vision of the origin, perhaps the future can be fixed. By saving the beginnings, the end can be achieved and the present can be transcended. African Hall offers a unique communion with nature at its highest and yet most vulnerable moment, the moment of the interface of the Age of Mammals with the Age of Man. This communion is offered through the sense of vision by the craft of taxidermy. Its most ecstatic and skillful moment joins ape and man in visual embrace.

Restoration of the origin, the task of genetic hygiene, is achieved in Carl Akeley's African Hall by an art that began for him in the 1880s with the crude stuffing of P. T. Barnum's elephant, Jumbo, who had been run down by a railroad train, the emblem of the Industrial Revolution. The end of his task came in the 1920s, with his exquisite mounting of the Giant of Karisimbi, the lone silverback male gorilla that dominates the diorama depicting the site of Akeley's own grave in the mountainous rain forest of the Congo, today's Zaire. So it could inhabit Akeley's monument to the purity of nature, this gorilla was killed in 1921, the same year the Museum hosted the Second International Congress of Eugenics. From the dead body of the primate, Akeley crafted something finer than the living organism; he achieved its true end, a new genesis. Decadence—the threat of the city, civilization, machine—was stayed in the politics of eugenics and the art of taxidermy. The Museum fulfilled its scientific purpose of conservation, preservation, and the production of permanence. Life was transfigured in the principal civic arena of western political theory—the natural body of man.[5]

Behind every mounted animal, bronze sculpture, or photograph lies a profusion of objects and social interactions among people and other animals, which can be recomposed to tell a biography embracing major themes for the twentieth-century United States. But the recomposition produces a story that is reticent, even mute, about Africa. H. F. Osborn, president of the American Museum from 1908–33, thought Akeley was Africa's biographer. But in a stronger sense, Akeley is America's biographer, at least for part of North America. Akeley thought in African Hall the visitor would experience nature at its moment of highest perfection. He did not dream that he crafted the means to experience a history of race, sex, and class in New York City that reached to Nairobi.

To enter the Theodore Roosevelt Memorial, the visitor must pass by a James Earle Fraser equestrian statue of Teddy majestically mounted as a father and protector between two "primitive" men, an American Indian and an African, both standing, dressed as "savages." The facade of the memorial, funded by the State of New York and awarded to the American Museum of Natural History on the basis of its competitive application in 1923, is classical, with four Ionic columns 54 feet high topped by statues of the great explorers Boone, Audubon, Lewis, and Clark. The coin-like, bas-relief seals of the United States and of the Liberty Bell are stamped on the front panels. Inscribed across the top are the words TRUTH, KNOWLEDGE, VISION and the dedication to Roosevelt as "a great leader of the youth of America, in energy and fortitude in the faith of our fathers, in defense of the rights of the people, in the love and conservation of nature and of the best in life and in man." Youth, paternal solicitude, virile defense of democracy, and intense emotional connection to nature are the unmistakable themes.[6]

The building presents itself in many visible faces. It is at once a Greek temple, a bank, a scientific research institution, a popular museum, a neoclassical theater. One is entering a space that sacralizes democracy, Protestant Christianity, adventure, science, and commerce. Entering this building, one knows that a drama will be enacted inside. Experience in this public monument will be intensely personal; this structure is one of North America's spaces for joining the duality of self and community.

Just inside the portals, the visitor enters the sacred space where transformation of consciousness and moral state will begin.[7] The walls are inscribed with Roosevelt's words under the headings Nature, Youth, Manhood, the State. The seeker begins in Nature: "There are no words that can tell the hidden spirit of the wilderness, that can reveal its mystery. . . . The nation behaves well if it treats its natural resources as assets which it must turn over to the next generation increased and not impaired in value." Nature is mystery and resource, a critical union in the history of civilization. The visitor—necessarily a white boy in moral state, no matter what accidents of biology or social gender and race might have pertained prior to the Museum excursion—progresses through Youth: "I want to see you game boys . . . and gentle and tender. . . . Courage, hard work, self mastery, and intelligent effort are essential to a successful life." Youth mirrors Nature, its pair across the room. The next stage is Manhood: "Only those are fit to live who do not fear to die and none are fit to die who have shrunk from the joy of life and the duty of life." Opposite is its spiritual pair, the State: "Aggressive fighting for the right is the noblest sport the world affords. . . . If I must choose between righteousness and peace, I choose righteousness." The walls of the atrium are full of murals depicting Roosevelt's life, the perfect illustration of his words. His life is inscribed in stone in a peculiarly literal way appropriate to this museum. One sees the man hunting big game in Africa, conducting diplomacy in the Philippines and China, helping boy and girl scouts, receiving academic honors, and presiding over the Panama Canal ("The land divided, the world united").

Finally, in the atrium stand the striking life-size bronze sculptures by Carl Akeley of the Nandi spearmen of East Africa on a lion hunt. These

African men and the lion they kill symbolize for Akeley the essence of the hunt, of what would later be named "man the hunter." Discussing the lion spearers, Akeley referred to them as men. In every other circumstance he referred to adult male Africans as boys. Roosevelt, the modern sportsman, and the "primitive" Nandi share in the spiritual truth of manhood. The noble sculptures express Akeley's great love for Roosevelt, his friend and hunting companion in Africa in 1910 for the killing of one of the elephants which Akeley mounted for the Museum. Akeley said he would follow Roosevelt anywhere because of his "sincerity and integrity" (Akeley 1923: 162).

In the Museum shop in the atrium in the 1980s, one may purchase *T.R.: Champion of the Strenuous Life*, a photographic biography of the 26th president. Every aspect of the fulfillment of manhood is depicted, even death is labeled "The Great Adventure." One learns that after defeat in the presidential campaign of 1912, Roosevelt undertook the exploration of the Amazonian tributary, the River of Doubt, under the auspices of the American Museum of Natural History and the Brazilian Government. It was a perfect trip. The explorers nearly died, the river had never before been seen by white men, and the great stream, no longer doubtful, was renamed Rio Roosevelt by the Brazilian State. In the picture biography, which includes a print of the adventurers paddling their primitive dugout canoe (one assumes before starvation and jungle fever attenuated the ardor of the photographer), the former president of a great industrial power explains his return to the wilderness: "I had to go. It was my last chance to be a boy" (Johnson 1958: 138, 126–7).[8]

The joining of life and death in these icons of Roosevelt's journeys and in the architecture of his stony memorial announces the central moral truth of the Museum. This is the effective truth of manhood, the state conferred on the visitor who successfully passes through the trial of the Museum. The body can be transcended. This is the lesson Simone de Beauvoir so painfully remembered in the *Second Sex*; man is the sex which risks life and in so doing, achieves his existence. In the upside down world of Teddy Bear Patriarchy, it is in the craft of killing that life is constructed, not in the accident of personal, material birth. Roosevelt is the perfect *locus genii* for the Museum's task of regeneration of a miscellaneous, incoherent urban public threatened with genetic and social decadence, threatened with the prolific bodies of the new immigrants, threatened with the failure of manhood.[9]

The Akeley African Hall itself is simultaneously a very strange place and an ordinary experience for literally millions of North Americans over more than five decades. The types of display in this hall are spread all over the country, and even the world, partly due to the craftspeople

Akeley himself trained. In the 1980s sacrilege is perhaps more evident than liminal experience of nature. What is the experience of New York streetwise kids wired to Walkman radios and passing the Friday afternoon cocktail bar by the lion diorama? These are the kids who came to the Museum to see the high tech Nature-Max films. But soon, for those not physically wired into the communication system of the late twentieth century, another time begins to take form. The African Hall was meant to be a time machine, and it is (Fabian 1983: 144). The individual enters the age of Mammals. But one enters alone, each individual soul, as part of no stable prior community and without confidence in the substance of one's body, in order to be received into a saved community. One begins in the threatening chaos of the industrial city, part of a horde, but here one will come to belong, to find substance. No matter how many people crowd the Great Hall, the experience is of individual communion with nature. The sacrament will be enacted for each worshipper. This nature is not constituted from a probability calculus. This is not a random world, populated by late twentieth-century cyborgs, for whom the threat of decadence is a nostalgic memory of a dim organic past, but the moment of origin where nature and culture, private and public, profane and sacred meet—a moment of incarnation in the encounter of man and animal.

The Hall is darkened, lit only from the display cases which line the sides of the spacious room. In the center of the Hall is a group of elephants so lifelike that a moment's fantasy suffices for awakening a premonition of their movement, perhaps an angry charge at one's personal intrusion. The elephants stand like a high altar in the nave of a great cathedral. That impression is strengthened by one's growing consciousness of the dioramas that line both sides of the main Hall and the spacious gallery above. Lit from within, the dioramas contain detailed and lifelike groups of large African mammals—game for the wealthy New York hunters who financed this experience. Called habitat groups, they are the culmination of the taxidermist's art. Called by Akeley a "peep-hole into the jungle,"[10] each diorama presents itself as a side altar, a stage, an unspoiled garden in nature, a hearth for home and family. As an altar, each diorama tells a part of the story of salvation history; each has its special emblems indicating particular virtues. Above all, inviting the visitor to share in its revelation, each tells the truth. Each offers a vision. Each is a window onto knowledge.

A diorama is eminently a story, a part of natural history. The story is told in the pages of nature, read by the naked eye. The animals in the habitat groups are captured in a photographer's and sculptor's vision. They are actors in a morality play on the stage of nature, and the eye is the critical organ. Each diorama contains a small group of animals in the foreground, in the midst of exact reproductions of plants, insects, rocks, soil. Paintings reminiscent of Hollywood movie set art curve in back of the group and up to the ceiling, creating a great panoramic vision of a scene on the African continent. Each painting is minutely appropriate to the particular animals in the foreground. Among the 28 dioramas in the Hall, all the major geographic areas of the African continent and most of the large mammals are represented.

Gradually, the viewer begins to articulate the content of the story. Most groups are made up of only a few animals, usually a large and vigilant male, a female or two, and one baby. Perhaps there are some other animals—a male adolescent maybe, never an aged or deformed beast. The animals in the group form a developmental series, such that the group can represent the essence of the species as a dynamic, living whole. The principles of organicism, that is, of the laws of organic form, rule the composition.[11] There is no need for the multiplication of specimens because the series is a true biography. Each animal is an organism, and the group is an organism. Each organism is a vital moment in the narrative of natural history, condensing the flow of time into the harmony of developmental form. The groups are peaceful, composed, illuminated—in "brightest Africa."[12] Each group forms a community structured by a natural division of function; the whole animal in the whole group is nature's truth. The physiological division of labor that has informed the history of biology is embodied in these habitat groups which tell of communities and families, peacefully and hierarchically ordered. Sexual specialization of function—the organic bodily and social sexual division of labor—is unobtrusively ubiquitous, unquestionable, right. The African buffalo, the white and black rhinos, the lion, the zebra, the mountain nyala, the okapi, all find their place in the differentiated developmental harmony of nature. The racial division of labor, the familial progress from youthful native to adult white man, was announced at the steps leading to the building itself; Akeley's original plan for African Hall included bas-relief sculptures of all the "primitive" tribes of Africa complementing the other stories of natural wild life in the Hall. Organic hierarchies are embodied in every organ in the articulation of natural order in the Museum.[13]

But there is a curious note in the story; it begins to dominate as scene after scene draws the visitor into itself through the eyes of the animals in the tableaux.[14] Each diorama has at least one animal that catches the viewer's gaze and holds it in communion. The animal is vigilant, ready to sound an alarm at the intrusion of man, but ready also to hold forever the gaze of meeting, the moment of truth, the original encounter. The moment seems fragile, the animals about to disappear, the communion about to break; the Hall threatens to dissolve into the chaos of the Age of Man. But it does not. The gaze holds, and the wary animal heals those who will look. There is no impediment to this vision, no mediation. The glass front of the diorama forbids the body's entry, but the gaze invites his visual penetration. The animal is frozen in a moment of supreme life, and man is transfixed. No merely living organism could accomplish this act. The specular commerce between man and animal at the interface of two evolutionary ages is completed. The animals in the dioramas have transcended mortal life, and hold their pose forever, with muscles tensed, noses aquiver, veins in the face and delicate ankles and folds in the supple skin all prominent. No visitor to a merely physical Africa could see these animals. This is a spiritual vision made possible only by their death and literal re-presentation. Only then could the essence of their life be present. Only then could the hygiene of nature cure the sick vision of civilized man. Taxidermy fulfills the fatal desire to represent, to be whole; it is a politics of reproduction.

There is one diorama that stands out from all the others, the gorilla group. It is not simply that this group is one of the four large corner displays. There is something special in the painting with the steaming volcano in the background and Lake Kivu below, in the pose of the enigmatic large silverback rising above the group in a chest-beating gesture of alarm and an unforgettable gaze in spite of the handicap of glass eyes. The painter's art was particularly successful in conveying the sense of limitless vision, of a panorama without end around the focal lush green garden. This is the scene that Akeley longed to return to. It is where he died, feeling he was at home as in no other place on earth. It is where he first killed a gorilla and felt the enchantment of a perfect garden. After his first visit in 1921, he was motivated to convince the Belgian government to make this area the first African national park to ensure a sanctuary for the gorilla. But the viewer does not know these things when he sees the five animals in a naturalistic setting. It is plain that he is looking at a natural family of close human relatives, but that is not the essence of this diorama. The viewer sees that the elephants, the lion, the rhino, and the water hole group—with its peaceful panorama of all the grassland species, including the carnivores, caught in a moment outside the Fall— all these have been a kind of preparation, not so much for the gorilla group, as for the Giant of Karisimbi. This double for man stands in a unique personal individuality, his fixed face molded forever from the death mask cast from his corpse by a taxidermist in the Kivu Mountains. Here is natural man, immediately known. His image may be purchased on a picture postcard at the desk in the Roosevelt atrium. (Figure 5.1)

It would have been inappropriate to meet the gorilla anywhere else but on the mountain. Frankenstein and his monster had Mont Blanc for

Fig. 5.1. The Giant of Karisimbi. Negative no. 315077. Published with permission of the Department of Library Services, American Museum of Natural History.

their encounter; Akeley and the gorilla first saw each other on the lush volcanoes of central Africa. The glance proved deadly for them both, just as the exchange between Victor Frankenstein and his creature froze each of them into a dialectic of immolation. But Frankenstein tasted the bitter failure of his fatherhood in his own and his creature's death; Akeley resurrected his creature and his authorship in both the sanctuary of Parc Albert and the African Hall of the American Museum of Natural History. Mary Shelley's story may be read as a dissection of the deadly logic of birthing in patriarchy at the dawn of the age of biology; her tale is a nightmare about the crushing failure of the project of man. But the taxidermist labored to restore manhood at the interface of the Age of Mammals and the Age of Man. Akeley achieved the fulfillment of a sportsman in Teddy Bear Patriarchy—he died a father to the game, and their sepulcher is named after him, the Akeley African Hall.

The gorilla was the highest quarry of Akeley's life as artist, scientist, and hunter, but why? He said himself (through his ghostwriter, the invisible Dorothy Greene), "To me the gorilla made a much more interesting quarry than lions, elephants, or any other African game, for the gorilla is still comparatively unknown" (Akeley 1923: 190). But so was the colobus monkey or any of a long list of animals. What qualities did it take to make an animal "game"? One answer is similarity to man, the ultimate quarry, a worthy opponent. The ideal quarry is the "other," the natural self. That is one reason Frankenstein needed to hunt down his creature. Hunter, scientist, and artist all sought the gorilla for his revelation about the nature and future of manhood. Akeley compared and contrasted his quest for the gorilla with the French-American Paul du Chaillu's, the first white man to kill a gorilla, in 1855, eight years after it was "discovered" to science. Du Chaillu's account of the encounter stands as the classic portrayal of a depraved and vicious beast killed in the heroic, dangerous encounter. Disbelieving du Chaillu, Akeley told his own readers how many times du Chaillu's publishers made him rewrite until the beast was fierce enough. Frankenstein plugged up his ears rather than listen to his awful son claim a gentle and peace loving soul. Akeley was certain he would find a noble and peaceful beast; so he brought his guns, cameras, and white women into the garden to hunt, wondering what distance measured courage in the face of a charging alter-ego.

Like du Chaillu, Akeley came upon a sign of the animal, a footprint, or in Akeley's case a handprint, before meeting face to face. "I'll never forget it. In that mud hole were the marks of four great knuckles where the gorilla had placed his hand on the ground. There is no other track like this in the world—there is no other hand in the world so large. . . . As I looked at that track I lost the faith on which I had brought my party to Africa. Instinctively I took my gun from the gun boy" (Akeley 1923: 203). Later, Akeley told that the handprint, not the face, gave him his greatest thrill. In the hand the trace of kinship writ large and terrible struck the craftsman.

But then, on the first day out from camp in gorilla country, Akeley did meet a gorilla face to face, the creature he had sought for decades, prevented from earlier success by mauling elephants, stingy millionaires, and world war. Within minutes of his first glimpse of the features of the face of an animal he longed more than anything to see, Akeley had killed him, not in the face of a charge, but through a dense forest screen within which the animal hid, rushed, and shook branches. Surely the taxidermist did not want to risk losing his specimen, for perhaps there would be no more. He knew the Prince of Sweden was just then leaving Africa after having shot fourteen of the great apes in the same region. The animals must be wary of new hunters; collecting might be very difficult.

Whatever the rational or fantastic logic that ruled the first shot, precisely placed into the aorta, the task that followed was arduous indeed—skinning the animal and transporting various remains back to camp. The corpse had nearly miraculously lodged itself against the trunk of a tree above a deep chasm. As a result of Herculean labors, which included casting the death mask pictured in *Lions, Gorillas, and their Neighbors* (Akeley and Akeley, 1922), Akeley was ready for his next gorilla hunt on the second day after shooting the first ape. The pace he was setting himself was grueling, dangerous for a man ominously weakened by tropical fevers. "But science is a jealous mistress and takes little account of a man's feelings."[15] The second quest resulted in two missed males, a dead female, and her frightened baby speared by the porters and guides. Akeley and his party had killed or attempted to kill every ape they had seen since arriving in the area.

On his third day out, Akeley took his cameras and ordered his guides to lead toward easier country. With a baby, female, and male, he could do a group even if he got no more specimens. Now it was time to hunt with the camera.[16] "Almost before I knew it I was turning the crank of the camera on two gorillas in full view with a beautiful setting behind them. I do not think at the time I appreciated the fact that I was doing a thing that had never been done before" (Akeley 1923: 221). But the photogenic baby and mother and the accompanying small group of other gorillas had become boring after two hundred feet of film, so Akeley provoked an action shot by standing up. That was interesting for a bit. "So finally, feeling that I had about all I could expect from that band, I picked out one that I thought to be an immature male. I shot and killed it and found, much to my regret, that it was a female. As it turned out, however, she was such a splendid large specimen that the feeling of regret was considerably lessened" (Akeley 1923: 222).

Satisfied with the triumphs of his gun and camera, Akeley decided it was time to ask the rest of the party waiting in a camp below to come up to hunt gorillas. He was getting considerably sicker and feared he would not fulfill his promise to his friends to give them gorilla. His whole purpose in taking white women into gorilla country depended on meeting this commitment: "As a naturalist interested in preserving wild life, I was glad to do anything that might make killing animals less attractive."[17] The best thing to reduce the potency of game for heroic hunting is to demonstrate that inexperienced women could safely do the same thing. Science had already penetrated; women could follow.

Two days of hunting resulted in Herbert Bradley's shooting a large silverback, the one Akeley compared to Jack Dempsey and mounted as the lone male of Karisimbi in African Hall. It was now possible to admit another level of feeling: "As he lay at the base of the tree, it took all one's scientific ardour to keep from feeling like a murderer. He was a magnificent creature with the face of an amiable giant who would do no harm except perhaps in self defense or in defense of his family" (Akeley 1923: 230). If he had succeeded in his aborted hunt, Victor Frankenstein could have spoken those lines.

The photograph in the American Museum film archive of Carl Akeley, Herbert Bradley, and Mary Hastings Bradley holding up the gorilla head and corpse to be recorded by the camera is an unforgettable image.[18] The face of the dead giant evokes Bosch's conception of pain, and the lower jaw hangs slack, held up by Akeley's hand. The body looks bloated and utterly heavy. Mary Bradley gazes smilingly at the faces of the male hunters, her own eyes averted from the camera. Akeley and Herbert Bradley look directly at the camera in unshuttered acceptance of their act. Two Africans, a young boy and a young man, perch in a tree above the scene, one looking at the camera, one at the hunting party. The contrast of this scene of death with the diorama framing the giant of Karisimbi mounted in New York is total; the animal came to life again, this time immortal.

There was no more need to kill, so the last capture was with the camera. "The guns were put behind and the camera pushed forward and we had the extreme satisfaction of seeing the band of gorillas disappear over the crest of the opposite ridge none the worse for having met with white men that morning. It was a wonderful finish to a wonderful gorilla hunt" (Akeley 1923: 235). Once domination is complete, conservation is urgent. But perhaps preservation comes too late.

What followed was the return to the United States and active work for an absolute gorilla sanctuary providing facilities for scientific research. Akeley feared the gorilla would be driven to extinction before it was adequately known to science (Akeley 1923: 248). Scientific knowledge canceled death; only death before knowledge was final, an abortive act in the natural history of progress. His health weakened but his spirit at its height, Akeley lived to return to Kivu to prepare paintings and other material for the gorilla group diorama. Between 1921 and 1926, he mounted his precious gorilla specimens, producing that extraordinary silverback whose gaze dominates African Hall. When he did return to Kivu in 1926, he was so exhausted from his exertions to reach his goal that he died on November 17, 1926, almost immediately after he and his party arrived on the slopes of Mt. Mikena, "in the land of his dreams" (M. J. Akeley 1929b: Chpt. XV).

Akeley's was a literal science dedicated to the prevention of decadence, of biological decay. His grave was built in the heart of the rain forest on the volcano, where "all the free wild things of the forest have perpetual sanctuary" (M. J. Akeley 1940: 341). Mary Jobe Akeley directed the digging of an eight-foot vault in lava gravel and rock. The hole was lined with closely set wooden beams. The coffin was crafted on the site out of solid native mahogany and lined with heavy galvanized steel salvaged from the boxes used to pack specimens to protect them from insect and other damage. Then the coffin was upholstered with camp blankets. A slab of cement ten by twelve feet and five inches thick was poured on top of the grave and inscribed with the name and date of death of the father of the game. The cement had been carried on porters' backs all the way from the nearest source in Kibale, Uganda. The men ditched the first load in the face of the difficult trails; they were sent back for a second effort. An eight-foot stockade fence was built around the grave to deter buffalo and elephant from desecrating the site. "Derscheid, Raddatz, Bill and I worked five days and five nights to give him the best home we could build, and he was buried as I think he would have liked with a simple reading service and a prayer" (M. J. Akeley 1929b: 189–90). The grave was inviolate, and reincarnation of the natural self would be immortal in African Hall. In 1979, "grave robbers, Zairoise poachers, violated the site and carried off [Akeley's] skeleton" (Fossey 1983: 3).

BIOGRAPHY

For this untruthful picture Akeley substitutes a real gorilla. (Osborn, in Akeley 1923: xii)

Of the two I was the savage and the aggressor. (Akeley 1923: 216)

Akeley sought to craft a true life, a unique life. The life of Africa became his life, his telos. But it is not possible to tell his life from a single point of view. There is a polyphony of stories, and they do not harmonize. Each source for telling the story of Akeley's life speaks in an authoritative mode, but I felt compelled to compare the versions, and then to cast Akeley's story in an ironic mode, the register most avoided by my subject. Akeley wanted to present an immediate vision; I would like to dissect and make visible layer after layer of mediation. I want to show the reader how the experience of the diorama grew from the safari in specific times and places, how the camera and the gun together are the conduits for the spiritual commerce of man and nature, how biography is woven into and from a social and political tissue. I want to show how the stunning animals of Akeley's achieved dream in African Hall are the product of particular technologies, i.e., the techniques of effecting meanings.

Life Stories

In harmony with the available plots in U.S. history, it is necessary that Carl Akeley (1864–1926) was born on a farm in New York of poor, but vigorous, old, (white—the only trait that didn't need to be named), American stock. The time of his birth, near the end of the Civil War, was an end and a beginning for so much in North America, including the history of biology and the structure of wealth and social class. In a boyhood full of hard farm labor, he learned self-reliance and skill with tools and machines. He passed long hours alone watching and hunting the wildlife of New York. By the age of 13, aroused by a borrowed book on the subject, Akeley was committed to the vocation of taxidermy. His vocation's bibliogenesis seems also ordained by the plot. At that age (or age 16 in some versions), he had a business card printed up. No Yankee boy could miss the connection of life's purpose with business, although young Carl scarcely believed he could make his living at such a craft. He took lessons in painting, so that he might provide realistic backgrounds for the birds he ceaselessly mounted. From the beginning Akeley's life had a single focus: the recapturing and representation of the nature he saw. On this point all the versions of Akeley's life concur.

After the crops were in, at the age of 19, Akeley set off from his father's farm "to get a wider field for my efforts" (Akeley 1923: 1). First he tried to get a job with a local painter and interior decorator whose hobby was taxidermy, but this man directed the boy to an institution which changed his life—Ward's Natural Science Establishment in Rochester, where Akeley would spend four years and form a friendship pregnant with consequences for the nascent science of ecology as it came to be practiced in museum exhibition. Ward's provided mounted specimens and natural history collections for practically all the museums in the nation. Several important men in the history of biology and museology in the United States passed through this curious institution, including Akeley's friend, William Morton Wheeler. Wheeler completed his career in entymology at Harvard, a founder of the science of animal ecology (which he called ethology—the science of the character of nature) and a mentor to the great organicists and conservative social philosophers in Harvard's biological and medical establishment (Russett 1966; Evans and Evans 1970; Cross and Albury 1987). Wheeler was then a young Milwaukee naturalist steeped in German "Kultur" who began tutoring the rustic Akeley for entry into Yale's Sheffield Scientific School. However, eleven hours of taxidermy in the day and long hours of study proved too much; so higher education was postponed, later permanently, in order to follow the truer vocation of reading nature's book directly.

Akeley was disappointed at Ward's because business imperatives al-lowed no room for improvement of taxidermy. He felt animals were "upholstered." Developing his own skill and technique in spite of the lack of encouragement, and the lack of money, he got a chance for public recognition when P. T. Barnum's famous elephant was run down by a locomotive in Canada in 1885. Barnum did not want to forego the fame and profit from continuing to display the giant (who had died trying to save a baby elephant, we are told), so Akeley and a companion were dispatched to Canada from Rochester to save the situation. Six butchers from a nearby town helped with the rapidly rotting carcass. What Akeley learned about very large mammal taxidermy from this experience laid the foundation for his later revolutionary innovations in producing light, strong, life-like pachyderms. The popular press followed the monumental mounting, and the day Jumbo was launched in his own railroad car into his post-mortem career, half the population of Rochester witnessed the resurrection.

In 1885, Wheeler returned to Milwaukee to teach high school and soon took up a curatorship in the Milwaukee Museum of Natural History. Wheeler urged his friend to follow, hoping to continue his tutoring and to secure Akeley commissions for specimens from the museum. Museums did not then generally have their own taxidermy departments, although around 1890 taxidermic technique flowered in Britain and the United States. Akeley opened his business shop on the Wheeler family property, and he and the naturalist spent long hours discussing natural history, finding themselves in agreement about museum display and about the character of nature. The most important credo for them both was the need to develop scientific knowledge of the whole animal in the whole group in nature—i.e., they were committed organicists. Wheeler soon became director of the Milwaukee Museum and gave Akeley significant support. Akeley had conceived the idea for habitat groups and wished to mount a series illustrating the fur-bearing animals of Wisconsin. His completed muskrat group (1889), minus the painted backgrounds, was probably the first mammalian habitat group anywhere.

As a result of a recommendation from Wheeler, in 1894 the British Museum invited Akeley to practice his trade in that world-famous institution. On the way to London, Akeley visited the Field Museum in Chicago, met Daniel Giraud Elliot and accepted his offer of preparing the large collection of specimens the Museum had bought from Ward's. In 1896, Akeley made his first collecting expedition to Africa, to British Somaliland, a trip that opened a new world to him. This was the first of five safaris to Africa, each escalating his sense of the purity of the continent's vanishing wildlife and the conviction that the meaning of his life was its preservation through transforming taxidermy into an art. He was next in Africa for the Field Museum in 1905, with his explorer/adventurer/author wife, Delia, to collect elephants in British East Africa. On this trip Akeley escaped with his life after killing a leopard in hand-to-fang combat.

In Chicago Akeley spent four years largely at his own expense preparing the justly famous Four Seasons deer dioramas. In 1908, at the invitation of the new president, H. F. Osborn, who was anxious to mark his office with the discovery of major new scientific laws and departures in museum exhibition and public education, Akeley moved to New York and the American Museum of Natural History in hope of preparing a major collection of large African mammals. From 1909–11 Carl and Delia collected in British East Africa, a trip marked by a hunt with Theodore Roosevelt and his son Kermit, who were collecting for the Washington National Museum. The safari was brought to a limping conclusion by Carl's being mauled by an elephant, delaying fulfillment of his dream of collecting gorillas. His plan for the African Hall took shape by 1911 and ruled his behavior thereafter. In World War I he was a civilian Assistant Engineer to the Mechanical and Devices Section of the Army. He is said to have refused a commission in order to keep his freedom to speak freely to anyone in the hierarchy.

During the war, his work resulted in several patents in his name. The theme of Akeley the inventor recurs constantly in his life story. Included in his roster of inventions, several of which involved subsequent business development, were a motion picture camera, a cement gun, and new taxidermic processes.

With the close of war, Akeley focused his energy on getting backing for the African Hall. He needed more than a million dollars. Lecture tours, articles, a book, and endless promotion brought him into touch with the major wealthy sportsmen of New York, but sufficient financial commitment eluded him. In 1921, financing half the expense himself, Akeley left for Africa, this time accompanied by a married couple, their 5-year-old daughter, their governess, and Akeley's adult niece whom he had promised to take hunting in Africa. In 1923 in New York, Carl and Delia divorced—an event unrecorded in versions of his life; Delia just disappears from the narratives. In 1924 Akeley married Mary L. Jobe, the explorer/adventurer/author who accompanied him on his last adventure, the Akeley-Eastman-Pomeroy African Hall Expedition, that collected for ten dioramas of the Great Hall. George Eastman, of Eastman Kodak fortunes, and Daniel Pomeroy, the benefactors, accompanied the taxidermist-hunter to collect specimens. Eastman, then 71 years old, went with his own physician and commanded his own railroad train for part of the excursion.

En route to Africa the Akeleys were received by the conservationist and war hero Belgian king, Albert. He was the son of the infamous Leopold II, whose personal rapacious control of the Congo for profit was wrested away and given to the Belgian government by other European powers in 1908. Leopold II had financed Henry Stanley's explorations of the Congo. Akeley is narrated as a man like the great explorers, Stanley and

Livingstone, but also as the man who witnessed, and helped birth, a new "bright" Africa. The "enlightened" Albert, led to his views on national parks by a visit to Yosemite, confirmed plans for the Parc Albert and commissioned the Akeleys to prepare topographical maps and descriptions of the area in cooperation with the Belgian naturalist, Jean Derscheid. There was no room for a great park for the Belgians in Europe, so "naturally" one was established in the Congo. Mandating protection for the Pygmies within park boundaries, the park was to provide sanctuary for "natural primitives," as well as foster scientific study by establishing permanent research facilities. After ten months of collecting, Carl and Mary Jobe set off for the Kivu forest, the heart of remaining unspoiled Africa, where he died and was buried "in ground the hand of man can never alter or profane" (M. J. Akeley 1940: 340).

Taxidermy: From Upholstery to Epiphany

Transplanted Africa stands before him—a result of Akeley's dream. (Clark 1936: 73)

The vision Carl Akeley had seen was one of jungle peace. His quest to *embody* this vision justified to himself his hunting, turned it into a tool of science and art, the scalpel that revealed the harmony of an organic, articulate world. Let us follow Akeley briefly through his technical contributions to taxidermy in order to grasp more fully the stories he needed to tell about the biography of Africa, the life history of nature.

It is a simple tale: Taxidermy was made into the servant of the "real." Artifactual children, better than life, were birthed from dead matter (Sofoulis 1988). Akeley's vocation, and his achievement, was the production of an organized craft for eliciting unambiguous experience of organic perfection. Literally, Akeley "typified" nature, made nature true to type. Taxidermy was about the single story, about nature's unity, the unblemished type specimen. Taxidermy became the art most suited to the epistemological and aesthetic stance of realism. The power of this stance is in its magical effects: what is so painfully constructed appears effortlessly, spontaneously found, discovered, simply there if one will only look. Realism does not appear to be a point of view, but appears as a "peephole into the jungle" where peace may be witnessed. Epiphany comes as a gift, not as the fruit of merit and toil, soiled by the hand of man. Realistic art at its most deeply magical issues in revelation. This art repays labor with transcendence. Small wonder that artistic realism and biological science were twin brothers in the founding of the civic order of nature at the American Museum of Natural History. It is also natural that taxidermy and biology depend fundamentally upon vision in a hierarchy of the senses; they are tools for the construction, discovery of form.

Akeley's eight years in Milwaukee from 1886 to 1894 were crucial for his working out techniques that served him the rest of his life. The culmination of that period was a head of a male Virginia deer that won first place in the first Sportsman's Show, in New York City in 1895. The judge in that national competition was Theodore Roosevelt, whom Akeley did not meet until they befriended each other on safari in Africa in 1906. The head, entitled "The Challenge," displayed a buck "in the full frenzy of his virility as he gave the defiant roar of the rutting season—the call to fierce combat" (M. J. Akeley 1940: 38). Jungle peace was not a passive affair, nor one unmarked by gender.

The head was done in a period of experimentation leading to the production of the Four Seasons group in Chicago, installed in 1902.[19] In crafting those groups over four years, Akeley worked out his manikin method, clay modeling, plaster casting, vegetation molding techniques, and the organized production system. He hired women and men workers by the hour to turn out the thousands of individual leaves needed to clothe the trees in the scenes. Charles Abel Corwin painted background canvases from studies in the Michigan Iron Mountains where the animals were collected. Akeley patented his vegetation process, but gave rights for its use free of charge to the Field Museum in Chicago. He allowed free, worldwide use of his patented methods of producing light, strong papier-mâché manikins from exact clay models and plaster casts. Cooperation in museum development was a fundamental value for Akeley, who did not make much money at his craft and whose inventions were significant for economic survival.

Akeley continued to make improvements in his taxidermic technique throughout his life, and he taught several other key workers, including James Lipsitt Clark, who was the Director of Arts, Preparation, and Installation at the American Museum after Akeley's death when African Hall was actually constructed. While Akeley worked long hours alone, taxidermy as he helped to develop it was not a solitary art. Taxidermy requires a complex system of coordination and division of labor, beginning in the field during the hunting of the animals and culminating in a finished diorama. A minimum list of workers on one of Akeley's projects includes taxidermists, collectors, artists, anatomists, and "accessory men" (M. J. Akeley 1940: 217). Pictures of work in the Museum taxidermy studios show men (males, usually white) tanning hides, working on clay models of sizable mammals (including elephants) or on plaster casts, assembling skeleton and wood frames, consulting scale models of the planned display, doing carpentry, making vegetation, sketching, etc. Clark reports that between 1926 and 1936, when African Hall opened, still unfinished, the staff of the project usually employed about 45 men. Painting the backgrounds was a major artistic specialization, and the artists based their final panoramas on numerous studies done at the site of collection. In the field, the entire operation rested on the organization of the safari, a complex social institution where race, sex, and class came together intensely. Skinning a large animal could employ 50 workers for several hours. Photographs, moving picture records, death masks, extensive anatomical measurements, initial treatment of skins, and sketches occupied the field workers. The production of a modern diorama involved the work of hundreds of people in a social system embracing structures of skill and authority on a worldwide scale.

How can such a system produce a unified biography of nature? How is it possible to refer to Akeley's African Hall when it was constructed after he died? On an ideological level, the answer to these questions connects to the ruling conception of organicism, an organic hierarchy, conceived as nature's principle of organization. Clark stressed the importance of "artistic composition" and described the process as a "recreation" of nature based on the principles of organic form. This process required a base of "personal experience," ideally actual presence in Africa, at the site of the animal's life and death. Technical crafts are always imagined to be subordinated by the ruling artistic idea, itself rooted authoritatively in nature's own life. "Such things must be felt, must be absorbed and assimilated, and then in turn, with understanding and enthusiasm, given out by the creator . . . Therefore, our groups are very often conceived in the very lair of the animals" (Clark 1936: 71).

The credos of realism and organicism interdigitate; both are systematizations of organization by a hierarchical division of labor, perceived as natural and so productive of unity. Unity must be *authored* in the Judeo-Christian myth system; just as nature has an Author, so does the organism or the realistic diorama. The author must be imagined with the aspects of mind, in relation to the body which executes. Akeley was intent on avoiding lying in his work; his craft was to tell the truth of nature. There was only one way to achieve such truth—the rule of mind rooted in the claim to experience. All the work must be done by men who did their collecting and studies on the spot because "[o]therwise, the exhibit is a lie and it would be nothing short of a crime to place it in one of the leading educational institutions of the country" (Akeley 1923: 265). A single mind infused collective experience: "If an exhibition hall is to approach its ideal, its plan must be that of a master mind, while in actuality it is the product of the correlation of many minds and hands" (Akeley 1923: 261). The "mind" is spermatic.

But above all, this sense of telling a true story rested on the selection of individual animals, the formation of groups of "typical" specimens. What was the meaning of "typical" for Akeley and his contemporaries in the biological departments of the American Museum of Natural History? What are the contents of these stories, and what must one do to see these contents? To respond, we must follow Carl Akeley into the field and watch him select an animal to mount. Akeley's concentration on finding the typical specimen, group, or scene cannot be overemphasized. But how could he know what was typical, or that such a state of being existed? This problem has been fundamental in the history of biology; one effort at solution is embodied in African Hall.

First, the concept includes the notion of perfection. The large bull giraffe in the water hole group in African Hall was the object of a hunt over many days in 1921. Several animals were passed over because they were too small or not colored beautifully enough. Remembering record trophies from earlier hunters undermined satisfaction with a modern, smaller specimen taken from the depleted herds of vanishing African nature. When at last the bull was taken as the result of great skill and daring, the minute details of its preservation and recreation were lovingly described.

Similarly, in 1910–11, the hunt for a large bull elephant provided the central drama of the safari for the entire two years. An animal with asymmetrical tusks was rejected, despite his imposing size. Character, as well as mere physical appearance, was important in judging an animal to be perfect. Cowardice would disqualify the most lovely and properly proportioned beast. Ideally, the killing itself had to be accomplished as a sportsmanlike act. Perfection was heightened if the hunt were a meeting of equals. So there was a hierarchy of game according to species: lions, elephants, and giraffes far outranked wild asses or antelope. The gorilla was the supreme achievement, almost a definition of perfection in the heart of the garden at the moment of origin. Perfection inhered in the animal itself, but the fullest meanings of perfection inhered in the meeting of animal and man, the moment of perfect vision, of rebirth. Taxidermy was the craft of remembering this perfect experience. Realism was a supreme achievement of the artifactual art of memory, a rhetorical achievement crucial to the foundations of Western science (Fabian 1983: 105–41). Memory was an art of reproduction.

There is one other essential quality for the typical animal in its perfect expression: it must be an adult male. Akeley describes hunting many fine females, and he cared for their hides and other details of reconstruction with all his skill. But never was it necessary to take weeks and risk the success of the entire enterprise to find the perfect female. There existed an image of an animal which was somehow *the* gorilla or *the* elephant incarnate. That particular tone of perfection could only be heard in the male mode. It was a compound of physical and spiritual quality judged truthfully by the artist-scientist in the fullness of direct experience. Perfection was marked by exact quantitative measurement, but even more by virile vitality known by the hunter-scientist from visual communion. Perfection was known by natural kinship; type, kind, and kin mutually and seminally defined each other.

Akeley hunted for a series or a group, not just for individuals. How did he know when to stop the hunt? Two groups give his criterion of wholeness, the gorilla group collected in 1921 and the original group of four elephants mounted by Akeley himself after the 1910–11 safari. Akeley once shot a gorilla, believing it to be a female, but found it to be a young male. He was disturbed because he wished to kill as few animals as possible and he believed the natural family of the gorilla did not contain more than one male. When he later saw a group made up of several males and females, he stopped his hunt with relief, confident that he could tell the truth from his existing specimens. Also, the photograph of Akeley's original group of four elephants unmistakably shows a perfect family. Nature's biographical unit, the reproductive group had the moral and epistemological status of truth-tellers.

Akeley wanted to be an artist and a scientist. Giving up his early plan of obtaining a degree from Yale Sheffield Scientific School and then of becoming a professional sculptor, he combined art and science in taxidermy. Since that art required that he also be a sculptor, he told some of his stories in bronzes as well as in dioramas. His criteria were similar; Akeley had many stories to tell, but they all expressed the same fundamental vision of a vanishing, threatened scene. In his determination to sculpt "typical" Nandi lion spearmen, Akeley used as models extensive photographs, drawings, and "selected types of American negroes which he was using to make sure of perfect figures" (Johnson 1936: 47). The variety of nature had a purpose—to lead to discovery of the highest type of each species of wildlife, including human beings outside "civilization."

Besides sculpture and taxidermy, Akeley perfected another narrative tool, photography. All of his story-telling instruments relied primarily on vision, but each caught and held slightly different manifestations of natural history. As a visual art, taxidermy occupied for Akeley a middle ground between sculpture and photography. Both sculpture and photography were subordinate means to accomplishing the final taxidermic scene. But photography also represented the future and sculpture the past. Akeley's practice of photography was suspended between the manual touch of sculpture, which produced knowledge of life in the fraternal discourses of organicist biology and realist art, and the virtual touch of the camera, which has dominated our understanding of nature since World War II. The nineteenth century produced the masterpieces of animal bronzes inhabiting the world's museums. Akeley's early twentieth-century taxidermy, seemingly so solid and material, appears as a brief frozen temporal section in the incarnation of art and science, before the camera technically could pervert his single dream into the polymorphous, absurdly intimate filmic reality we now take for granted. Critics accuse Akeley's taxidermy and the American Museum's expensive policy of building the great display halls in the years before World War II of being armature against the future, of having literally locked in stone one historical moment's way of seeing, while calling this vision the whole (Kennedy 1968: 204). But Akeley was a leader technically and spiritually in the perfection of the camera's eye. Taxidermy was not armed against the filmic future, but froze one frame of a far more intense visual communion to be consummated in virtual images. Akeley helped produce the armature—and armament—that would advance into the future.

Photography: Hunting with the Camera

> Guns have metamorphosed into cameras in this earnest comedy, the ecology safari, because nature has ceased to be what it had always been—what people needed protection from. Now nature—tamed, endangered, mortal—needs to be protected from people. When we are afraid, we shoot. But when we are nostalgic, we take pictures. (Sontag 1977: 15)

Akeley and his peers feared the disappearance of their world, of their social world in the new immigrations after 1890 amd the resulting dissolution of the old imagined hygienic, pre-industrial America. Civilization appeared to be a disease in the form of technological progress and the vast accumulation of wealth in the practice of monopoly capitalism by the very wealthy sportsmen who were trustees of the Museum and the backers of Akeley's African Hall. The leaders of the American Museum were afraid for their health; that is, their manhood was endangered. Theodore Roosevelt knew the prophylaxis for this specific historical malaise: the true man is the true sportsman. Any human being, regardless of race, class, and gender, could spiritually participate in the moral status of healthy manhood in democracy, even if only a few (Anglo-Saxon, male, heterosexual, Protestant, physically robust, and economically comfortable) could express manhood's highest forms. From about 1890 to the 1930s, the Museum was a vast public education and research program for producing experience potent to induce the fertile state of manhood. The Museum, in turn, was the ideological and material product of the sporting life. As Mary Jobe Akeley realized, "[the true sportsman] loves the game as if he were the father of it" (M. J. Akeley 1929b: 116). Akeley believed that the highest expression of sportsmanship was hunting with the camera: "Moreover, according to any true conception of sport—the use of skill, daring, and endurance in overcoming difficulties—camera hunting takes twice the man that gun hunting takes" (Akeley 1923: 155). The true father of the game loves nature with the camera; it takes twice the man, and the children are in his perfect image. The eye is infinitely more potent than the gun. Both put a woman to shame—reproductively.

At the time of Akeley's first collecting safari in 1896, cameras were a nearly useless encumbrance, incapable of capturing the goal of the hunt—life. According to Akeley, the first notable camera hunters in Africa appeared around 1902, beginning with Edward North Burton. The early books like Burton's were based on still photographs; moving picture wildlife photography, owing much to Akeley's own camera, did not achieve anything before the 1920s. On his 1910–11 safari to east Africa, with the best available equipment, Akeley tried to film the Nandi lion spearing. His failure due to inadequate cameras, described with great emotional intensity, led him during the next five years to design the Akeley camera, which was used extensively by the Army Signal Corps during World War I. Akeley formed the Akeley Camera Company to develop his invention, which received its civilian christening by filming Man o' War win the triplecrown races in 1920, and his camera's innovative telephoto lens caught the Dempsey-Carpentier heavyweight battle. Akeley's first taste of his own camera in the field was in 1921 in the Kivu forest. Within a few days, Akeley shot his first gorillas with both gun and camera: in these experiences he saw the culmination of his life. Awarded the John Price Wetherhill Medal at the Franklin Institute in 1926 for his invention, Akeley succeeded that year in filming to his satisfaction African lion spearing, on the same safari on which Rochester's George Eastman, of Eastman-Kodak fortunes, was both co-sponsor and hunter-collector.[20]

The ambiguity of the gun and camera runs throughout Akeley's work. He is a transitional figure from the western image of darkest to lightest Africa, from nature worthy of manly fear to nature in need of motherly nurture. The woman/scientist/mother of orphaned apes popularized by the National Geographic Society's magazine and films in the 1970s was still half a century away. With Akeley, manhood tested itself against fear, even as the lust for the image of jungle peace held the finger on the gun long enough to take the picture and even as the intellectual and mythic certainty grew that the savage beast in the jungle was human, in particular, industrial human. The industrialist in the field with Akeley, George Eastman, was an object lesson in the monopoly capitalist's greater fear of decadence than of death. The narrative has a septuagenarian Eastman getting a close-up photograph at 20 feet of a charging rhino, directing his white hunter when to shoot the gun, while his personal physician looks on. "With this adventure Mr. Eastman began to enjoy Africa thoroughly..." (M. J. Akeley 1940: 270).

Even at the literal level of physical appearance, "[t]o one familiar with the old types of camera the Akeley resembled a machine gun quite as much as it resembled a camera" (Akeley 1923: 166). Akeley said he set out to design a camera "that you can aim...with about the same ease that you can point a pistol" (Akeley 1923: 166). He enjoyed retelling the apocryphal story of seven Germans mistakenly surrendering to one American when they found themselves faced by an Akeley. "The fundamental difference between the Akeley motion-picture camera and the others is a panoramic device which enables one to swing it all about, much as one would swing a swivel gun, following the natural line of vision" (Akeley 1923: 167). Akeley semi-joked in knowing puns on the penetrating, deadly invasiveness of the camera, naming one of his image machines "The Gorilla." "'The Gorilla' had taken 300 feet of film of the animal that had never heretofore been taken alive in its native wilds by any camera...I was satisfied—more satisfied than a man ever should be—but I revelled in the feeling."[21]

The taxidermist, certain of the essential peacefulness of the gorilla, wondered how close he should let a charging male get before neglecting the camera for the gun. "I hope that I shall have the courage to allow an apparently charging gorilla to come within a reasonable distance before shooting. I hesitate to say just what I consider a reasonable distance at the present moment. I shall feel very gratified if I can get a photograph at twenty feet. I should be proud of my nerve if I were able to show a photograph of him at ten feet, but I do not expect to do this unless I am at the moment a victim of suicidal mania" (Akeley 1923: 197). Akeley wrote these words before he had ever seen a wild gorilla. What was the boundary of courage; how much did nature or man need protecting? What if the gorilla never charged, even when provoked? What if the gorilla were a coward (or a female)? Who, precisely, was threatened in the drama of natural history in the early decades of monopoly capitalism's presence in Africa and America?

Aware of a disturbing potential of the camera, Akeley set himself against faking. He stuffed Barnum's Jumbo, but he wanted no part of the great circus magnate's cultivation of the American popular art form, the hoax (Harris 1973). But hoax luxuriated in early wildlife photography (and anthropological photography). In particular, Akeley saw unscrupulous men manipulate nature to tell the story of a fierce and savage Africa that would sell in the motion picture emporia across America. Taxidermy had always threatened to lapse from art into deception, from life to upholstered death as a poor sportsman's trophy. Photography too was full of philistines who could debase the entire undertaking of nature work, the Museum's term for its educational work in the early 1900s. The Museum was for public entertainment (the point that kept its Presbyterian trustees resisting Sunday opening in the 1880s despite that day's fine potential for educating the new Catholic immigrants, who worked a six-day week); but entertainment only had value if it communicated the truth. Therefore, Akeley encouraged an association between the American Museum and the wildlife photographers, Martin and Osa Johnson, who seemed willing and able to produce popular motion pictures telling the story of jungle peace. Johnson claimed in his 1923 prospectus to the American Museum, "The camera cannot be deceived... [therefore, it has] enormous scientific value."[22]

Entertainment was interwoven with science, art, hunting, and education. Barnum's humbug tested the cleverness, the scientific acumen, of the observer in a republic where each citizen could discover the nakedness of the emperor and the sham of his rationality. This democracy of reason was always a bit dangerous. There is a tradition of active participation in the eye of science in America which makes the stories of nature ready to erupt into popular politics. Natural history can be—and has sometimes been—a means for millenial expectation and disorderly action. Akeley himself is an excellent example of a self-made man who made use of the mythic resources of the independent man's honest vision, the appeal to experience the testimony of one's own eyes. He *saw* the Giant of Karisimbi. The camera, an eminently democratic machine, has been crucial to crafting stories in biology. Its control has eluded the professional and the moralist, the official scientist. But in Martin Johnson, Akeley hoped he had the man who would tame specular entertainment for the social uplift promised by science.

In 1906 Martin Johnson shipped out with Jack London for a two-year south sea voyage. The ship, the *Snark,* was the photographer's *Beagle.* Its name could hardly have been better chosen for the ship carrying the two adventurers whose books and films complemented *Tarzan* for recording the dilemma of manhood in the early twentieth century. Lewis Carroll's *The Hunting of the Snark* parodically anticipates the revelation of men like Johnson, London, and Akeley:

> In one moment I've seen what has hitherto been
> Enveloped in absolute mystery,
> And without extra charge I will give you at large
> A Lesson in Natural History. (Carroll 1971: 225)

From 1908–13 Johnson ran five motion picture houses in Kansas. He and Osa traveled in the still mysterious, potent places to film "native life": Melanesia, Polynesia, Malekula, Borneo, Kenya Colony. In 1922 the Johnsons sought Akeley's opinion of their new film, *Trailing African Wild Animals*. Akeley was delighted, and the Museum set up a special corporation to fund the Johnsons on a five-year African film safari. They planned a film on "African Babies." "It will show elephant babies, lion babies, zebra babies, giraffe babies, and black babies ... showing the play of wild animals and the maternal care that is so strange and interesting a feature of wildlife."[23] African human life had the status of wildlife in the Age of Mammals. That was the logic for "protection"—the ultimate justification for domination. Here was a record of jungle peace.

The Johnsons also planned a big animal feature film. The museum lauded both the commercial and educational values. Osborn enthused, "The double message of such photography is, first, that it brings the aesthetic and ethical influence of nature within the reach of millions of people ... second, it spreads the idea that our generation has no right to destroy what future generations may enjoy."[24] Johnson was confident that their approach of combining truth and beauty without hoax would ultimately be commercially superior, as well as scientifically accurate. "[T]here is no limit to the money it can make ... My past training, my knowledge of showmanship, mixed with the scientific knowledge I have absorbed lately, and the wonderful photographic equipment ... make me certain that this Big Feature is going to be the biggest money maker ever placed on the market, as there is no doubt it will be the last big Africa Feature made, and it will be so spectacular that there will be no danger of another film of like nature competing with it. For these reasons it will produce an income as long as we live."[25] Africa had always promised gold.

The "naked eye" science advocated by the American Museum perfectly suited the camera, ultimately so superior to the gun for the possession, production, preservation, consumption, surveillance, appreciation, and control of nature. Akeley's aesthetic ideology of realism was part of his effort to bridge the yawning gaps in the endangered self. To make an exact image is to insure against disappearance, to cannibalize life until it is safely and permanently a specular image, a ghost. The image arrested decay. That is why nature photography is so beautiful and so religious—and such a powerful hint of an apocalyptic future. Akeley's aesthetic combined the instrumental and contemplative into a photographic technology providing a transfusion for a steadily depleted sense of reality. The image and the real define each other, as all of reality in late capitalist culture lusts to become an image for its own security. Reality is assured, insured, by the image, and there is no limit to the amount of money that can be made. The camera is superior to the gun for the control of time; and Akeley's dioramas with their photographic vision, sculptor's touch, and taxidermic solidity were about the end of time (Sontag 1977).

TELLING STORIES

The synthetic story told so far has had three major and many minor sources. Telling a life synthetically masks the tones emerging from inharmonious versions. The single biography, the achieved unity of African Hall, can be unraveled to tie its threads into an imagined heteroglossic narrative of nature yet to be written. A polyphonic natural history waits for its sustaining social history. To probe more deeply into the tissue of meanings and mediations making the specific structure of experience possible for the viewer of the dioramas of African Hall, I would like to tease apart the sources for a major event in Akeley's life, an elephant mauling in British East Africa in 1910. This event leavens my story of the structure and function of biography in the construction of a twentieth-century primate order, with its multiform hierarchies of race, sex, species, and class. Whose stories appear and disappear in the web of social practices that constitute Teddy Bear Patriarchy? Questions about authorized writing enforced by publishing practices and about labor that never issues in acknowledged authorship (never becomes father of the game) make up my story.[26]

Authors and Versions

She didn't write it.
She wrote it but she shouldn't have.
She wrote it, but look what she wrote about. (Russ 1983: 76)

In Brightest Africa appears to be written by Carl Akeley. But we learn from Mary Jobe Akeley (1940: 222), a prolific author, that the taxidermist "hated to wield a pen." She elaborates that Doubleday and Page (the men, not the company), were enthralled by Carl's stories told in their homes at dinner and so "determined to extract a book from him." So one evening after dinner Arthur W. Page "stationed a stenographer behind a screen, and without Carl's knowledge, she recorded everything he said while the guests lingered before the fire." Editing of this material is credited to Doubleday and Page, and the author is named as Carl. The stenographer is an unnamed hand. Her notes gave rise to articles in a journal called *World's Work,* but the publishers wanted a book. Then Akeley read a newspaper account of his Kivu journey that he liked; it had been written by Dorothy S. Greene while she worked for the director of the American Museum. Akeley hired her as his secretary, to record his stories while he talked with explorers and scientists or lectured to raise funds for African Hall. "She unobtrusively jotted down material which could be used in a book" (M. J. Akeley 1940: 223). Who wrote *In Brightest Africa*? To insist on that question troubles official versions of the relation of mind and body in western authorship.

The physical appearance of the books is itself an eloquent story. The stamp of approval from men like H. F. Osborn in the dignified prefaces, the presence of handsome photographs, a publishing house that catered to wealthy hunters: all compose the authority of the books. The frontispieces are like Orthodox icons; the entire story can be read from them. In *Lions, Gorillas and their Neighbors*, published for young people, the frontispiece shows an elderly Carl Akeley in his studio gazing intently into the eyes of the plaster death mask of the first gorilla he ever saw. Maturity in the encounter with nature is announced. *The Wilderness Lives Again*, the biography that resurrected Carl through his wife's vicarious authorship, displays in the front a young Carl, arm and hand bandaged heavily, standing outside a tent beside a dead leopard suspended by her hind legs. The caption reads: "Carl Akeley, when still in his twenties, choked this wounded infuriated leopard to death with his naked hands as it attacked him with intent to kill."

Carl Akeley's story of his encounter with the elephant that mauled him is in a chapter titled "Elephant Friends and Foes." Moral lessons pervade the chapter, prominently those of human ignorance of the great animals—partly because hunters are only after ivory and trophies, so that their knowledge is only of tracking and killing, not of the animals' lives—and of Akeley's difference because of his special closeness to nature embodied in the magnificent elephants. Akeley witnessed two elephants help a wounded comrade escape from the scene of slaughter, inspiring one of the taxidermist's bronzes. But, the reader also sees Akeley making a table to seat eight people out of elephant ears from a specimen which nearly killed him and Delia, despite each of them shooting into his head about 13 times. In this chapter, the taxidermist is hunting as an equal with his wife. He does not hide stories which might seem a bit seedy or full of personal bravado; yet his "natural nobility" pervaded all these anecdotes, particularly for an audience of potential donors to African Hall, who might find themselves shooting big game in Africa.

His near fatal encounter with an elephant occurred when Akeley had gone off without Delia to get photographs, taking "four days' rations, gun boys, porters, camera men, and so forth—about fifteen men in all" (Akeley 1923: 45). He was tracking an elephant whose trail was very fresh, when he suddenly became aware that the animal was bearing down on him directly:

> I have no knowledge of how the warning came . . . I only know that as I picked up my gun and wheeled about I tried to shove the safety catch forward. It refused to budge . . . My next mental record is of a tusk right at my chest. I grabbed it with my left hand, the other one with my right hand, and swinging in between them went to the ground on my back. This swinging in between the tusks was purely automatic. It was the result of many a time on the trails imagining myself caught by an elephant's rush and planning what to do, and a very profitable planning too: for I am convinced that if a man imagines such a crisis and plans what he would do, he will, when the occasion occurs, automatically do what he planned . . . He drove his tusks into the ground on either side of me. (Akeley 1923: 48–49)

Akeley tells that he lay unconscious and untouched for hours because his men felt he was dead, and they came from groups which refused to touch a dead man. When he came to, he shouted and got attention. He relates that word had been sent to Mrs. Akeley at base camp, who valiantly mounted a rescue party in the middle of the night against the wishes of her guides (because of the dangers of night travel through the bush), whom she pursued into their huts to force their cooperation. Sending word to the nearest government post to dispatch a doctor, she arrived at the scene of the injury by dawn. Akeley attributed his recovery to Delia's fast action, but more to the subsequent speedy arrival of a neophyte Scottish doctor, who sped through the jungle to help the injured man partly out of his ignorance of the foolishness of hurrying to help anyone mauled by an elephant—such men simply didn't survive to pay for one's haste. The more seasoned chief medical officer arrived considerably later.

The remainder of the chapter recounts Akeley's chat with other old hands in Africa about their experiences surviving elephant attacks. Like his thoughts as he swung between the giant tusks, the tone is reasoned, scientific, focused on the behavior and character of those interesting aspects of elephant behavior. The ubiquitous moral concludes the chapter:

> But although the elephant is a terrible fighter in his own defense when attacked by man, that is not his chief characteristic. The things that stick in my mind are his sagacity, his versatility, and a certain comradeship which I have never noticed to the same degree in other animals . . . I like to think back to the day I saw the group of baby elephants playing with a great ball of baked dirt . . . They have no enemy but man and are at peace amongst themselves. It is my friend the elephant that I hope to perpetuate in the central group in Roosevelt African Hall . . . In this, which we hope will be an everlasting monument to the Africa that was, the Africa that is fast disappearing, I hope to place the elephant on a pedestal in the centre of the hall—the rightful place for the first among them. (Akeley 1923: 54–5)

Akeley sees himself as an advocate for "nature" in which "man" is the enemy, the intruder, the dealer of death. His own exploits in the hunt stand in ironic juxtaposition only if the reader evades their true meaning—the tales of a pure man whose danger in pursuit of a noble cause brings him into communion with nature through the beasts he kills. This nature is a worthy brother of man, a worthy foil for his manhood. Akeley's elephant is profoundly male, singular, and representative of the possibility of nobility. The mauling was an exciting tale, with parts for many actors, including Delia, but the brush with death and the details of rescue are told with the cool humor of a man ready for his end dealt by such a noble friend and brother, his best enemy, the object of his scientific curiosity. The putative behavior of the "boys" underlines the confrontation between white manhood and the noble beast. "I never got much information out of the boys as to what did happen, for they were not proud of their part in the adventure . . . It is reasonable to assume that they had scattered through [the area which the elephant thoroughly trampled] like a covey of quail . . . " (1923: 49). Casual and institutional racism heightens the life story of the single adult man. The action in Akeley's stories focuses on the center of the stage, on the meeting of the singular man and animal. The entourage is inaudible, invisible, except for comic relief and anecdotes about native life. In Akeley's rendering, empowered by class and race, white woman stands without much comment in a similar moral position as white man—a hunter, an adult.

Mary Jobe Akeley published her biography of her husband, *The Wilderness Lives Again*, in 1940, four years after the Akeley African Hall opened to the public. Her purpose was to promote conservation and fulfill her life's purpose—accomplishing her husband's life work. She presents herself as the inspired scribe for her husband's story. Through her vicarious authorship and through African Hall and the Parc Albert, not only the wilderness, but Akeley himself, whose meaning was the wilderness, lives again.

Mary Jobe had not always lived for a husband.[27] An explorer since 1913, she had completed ten expeditions to explore and map British Columbian wilderness; and the Canadian government named a peak Mt. Jobe. She recounts the scene at Carl's death when she accepted his commission for her, that she would live thereafter to fulfill his work. The entire book is suffused with her joy in this task. Her self-construction as the other is breathtaking in its ecstasy. The story of the elephant mauling undergoes interesting emendations to facilitate her accomplishment. One must read this book with attention because Carl's words from his field diaries and publications are quoted at great length with no typographical differentiation from the rest of the text. At no point does the wife give a source for the husband's words; they may be from conversation, lectures, anywhere. It does not matter, because the two are one flesh. The stories of Carl and Mary Jobe blend imperceptibly—until the reader starts comparing other versions of the "same" incidents, even the ones written apparently in the direct words of the true, if absent, author-husband.

The key emendation is an absence; the entire biography of Carl Akeley by Mary Jobe Akeley does not mention the name or presence of Delia. Her role in the rescue is taken by the Kikuyu man Wimbia Gikungu, called "Bill," Akeley's gun bearer and companion on several safaris. Bill roused the recalcitrant guides and notified the government post, thus bringing on the Scotsman posthaste (M. J. Akeley, 1940: Chpt. IX). The long quotation from Carl in which the whole story is told simply lacks mention of his previous wife.

Mary Jobe tells a sequel to the mauling not in Akeley's published stories, and apparently taken from his field diaries or lectures. Because it is not uncommon for a man to lose his nerve after an elephant mauling and decline to hunt elephants again, it was necessary for Akeley to face elephants as soon as possible. Again, the first thing to notice is an absence; there is no question that such courage should be regained. But the explicit story does not ennoble Akeley. He tracked an elephant before he was really healthy, needing his "boys" to carry a chair on the trail for him to sit on as he tired; he wounded the elephant with unsportsmanlike hasty shots; and it was not found before dying. If Akeley's nobility is saved in this story, it is by his humility: "The whole thing had been stupid and unsportsmanlike" (M. J. Akeley 1940: 126).

Mary Jobe Akeley pictures herself as Carl's companion and soul mate, but not really as his co-adventurer and buddy hunter—with one exception. Mary Jobe fired two shots in Africa, and killed a magnificent male lion: "An hour later we came upon a fine old lion, a splendid beast, Carl said, and good enough for me to shoot. And so I shot. . . . Carl considered it a valuable specimen; but I was chiefly concerned that I fulfilled Carl's expectations and had killed the lion cleanly and without assistance" (M. J. Akerley 1940: 303). Mary Jobe's authority as a biographer does not depend on her being a hunter, but her status was enhanced by this most desirable transforming experience.

Delia Akeley pictures herself as a joyous and unrepentant hunter; but, by the publication of *Jungle Portraits* in 1930, her husband has some warts. Delia does not bear the authorial moral status of the artist-scientist, Carl Akeley, or his socially sure second wife. Delia's tales clarify the kind of biography that was to be suppressed in African Hall. In Delia's story of the rescue, "Bill" also appears, and he behaves well. But her own heroism in confronting the superstitions of the "boys" and in saving her endangered husband is the central tale: "Examining and cleansing

Mr. Akeley's wounds were my first consideration.... The fact that his wounds were cared for so promptly prevented infection, and without doubt saved his life" (D. Akeley 1930: 249).

Delia produced a biographical effect at odds with the official histories; she showed the messiness behind the "unified truth" of natural history museums. Delia dwelt on the sickness and injury of early collectors and explorers; she remarked pointedly on insects, weariness, and failure in the past and contrasted that with the experience provided the current (1930) traveler, the tourist, and museum visitor. She foregrounded the devoted and unrewarded wife who kept camp in the jungle and house at home. The wife-manager of Carl's safaris, aware of the material mediations in the quest for manhood and natural truth, showed pique at all the attention given her scientist-husband: "The thrilling story of the accident and his miraculous escape from a frightful death has been told many times by himself from the lecture platform. But a personal account of my equally thrilling night journey to his rescue through one of the densest, elephant-infested forests on the African continent is not nearly so well known" (D. Akeley 1930: 233). This is not the wife who devotes herself to her husband's authorship of wilderness. Indeed, she insisted on "darkest Africa" throughout her book.

Delia foregrounded her glory at the expense of her husband's official nobility. Delia's reader discovers Carl frequently sick in his tent, an invalid dangerously close to death while the courageous wife hunts not only for food for the entire camp, but also for scientific specimens so that he may hasten out of this dangerous continent before it claims him. In the elephant hunt following the mauling, Carl was still searching to restore his endangered "morale." But this time his wife was his companion in what is portrayed as a dangerous hunt terminating in a thrilling kill marked by a dangerous charge. Delia's story demurred on who fired the fatal shot, but "fatigue and a desire to be sure of his shot made Mr. Akeley slow in getting his gun in position" (D. Akeley 1930: 93).

Delia published an extraordinary photograph of a dashing Carl Akeley smoking a pipe and lounging on the body of a large fallen elephant; her caption reads, "Carl Akeley and the first elephant he shot after settling the question of his morale." A reader will not find that particular photograph of Akeley in any other publication than Delia's. Further, my hunt in the Museum's archive for the image of Akeley lounging astride his kill caught Delia in a lie (hoax?) about that elephant. But the lie reveals another truth. The accompanying photos in the archive suggest a version of reality, a biography of Africa, which the Museum and its official representatives did not want displayed in their Halls or educational publications.

Fig. 5.2. The Christening. Negative no. 211526. Published with permission of the Department of Library Services, American Museum of Natural History.

The images from the photo archive upstairs haunt the mind's eye as the viewer stands before the elephant group in African Hall. First, the particular elephant with the lounging Carl could not have been killed on the occasion Delia described. The cast of characters evidences a different year; a picture clearly taken on the same occasion shows the white hunter, the Scotsman Richard John Cunninghame, hired by Akeley in 1909 to teach him how to hunt elephants, lounging with Delia on the same carcass. The Museum archive labels the photo "Mrs. Akeley's first elephant." It is hard not to order the separate photos in the folder into a narrative series. The next snapshot shows the separated and still slightly bloody tusks of the elephant held in a gothic arch over a pleased, informal Delia. She is standing confidently under the arch, each arm reaching out to grasp a curve of the elephantine structure. But the real support for the ivory is elsewhere. Cut off at the edge of the picture are four black arms; the hands come from the framing peripheral space to encircle the tusks arching over the triumphant white woman. The museum archive labels this photo "Mrs. Akeley's ivory." The last photograph shows a smiling Cunninghame anointing Mrs. Akeley's forehead with the pulp from the tusk of the deceased elephant. She stands with her head bowed under the ivory arch, now supported by a single, solemn African man. The Museum's spare comment reads, "The Christening." (Figure 5.2)

Here is an image of a sacrament, a mark on the soul signing a spiritual transformation effected by the act of first killing. It is a sacred moment in the life of the hunter, a rebirth in the blood of the sacrifice, of conquered nature. This elephant stands a fixed witness in Akeley African Hall to its dismembered double in the photograph, whose bloody member signed the intersection of race, gender, and nature on the soul of the western hunter. In this garden, the camera captured a retelling of a Christian story of origins, a secularized Christian sacrament in a baptism of blood from the victim whose death brought spiritual adulthood, i.e., the status of hunter, the status of the fully human being who is reborn in risking life, in killing. Versions of this story proliferate in the history of American approaches to the sciences of life, especially primate life. With Delia, the story is near parody; with Carl it is near epiphany. His was authorized to achieve a fusion of science and art. Delia, the more prolific author, who neither had nor was a ghostwriter, was erased—by divorce and by duplicity.

Safari: A Life of Africa

> Now with few exceptions our Kivu savages, lower in the scale of intelligence than any others I had seen in Equatorial Africa, proved kindly men. . . . How deeply their sympathy affected me! As I think of them, I am reminded of the only playmate and companion of my early childhood, a collie dog. . . . (M. J. Akeley 1929b: 200)

The great halls of the American Museum of Natural History would not exist without the labor of Africans (or South Americans or the Irish and Negroes in North America). The Akeleys would be the first to acknowledge this fact; but they would claim the principle of organization came from the white safari managers, the scientist-collector and his camp-managing wife, the elements of mind overseeing the principle of execution. From the safari of 1895, dependent upon foot travel and the strong backs of "natives," to the motor safaris of the 1920s, the everyday survival of Euro-Americans in the field depended upon the knowledge, good sense, hard work, and enforced subordination of people the white folk insisted on seeing as perpetual children or as wildlife. If a black person accomplished some exceptional feat of intelligence or daring, the explanation was that he (or she?) was inspired, literally moved, by the spirit of the master. As Mary Jobe (1929b: 199) put it in her unself-conscious colonial voice, "It was as if the spirit of his master had descended upon him, activating him to transcendent effort." This explanation was all the more powerful if the body of the master was physically far removed, by death or trans-Atlantic residence. Aristotle was as present in the safari as he was in the taxidermic studios in New York or in the physiological bodies of organisms. Labor was not authorized as action, as mind, or as form. Labor was the marked body.

Carl and Mary Jobe Akeley's books elucidate safari organization over a thirty-year span. The photographs of solemn African people in a semicircle around the core of white personnel, with the cars, cameras, and abundant baggage in the background, are eloquent about race, gender, and colonialism. The chapters discuss the problems of cooks, the tasks of a headman, the profusion of languages which no white person on the journey spoke, numbers of porters (about thirty for most of the 1926 trip, many more in 1895) and problems in keeping them, the contradictory cooperation of local African leaders (often called "sultans"), the difficulty of providing white people coffee and brandy in an "unspoiled" wilderness, the hierarchy of pay scales and food rations for safari personnel, the behavior of gun bearers, and the punishment for perceived misdeeds. The chapters portray a social organism ordered by the principles of organic form: hierarchical division of labor called cooperation and coordination. The safari was an icon of the whole enterprise in its logic of mind and body, in its scientific marking of the body for functional efficiency (Sohn-Rethel 1978; Young 1977b; Rose 1983). In western inscriptions of race, Africans were written into the script of the story of life—and written out of authorship.

Few of the black personnel appear with individual biographies in the safari literature, but there are exceptions, object lessons or type life histories. Africans were imagined as either "spoiled" or "unspoiled," like the nature they signified. Spoiled nature could not relieve decadence, the malaise of the imperialist and city dweller, but only presented evidence of decay's contagion, the germ of civilization, the infection which was obliterating the Age of Mammals. And with the end of that time came the end of the essence of manhood, hunting. But unspoiled Africans, like the Kivu forest itself, were solid evidence of the resources for restoring manhood in the healthy activity of sportsmanlike hunting. Hinting at the complexity of the relation of master and servant in the pursuit of science on the safari, the life story is told from the point of view of the white person. Wimbia Gikungu, the Kikuyu known as Bill who joined Carl Akeley in British East Africa in 1905 at thirteen years of age, did not write—or ghost write—my sources. He was not the author of his body, but he was the Akeleys' favorite "native."

Bill began as an assistant to Delia Akeley's "tent boy," but is portrayed as rapidly learning everything there was to know about the safari through his unflagging industry and desire to please. He was said to have extraordinary intelligence and spirit, but suffered chronic difficulty with authority and from inability to save his earnings. "He has an independence that frequently gets him into trouble. He does not like to take orders from any one of his own color" (Akeley 1923: 143). He served with Akeley safaris in 1905, 1909–11, and 1926, increasing in authority and power over the years until there was no African whom Carl Akeley respected more for his trail knowledge and judgment. Bill got into trouble serving on the Roosevelt safari, was dismissed and blacklisted. Nonetheless, Akeley immediately rehired him, assuming he had had some largely innocent (i.e., not directed against a white person) eruption of his distaste for authority (Akeley 1923: 144).

Akeley describes three occasions on which he "punished" Gikungu; these episodes are icons of Akeley's paternal ideology. Once Bill refused to give the keys for Carl's trunk to other white people when they asked, "saying that he must have an order from his own Bwana. It was cheek, and he had to be punished; the punishment was not severe, but coming from me it went hard with him and I had to give him a fatherly talk to prevent his running away" (Akeley 1923: 134). The "father to the game" claimed the highest game of all in the history of colonialism—the submission of man. Later, the Kikuyu shot at an elephant he believed was charging an unsuspecting Akeley. Akeley had seen the animal, but did not know his "gun boy" did not know. Akeley slapped Gikungu "because he had broken one of the first rules of the game, which is that a black boy must never shoot without orders, unless his master is down and at the mercy of a beast." Realizing his mistake, "my apologies were prompt and as humble as the dignity of a white man would permit" (M. J. Akeley 1940: 132). The African could not be permitted to hunt independently with a gun in the presence of a white man. The entire logic of restoring threatened white manhood depended on that rule. Hunting was magic; Bill's well-meaning (and well-placed) shot was pollution, a usurpation of maturity. Finally, Akeley had Gikungu put in jail during the 1909–11 safari when "Bill" actively declined to submit when Carl "found it necessary to take him in hand for mild punishment" for another refusal of a white man's orders about baggage (Akeley 1923: 144). Gikungu spend two weeks in jail; the white man's paternal solicitude could be quite a problem.

Akeley relied on Gikungu's abilities and knowledge. Always, his performance was attributed to his loyalty for the master. Collecting the ivory of a wounded elephant, organizing the rescue after the elephant mauling, assisting Mary Jobe Akeley after Carl's death—these deeds were the manifestations of subordinate love. There is no hint that Gikungu might have had other motives—perhaps a non-subservient pity for a white widow in the rain forest, pleasure in his superb skills, complex political dealings with other African groups, or even a superior hatred for his masters. Attributing intentions to "Bill" is without shadow of doubt; the African played his role in the safari script as the never quite tame, permanently good boy. Bill was believed to be visible; other Africans largely remained invisible. The willed blindness of the white lover of nature remained characteristic of the scientists who went to the Garden to study primates, to study origins, until cracks began to show in this consciousness around 1970.

Institution

Speak to the Earth and It Shall Teach Thee. (Job 12:8)[28]

Every specimen is a permanent fact.[29]

From 1890 to 1930 the "Nature Movement" was at its height in the United States. Conventional western ambivalence about "civilization" was never higher than during the early decades of monopoly capital formation (Marx 1964; Nash 1982). The woes of "civilization" were often blamed on technology—fantasized as "the Machine." Nature is such a potent symbol of innocence partly because "she" is imagined to be without technology. Man is not *in* nature partly because he is not seen, is not the spectacle. A constitutive meaning of masculine gender for us is to be the unseen, the eye (I), the author, to be Linnaeus who fathers the primate order. That is part of the structure of experience in the Museum, one of the reasons one has, willy nilly, the moral status of a young boy undergoing initiation through visual experience. The Museum is a visual technology. It works through desire for communion, not separation, and one of its products is gender. Who needs infancy in the nuclear family when we have rebirth in the ritual spaces of Teddy Bear Patriarchy?

Social relations of domination are built into the hardware and logics of technology, producing the illusion of technological determinism. Nature is, in "fact," constructed as a technology through social praxis. And dioramas are meaning-machines. Machines are maps of power, arrested moments of social relations that in turn threaten to govern the living. The owners of the great machines of monopoly capital were, with excellent reason, at the forefront of nature work—because it was one of the means of production of race, gender, and class. For them, "naked eye science" could give direct vision of social peace and progress despite the appearances of class war and decadence. They required a science "instaurating" jungle peace; and so they bought it.

This scientific discourse on origins was not cheap; and the servants of science, human and animal, were not always docile. But the relations of knowledge and power at the American Museum of Natural History should not be narrated as a tale of evil capitalists in the sky conspiring to obscure the truth. Quite the opposite, the tale must be of committed Progressives struggling to dispel darkness through research, education, and reform. The capitalists were not in the sky; they were in the field, armed with the Gospel of Wealth.[30] They were also often armed with an elephant gun and an Akeley camera. Sciences are woven of social relations throughout their tissues. The concept of social relations must include the entire complex of interactions among people; objects, including books, buildings, and rocks; and animals.[31]

One band in the spectrum of social relations—the philanthropic activities of men in the American Museum of Natural History, which fostered exhibition (including public education and scientific collecting), conservation, and eugenics—is the optic tectum of naked eye science, i.e., the neural organs of integration and interpretation. After the immediacy of experience and the mediations of biography and story telling, we now must attend to the synthetic organs of social construction as they came together in an institution.[32]

Decadence was the threat against which exhibition, conservation, and eugenics were all directed as prophylaxis for an endangered body politic. The Museum was a medical technology, a hygienic intervention, and the pathology was a potentially fatal organic sickness of the individual and collective body. Decadence was a venereal disease proper to the organs of social and personal reproduction: sex, race, and class. From the point of view of Teddy Bear Patriarchy, race suicide was a clinical manifestation whose mechanism was the differential reproductive rates of anglo-saxon vs. "non-white" immigrant women. Class war, a pathological antagonism of functionally related groups in society, seemed imminent. And middle class white women undertaking higher education might imperil their health and reproductive function. Were they unsexed by diverting the limited store of organic energy to their heads at crucial organic moments? Lung disease (remember Teddy Roosevelt's asthma), sexual disease (what was *not* a sexual disease, when leprosy, masturbation, and Charlotte Perkins Gilman's need to write all qualified?), and social disease (like strikes and feminism) all disclosed ontologically and epistemologically similar disorders of the relations of nature and culture. Decadence threatened in two interconnected ways, both related to energy-limited, productive systems—one artificial, one organic. The machine threatened to consume and exhaust man. And the sexual economy of man seemed vulnerable both to exhaustion and to submergence in unruly and primitive excess. The trustees and officers of the Museum were charged with the task of promoting public health in these circumstances.

Three public activities of the Museum were dedicated to preserving a threatened manhood: exhibition, eugenics, and conservation. Exhibition was a practice to produce permanence, to arrest decay. Eugenics was a movement to preserve hereditary stock, to assure racial purity, to prevent race suicide. Conservation was a policy to preserve resources, not only for industry, but also for moral formation, for the achievement of manhood. All three activities were prescriptions against decadence, the dread disease of imperialist, capitalist, white culture. Forms of education and science, they were also very close to religious and medical practice. These three activities were about the transcendence of death, personal and collective. They attempted to insure preservation without fixation and paralysis, in the face of extraordinary change in the relations of sex, race, and class.

Exhibition

The American Museum of Natural History was (and is) a "private" institution, as private could only be defined in the United States. In Europe the natural history museums were organs of the state, intimately connected to the fates of national politics (Holton and Blanpied 1976). The development of U.S. natural history museums was tied to the origins of the great class of capitalists after the Civil War (Kennedy 1968). The social fate of that class was also the fate of the Museum; its rearrangements and weaknesses in the 1930s were reproduced in crises in the Museum, ideologically and organizationally. The American Museum, relatively unbuffered from intimate reliance on the personal beneficence of a few wealthy men, is a peephole for spying on the wealthy in their ideal incarnation. They made dioramas of themselves.

The great scientific collecting expeditions from the American Museum began in 1888 and stretched to the 1930s. By 1910, they had gained the Museum scientific prestige in selected fields, especially paleontology, ornithology, and mammalogy. The Museum in 1910 boasted nine scientific departments and twenty-five scientists. Anthropology also benefited, and the largest collecting expedition ever mounted by the Museum was the 1890s' Jesup North Pacific Expedition so important to Franz Boas's career (Kennedy 1968: 141ff.). The sponsors of the Museum liked a science that stored facts safely; and they liked the public popularity of the new exhibitions. Many people among the white, protestant, middle and upper classes in the United States were committed to nature, camping, and the outdoor life; Teddy Roosevelt embodied their politics and their ethos. Theodore Roosevelt's father was one of the incorporators of the Museum in 1868. His son, Kermit, was a trustee during the building of African Hall. Others in that cohort of trustees were J. P. Morgan, William K. Vanderbilt, Henry W. Sage, H. F. Osborn, Daniel Pomeroy, E. Roland Harriman, Childs Frick, John D. Rockefeller III, and Madison Grant. Patrons of science, these are leaders of movements for eugenics, conservation, and the rational management of capitalist society.

The first hall of dioramas was Frank Chapman's Hall of North American Birds, opened in 1903. Akeley, hired to prepare African game, especially elephants, conceived the idea for African Hall on his first collecting trip for the American Museum. Osborn hoped for—and got—a North American and Asian Mammal Hall after the African one. The younger trustees in the 1920s formed an African Big Game Club that invited wealthy sportsmen to join in contributing specimens and money to African Hall. The 1920s were prosperous for these men, and they gave generously. There were over one hundred expeditions in the field for the American Museum in the 1920s discovering facts (Kennedy 1968: 192).

There was also a significant expansion of the museum's educational endeavors. Over a million children per year in New York were looking at the Museum's "nature cabinets" and food exhibits circulated through the city public health department. Radio talks, magazine articles, and books covered the Museum's popular activities, which appeared in many ways to be a science for the people, like that of the *National Geographic,* which taught republican Americans their responsibilities in empire after 1888. Both *Natural History,* the Museum's publication, and *National Geographic* relied heavily on photographs. There was a big building program from 1909 to 1929; and the Annual Report of the Museum for 1921 quoted the estimate by its director that 2 1/2 million people were reached by the Museum and its education extension program.

Osborn summarized the fond hopes of educators like himself in his claim that children passing through the Museum's halls "become more reverent, more truthful, and more interested in the simple and natural laws of their being and better citizens of the future through each visit." He maintained that the book of nature, written only in facts, was proof against the failing of other books: "The French and Russian anarchies were based in books and in oratory in defiance of every law of nature."[33] Going beyond pious hopes, Osborn had the power to construct a Hall of the Age of Man to make the moral lessons of racial hierarchy and progress explicit, lest they be missed in gazing at elephants. He countered those who criticized the halls and educational work for requiring too much time and money better spent on science itself. "The exhibits in these Halls have been criticized only by those who speak without knowledge. They all tend to demonstrate the slow upward ascent and the struggle of man from the lower to the higher stages, physically, morally, intellectually, and spiritually. Reverently and carefully examined, they put man upwards towards a higher and better future and away from the purely animal stage of life."[34] This is the Gospel of Wealth, reverently examined.

Prophylaxis

Eugenics and conservation were closely linked in philosophy and in personnel at the Museum, and they tied in closely with exhibition and research. For example, the white-supremacist author of *The Passing of the Great Race,* Madison Grant, was a successful corporation lawyer, a trustee of the American Museum, an organizer of support for the North American Hall, a co-founder of the California Save-the-Redwoods League, activist for making Mt. McKinley and adjacent lands a national park, and the powerful secretary of the New York Zoological Society. His preservation of nature and germ plasm all seemed the same sort of work. Grant was not a quack or an extremist. He represented a band of Progressive opinion terrified of the consequences of unregulated monopoly capitalism, including failure to regulate the importation of non-white (which included Jewish and southern European) working classes, who invariably had more prolific women than the "old American stock." Powerful men in the American scientific establishment were involved in establishing Parc Albert in the Congo, a significant venture in international scientific cooperation: John C. Merriam of the Carnegie Institution of Washington, George Vincent of the Rockefeller Foundation, Osborn at the American Museum. The first significant user of the sanctuary would be sent by the "father" of primatology in America, Robert Yerkes, for a study of the psychobiology of wild gorillas. Yerkes was a leader in the movements for social hygiene, the category in which eugenics and conservation also fit. It was all in the service of science.

The Second International Congress of Eugenics was held at the American Museum of Natural History in 1921 while Akeley was in the field collecting gorillas and initiating plans for Parc Albert. Osborn, an ardent eugenicist, believed that it was "[p]erhaps the most important scientific meeting ever held in the Museum." Leading U.S. universities and state institutions sent representatives, and there were many eminent foreign delegates. The collected proceedings were titled "Eugenics in Family, Race, and State." U.S. lawmakers were one intended audience. "The section of the exhibit bearing on immigration was then sent to Washington by the Committee on Immigration of the Congress, members of which made several visits to the Museum to study the exhibit. The press was at first inclined to treat the work of the Congress [of Eugenics] lightly ... but the influence of the Congress grew and found its way into news and editorial columns of the entire press of the United States."[35] In 1923 the United States Congress passed immigration restriction laws, to protect the Race, the only race needing a capital letter.

The 1930s were a hiatus for the Museum. The Depression led to reduced contributions, and basic ideologies and politics shifted. The changes were not abrupt; but even the racial doctrines so openly championed by the Museum were publicly criticized in the 1940s, though not until then. Conservation was pursued with different political and spiritual justifications. A different biology was being born, more in the hands of the Rockefeller Foundation and in a different social womb. The issue would be molecular biology and other forms of post-organismic cyborg biology. The threat of decadence gave way to the catastrophes of the obsolescence of man (and of all organic nature) and the disease of stress, realities announced vigorously after World War II. Different forms of capitalist patriarchy and racism would emerge, embodied in a retooled nature. Decadence is a disease of organisms; obsolescence and stress are conditions of technological systems. Hygiene would give way to systems engineering as the basis of medical, religious, political, and scientific story-telling practices.

The early leaders of the American Museum of Natural History would insist that they were trying to know and to save nature, reality. And the real was one. The explicit ontology was holism, organicism. The aesthetic appropriate to exhibition, conservation, and eugenics from 1890 to 1930 was realism. But in the 1920s the surrealists knew that behind the day lay the night of sexual terror, disembodiment, failure of order; in short, castration and impotence of the seminal body which had spoken all the important words for centuries, the great white father, the white hunter in the heart of Africa. The strongest evidence in this chapter for the correctness of their judgment has been a literal reading of the realist, organicist artifacts and practices of the American Museum of Natural History. Their practice and mine have been literal, dead literal.

NOTES

1. H. F. Osborn (1908, in Kennedy 1968: 347).
2. Edgar Rice Burroughs, in Porges (1975: 129).
3. A plaque at the Deauvereaux or Hotel Colorado in Glenwood Springs inscribes one version of the origin of the Teddy Bear, emblem of Theodore Roosevelt: T.R. returned empty-handed from a hunting trip to the hotel, and so a hotel maid created a little stuffed bear and gave it to him. Word spread, and the Bear was soon manufactured in Germany. It is a pleasure to compose a chapter in feminist theory on the subject of stuffed animals.

4. Believing "man" arose in Asia, H. F. Osborn presided over Museum expeditions into the Gobi desert in the 1920s in an attempt to prove this position. However, Africa still had special meaning as the core of primitive nature, and so as origin in the sense of potential restoration, a reservoir of original conditions where "true primitives" survived. Africa was not established as the scene of the original emergence of our species until well after the 1930s. Pietz (1983) theorizes Africa as the locus for the inscription of capitalist desire in history.

5. Feminist theory emphasizes the body as generative political construction (Hartsock 1983; Moraga 1983; de Lauretis 1984, 1987; Martin 1987; Moraga and Anzaldua 1981; Hartouni 1987; Spillers 1987). See also *Social Research*, Winter 1974, essays from the New School for Social Research, "Conference on the Meaning of Citizenship."

6. Visual communion, a form of erotic fusion in themes of heroic action, especially death, infuses modern scientific ideologies. Its role in masculinist epistemology in science, with its politics of rebirth, is fundamental. Feminist theory so far has paid more attention to gendered subject/object splitting and not enough to love in specular domination's construction of nature and her sisters (Merchant 1980; Keller 1985; Keller and Grontkowski 1983; Sofoulis 1988).

7. William Pietz's 1983 UCSC slide lecture on the Chicago Field Museum analyzed museums as scenes of ritual transformation.

8. See also McCullough (1981); Cutright (1956). On travel and the modern Western self, especially the penetration of Brazil, see Defert (1982). Travel as science and as heroic quest interdigitate.

9. Women had all the frightening babies, a detail basic to their immigrant life in a racist society (Gordon 1976; Reed 1978; McCann 1987). Roosevelt's 1905 speech popularized the term "race suicide."

10. Akeley to Osborn, 29 March 1911, in Kennedy (1968: 186).

11. On principles of composition: Clark (1936); *The Mentor,* January 1926; *Natural History,* January 1936. Lowe (1982) illuminates the production of the transcendental subject from the structured relations of human eye/subject/technical apparatus.

12. For a genealogy of darkest Africa, see Brantlinger (1985).

13. Malvina Hoffman's bronzes of African men and women in this hall and her heads of Africans at the entrance to the hall testify to a crafted human beauty, not a story of natural primitives. On Osborn's failed effort to enlist Hoffman in his projects, see Porter (n.d.) and Taylor (1979).

14. James Clifford's sharp eye supplied this perception. See Landau (1981, 1984) on evolutionary texts as narrative.

15. Akeley (1923: 211). The jealous mistress trope is a ubiquitous element of the heterosexist gender anxieties in male scientists' writing about their endeavors (Keller 1985).

16. Nesbit (1926); Guggisberg (1977).

17. Akeley (1923: 226). Bradley (1922) wrote the white woman's account of this trip. The white child, daughter of Mary Hastings and Herbert Bradley, became James Tiptree, Jr., the science fiction writer. Introducing *Warm Worlds and Otherwise* (Tiptree 1975), Robert Silverberg used Tiptree, whom he later learned was Dr. Alice B. Sheldon, as an example of fine masculine writing that must have been produced by a "real" man. Sheldon earned her doctorate in experimental psychology at age 52 and then began a career as a science fiction writer. Silverberg compared Tiptree to Hemingway—citing "that prevailing masculinity about both of them" (Silverberg, in Tiptree 1975: xv). Tiptree's fiction drew deeply from her travels with her naturalist parents to Africa and Indonesia. Writing as Racoona Sheldon when she wished a female-identified persona, Tiptree kept her "real" gender identity obscure until 1976, near the time of publication of her first novel, *Up the Walls of the World* (1978), which explores an alien species

in which males mother the young. Tiptree's fiction and her publishing practices both interrogate gender. A man and a mother, a scientist and a writer of science fiction, a woman and a masculine author, Tiptree is an oxymoronic figure reconstructing social subjectivities out of a childhood colonial past and into a post-colonial world of other possibilities. In ill health, Tiptree committed suicide with her aged husband in 1987 in their Virginia home.

18. Reserving it for internal use only, the Museum refused permission to publish this photograph. Is it still so sensitive after 68 years?

19. Frank Chapman of the Department of Mammalogy and Ornithology was working on North American bird habitat groups, installed for the public in a large hall in 1903. In the 1880s, British Museum workers innovated methods for mounting birds, including making extremely lifelike vegetation. The American Museum founded its own department of taxidermy in 1885 and hired two London taxidermists, the brother and sister Henry Minturn and Mrs. E. S. Mogridge, to teach how to mount the groups. Joel Asaph Abel, Head of Mammalogy and Ornithology, hired Frank Chapman in 1887. Chapman, a major figure in the history of American ornithology, influenced the start of field primatology in the 1930s. American Museum bird groups from about 1886 were very popular. "Wealthy sportsmen, in particular, began to give to the museum." This turning point in fortunes was critical to the U.S. conservation movement. Department of Mammalogy and Ornithology scientists significantly enhanced the scientific reputation of the American Museum in the late 1800s (Kennedy 1968: 97–104; Chapman 1929, 1933; Chapman and Palmer 1933; pamphlet of Chicago Field Columbia Museum, 1902, "The Four Seasons"; "The Work of Carl E. Akeley in the Field Museum of Natural History," Chicago: Field Museum, 1927).

20. M. J. Akeley (1929b: 127–30, 1940: 115).

21. Akeley (1923: 223–24). Akeley recognized the utility of his camera's telephoto feature to anthropologists for making "motion pictures of natives of uncivilized countries without their knowledge" (Akeley 1923: 166).

22. October 1923, prospectus, AMNH; Johnson (1936); M. J. Akeley (1929b: 129); July 26, 1923, Akeley memorandum on Martin Johnson Film Expedition, microfilm 1114a and 1114b. The Johnsons' films were *Simba*, made on the Eastman-Pomeroy expedition, and *Trailing African Wild Animals*. *Cannibal of the South Seas* was earlier. See Osa Johnson's (1940) thriller about their lives.

23. October 1923, prospectus to the AMNH, microfilm 1114a.

24. October 1923, Osborn endorsement, AMNH microfilm 1114a.

25. Martin Johnson, July 26, 1923, prospectus draft, microfilm 1114a. The expectation that a film (*Simba*) made in the 1920s would be the last wildlife extravaganza was a wonderful statement of the belief that nature existed in essentially one form and could be captured in one vision, if only the visualizing technology were adequate.

26. The principal sources for this section are correspondence, annual reports, photographic archives, and artifacts in the AMNH; Akeley (1923); M. J. Akeley (1940); Akeley and Akeley (1922); Mary Jobe and Carl Akeley's articles in *The World's Work*; and Delia Akeley's adventure book (1930). Delia is Delia Denning/Akeley/Howe. See *N.Y. Times*, 23 May 1970, 23. The buoyant racism in the books and articles of this contemporary of Margaret Mead makes Mary Jobe and Carl look cautious. Olds (1985: 71–154) provides a biographic portrait of Delia Akeley. Not sharing the elite social origins of most women explorers, Delia was born about 1875, on a farm near a small Wisconsin town, the youngest of nine children of devout Catholic Irish immigrant parents. She ran away from home at 13 and married a barber a year later. Nothing is known about the end of that marriage. Probably meeting him on hunting trips with her husband in Wisconsin, she married Carl Akeley in 1902, when he was still a taxidermist-sculptor at

the Milwaukee Public Museum. Without hint of irony, Olds comments on the 1905–6 Akeley trip in Kenya: "The indispensable 'boys' took the place of horses, mules, or donkeys, because the tsetse fly made use of beasts of burden impossible" (Olds 1985: 87). In the project of recovering great white foremothers, Olds writes in 1985 in the same colonialist tones that permeate Akeley's work 60 years earlier. Olds's book is appropriately endorsed on the back cover by a NASA administrator. Olds makes a convincing case for the official scientific community's covering up Delia's role in Carl Akeley's explorations in favor of the story of Mary Jobe (Olds 1985: 150). Delia's bull elephant kill from 1906 is mounted in the Chicago Field Museum. She collected 19 mammalian species listed in the Field Museum catalogue, in addition to a large bird collection. Six weeks after her divorce from Carl in 1923, Delia was commissioned by the Brooklyn Museum of Arts and Sciences to lead an expedition to East and Central Africa. The museum director reported that it would be a "one-woman expedition"; i.e., "her sole companions on trips into the interior will be natives selected and trained by her" (Olds 1985: 114). To be with "natives" was to be "alone" epistemically. This scientific expedition was the first such venture led by a woman. Including a Dutch heiress, Alexine Tinne in 1862, an American feminist, May French Seldon, in 1891, and the British intellectual Mary Kingsley, who traded throughout the Congo Free State in the 1890s, the theme of adventurer-white women "alone" in the "interior" of Africa does not begin with Jane Goodall and the *National Geographic* sagas. But the later coding of the woman as scientist is different. Contrast the popular reporting of the Goodall story in the 1960s with the 1923 headline in the *New York World*: "Woman to Forget Marital Woe by Fighting African Jungle Beasts." For the world in which Delia and Mary Jobe worked, see Rossiter (1982).

27. From English extraction on both sides, Mary Jobe Akeley (1878–1966) was born and went to college in Ohio. Her father's family had been in America since colonial times. Mary Jobe studied English and history for two years in graduate school at Bryn Mawr; earned a Master's degree at Columbia in 1909; and was on the Hunter College faculty until 1916. She owned and ran a summer camp for upper class girls in Mystic, Connecticut, from 1916–30, where Martin and Osa Johnson talked of their adventures. Married to Carl Akeley in 1924, she led her own expeditions in 1935 and 1947. Her wildlife photography dates from about 1914 (McKay 1980).

28. Engraved on a plaque at the entrance to Earth History Hall, AMNH.

29. H. F. Osborn, 54th Annual Report to the Trustees, p. 2, AMNH.

30. Carnegie (1889); Domhoff (1967); Kolko (1977); Weinstein (1969); Wiebe (1966); Hofstadter (1955); Starr (1982); Oleson and Voss (1979); Nielson (1972).

31. Latour (1988); Latour and Woolgar (1979); Knorr-Cetina and Mulkay (1983).

32. On decadence and the crisis of white manhood: F. Scott Fitgerald, *The Great Gatsby* (1925); Henry Adams, *The Education of Henry Adams* (privately printed 1907); Ernest Hemingway, *Green Hills of Africa* (1935). On the history of conservation: Nash (1977, 1982); Hays (1959). On eugenics, race doctrines, and immigration: Higham (1975); Haller (1971); Chase (1977); Ludmerer (1972); Pickens (1968); Gould (1981); Chorover (1979); Cravens (1978); Kevles (1985). On sexuality, hygiene, decadence, birth control, and sex research in the early 1900s in life and social sciences: Rosenberg (1982); McCann (1987); Sayers (1982).

33. AMNH: Osborn, "The American Museum and Citizenship," 53rd Annual Report, 1922, p. 2. For the sweep of his work, see Osborn (1930).

34. Osborn, "Citizenship," p. 2.

35. Osborn, 53rd Annual Report, 1921, pp. 31–32. Ethel Tobach helped me find material on AMNH social networks, eugenics, racism, and sexism. Galton Society organizing meetings were in Osborn's home.

REFERENCES

Akeley, Carl E. 1923. *In Brightest Africa*. New York: Douleday, Page & Co.

Akeley, Delia. 1930. *Jungle Portraits*. New York: Macmillan.

Akeley, Mary Jobe. 1929a. "Africa's Great National Park. The Formal Inauguration of Parc Albert at Brussels." *Natural History* 29:638–50.

———. 1929b. *Carl Akeley's Africa*. New York: Dodd & Mead.

———. 1940. *The Wilderness Lives Again. Carl Akeley and the Great Adventure*. New York: Dodd & Mead.

Akeley, Carl E., and Mary Jobe Akeley. 1922. *Lions, Gorillas, and Their Neighbors*. New York: Dodd & Mead.

Bradley, Mary Hastings. 1922. *On the Gorilla Trail*. New York: Appleton.

Brantlinger, Patrick. 1985. "Victorians and Africans: The Genealogy of the Myth of the Dark Continent." *Critical Inquiry* 12(1): 166–203.

de Beauvoir, Simone. 1954. *The Second Sex*. Translated and edited by H. M. Parshley. New York: Vintage.

Carnegie, Andrew. 1889. "The Gospel of Wealth." *North American Review*, vol. 149–150.

Carroll, Lewis. 1971 [1876]. "The Hunting of the Snark." In *Alice in Wonderland*. New York: Norton Critical Edition.

Chapman, Frank M. 1929. *My Tropical Air Castle*. New York: Appleton-Century.

———. 1933. *Autobiography of a Bird Lover*. New York: Appleton-Century.

Chapman, Frank M., and T. S. Palmer, eds. 1933. *Fifty Years of Progress in American Ornithology, 1883–1933*. Lancaster, PA: American Ornithologists Union.

Chase, Allan. 1977. *The Legacy of Malthus: The Social Costs of the New Scientific Racism*. New York: Knopf.

Chorover, Stephen L. 1979. *From Genesis to Genocide*. Cambridge, MA: M.I.T. Press.

Clark, James. 1936. "The Image of Africa." In *The Complete Book of Africa Hall*. New York: American Museum of Natural History.

Cravens, Hamilton. 1978. *The Triumph of Evolution: American Scientists and the Heredity Environment Controversy, 1900–41*. Philadelphia: University of Pennsylvania Press.

Cross, Stephen J., and William R. Albury. 1987. "Walter B. Cannon, L. J. Henderson, and the Organic Analogy." *Osiris*, 2nd series, 3: 165–92.

Cutright, Paul Russell. 1956. *Theodore Roosevelt the Naturalist*. New York: Harper & Row.

de Lauretis, Teresa. 1984. *Alice Doesn't: Feminism, Semiotics, and Cinema*. Bloomington: Indiana University Press.

———. 1987. *Technologies of Gender: Essays on Theory, Film, and Fiction*. Bloomington: Indiana University Press.

Defert, Daniel. 1982. "The Collection of the World: Accounts of Voyages from the Sixteenth to the Eighteenth Centuries." *Dialectical Anthropology* 7: 11–20.

Domhoff, G. William. 1967. *Who Rules America?* Englewood Cliffs, NJ: Prentice-Hall.

Du Chaillu, Paul. 1861. *Explorations and Adventures in Equatorial Africa; with Accounts of the Manners and Customs of the People, and the Chase of the Gorilla, Crocodile, Leopard, Elephant, Hippopotamus, and Other Animals*. London: Murray.

Evans, Mary Alice, and Howard Ensign Evans. 1970. *William Morton Wheeler, Biologist*. Cambridge, MA: Harvard University Press.

Fabian, Johannes. 1983. *Time and the Other: How Anthropology Makes Its Object*. New York: Columbia University Press.

Fossey, Dian. 1983. *Gorillas in the Mist*. Boston: Houghton-Mifflin.

Gordon, Linda. 1976. *Woman's Body, Woman's Right: A Social History of Birth Control in America*. New York: Viking.

Gould, Stephen Jay. 1981. *The Mismeasure of Man*. New York: Norton.

Guggisberg, G. A. 1977. *Early Wildlife Photographers*. New York: Talpinger.

Haller, John. 1971. *Outcasts from Evolution*. Urbana, IL: Illinois University Press.

Harris, Neil. 1973. *Humbug: The Art of P. T. Barnum*. Boston: Little, Brown.

Hartouni, Val. 1987. "Personhood, Membership, and Community: Abortion Politics and the Negotiation of Public Meanings." Ph.D. thesis, University of California Santa Cruz.

Hartsock, Nancy. 1983. *Money, Sex, and Power*. New York: Longman.

Hays, Samuel. 1959. *Conservation and the Gospel of Efficiency: The Progressive Conservation Movement, 1890–1920*. Cambridge, MA: Harvard University Press.

Higham, John. 1975. *Send These to Me: Jews and Other Immigrants in Urban America*. New York: Atheneum.

Hofstadter, Richard. 1955. *The Age of Reform*. New York: Knopf.

Holton, Gerald, and William Blanpied, eds. 1976. *Science and Its Public: The Changing Relation*. Dordrecht: Holland.

Johnson, Martin. 1936. "Camera Safaris." In *The Complete Book of African Hall*. New York: American Museum of Natural History.

Johnson, Osa. 1940. *I Married an Adventurer: The Lives and Adventures of Martin and Osa Johnson*. Garden City, NY: Garden City Publishers.

Johnson, William Davidson. 1958. *T.R: Champion of the Strenuous Life*. New York: Theodore Roosevelt Association.

Keller, Evelyn Fox. 1985. *Reflections on Gender and Science*. New Haven: Yale University Press.

Keller, Evelyn Fox, and Christine Grontkowski. 1983. "The Mind's Eye." In *Discovering Reality: Feminist Perspectives on Epistemology, Metaphysics, Methodology, and Philosophy of Science*, edited by S. Harding and M. Hintikka. Dordrecht: Reidel.

Kennedy, John Michael. 1968. "Philanthropy and Science in New York City: The American Museum of Natural History, 1868–1968." Ph.D. thesis, Yale University.

Kevles, Daniel J. 1985. *In the Name of Eugenics: Genetics and the Uses of Human Heredity*. New York: Knopf.

Knorr-Cetina, Karin D., and Michael Mulkay, eds. 1983. *Science Observed: Perspectives on the Social Study of Science*. London: Sage.

Kolko, Gabriel. 1977. *The Triumph of Conservatism*. New York: Free Press.

Landau, Misia. 1981. "The Anthropogenic: Paleoanthropological Writing as a Genre of Literature." Ph.D. thesis, Yale University.

———. 1984. "Human Evolution as Narrative." *American Scientist* 72: 362–8.

Latour, Bruno. 1988. *The Pasteurization of France*. Translated by Alan Sheridan and John Law. Cambridge, MA: Harvard University Press.

Latour, Bruno, and Stephen Woolgar. 1979. *Laboratory Life: The Social Construction of Scientific Facts*. London: Sage.

Lowe, Donald. 1982. *The History of Bourgeois Perception*. Chicago: University of Chicago Press.

Ludmerer, Kenneth. 1972. *Genetics and American Society*. Baltimore: Johns Hopkins University Press.

Martin, Emily. 1987. *The Woman in the Body: A Cultural Analysis of Reproduction*. Boston: Beacon Press.

Marx, Leo. 1964. *The Machine in the Garden*. London: Oxford University Press.

McCann, Carole Ruth. 1987. "Race, Class, and Gender in U.S. Birth Control Politics." Ph.D. thesis, University of California Santa Cruz.

McCullough, David. 1981. *Mornings on Horseback*. New York: Simon & Schuster.

McKay, Mary. 1980. "Akeley, Mary Lee Jobe 1878–1966," *Notable American Women: The Modern Period*. Cambridge, MA: Harvard University Press, pp. 8–10.

Merchant, Carolyn. 1980. *The Death of Nature: Women, Ecology, and the Scientific Revolution*. New York: Harper & Row.

Moraga, Cherrie. 1983. *Loving in the War Years*. Boston: Southend.

Moraga, Cherrie, and Gloria Anzaldúa. 1981. *This Bridge Called My Back*. Watertown, MA: Persephone.

Nash, Roderick. 1977. "The Exporting and Importing of Nature: Nature-Appreciation as a Commodity, 1850–1980." *Perspectives in American History* XII: 517–60.

———. 1982. *Wilderness and the American Mind,* 3rd edition. New Haven: Yale University Press.

Nesbit, William. 1926. *How to Hunt with the Camera.* New York: Dutton.

Nielson, Waldemar A. 1972. *The Big Foundations.* New York: Columbia University Press.

Olds, Elizabeth Fagg. 1985. *Women of the Four Winds: The Adventures of Four of America's First Women Explorers.* Boston: Houghton Mifflin.

Oleson, Alexandra, and John Voss, eds. 1979. *The Social Organization of Knowledge in Modern America, 1860–1920.* Baltimore: Johns Hopkins University Press.

Osborn, Henry Fairfield. 1930. *Fifty-two Years of Research, Observation, and Publication.* New York: American Museum of Natural History.

Pickens, Donald. 1968. *Eugenics and the Progressives.* Nashville: Vanderbilt University Press.

Pietz, William. 1983. "The Phonograph in Africa: International Phonocentrism from Stanley to Sarnoff." Unpublished paper from the Second International Theory and Text Conference, Southampton, England.

Porges, Irwin. 1975. *Edgar Rice Burroughs: The Man Who Created Tarzan.* Provo, UT: Brigham Young University Press.

Porter, Charlotte. n.d. "The Rise to Parnassus: Henry Fairfield Osborn and the Hall of the Age of Man." Unpublished manuscript, Smithsonian Institution.

Reed, James. 1978. *From Private Vice to Public Virtue: The Birth Control Movement and American Society since 1830.* New York: Basic Books.

Rose, Hilary. 1983. "Hand, Brain, and Heart: A Feminist Epistemology for the Natural Sciences." *Signs* 9:73–90.

Rosenberg, Rosalind. 1982. *Beyond Separate Spheres: Intellectual Roots of Modern Feminism.* New Haven: Yale University Press.

Rossiter, Margaret. 1982. *Women Scientists in America: Struggles and Strategies to 1914.* Baltimore: Johns Hopkins University Press.

Russ, Joanna. 1983. *How to Suppress Women's Writing.* Austin: University of Texas Press.

Russett, Cynthia. 1966. *The Concept of Equilibrium in American Social Thought.* New Haven: Yale University Press.

Sayers, Janet. 1982. *Biological Politics: Feminist and Anti-Feminist Approaches.* London: Tavistock.

Shelley, Mary. 1818. *Frankenstein.* London: Lackington, Hughes, Harding, Mavor, and Jones.

Sofoulis, Zoe. 1988. "Through the Lumen: Frankenstein and the Optics of Re-origination." Ph.D. thesis, University of California Santa Cruz.

Sohn-Rethel., Alfred. 1978. *Intellectual and Manual Labor.* London: Macmillan.

Sontag, Susan. 1977. *On Photography.* New York: Delta.

Spillers, Hortense. 1987. "Mama's Baby, Papa's Maybe: An American Grammar Book." *Diacritics* 17(2): 65–81.

Starr, Paul. 1982. *The Social Transformation of American Medicine.* New York: Basic Books

Taylor, Joshua. 1979. "Malvina Hoffman." *American Art and Antiques* 2 (July/August): 96–103.

Tiptree, James, Jr. 1975. *Warm Worlds and Otherwise.* Introduction by Robert Silverberg. New York: Ballantine.

———. 1978. *Up the Walls of the World.* New York: Macmillan.

Weinstein, James. 1969. *The Corporate Ideal in the Liberal State, 1900–1918.* Boston: Beacon.

Wiebe, Robert. 1966. *The Search for Order, 1877–1920.* New York: Hill & Wang.

Young, Robert M. 1977. "Science *Is* Social Relations." *Radical Science Journal* 5:65–129.

6

MORPHING IN THE ORDER: FLEXIBLE STRATEGIES, FEMINIST SCIENCE STUDIES, AND PRIMATE REVISIONS

SCIENTIFIC PRACTICE AND TOUGH LOVE

"For thus all things must begin, with an act of love." In a 1980 broadcast on South Africa Radio, "The Soul of the White Ant," Eugene Marais, a naturalist who published his observations on baboons in 1926, enunciated a central ethical and epistemological point about where to start a scientific account. Marais's comment applies to more than the sexual and reproductive capers of the animals he observed. I understand Marais's statement, which also opened my book *Primate Visions* (1989), to state the relationship between scientists and their subjects and between science studies scholars and their subjects. The greatest origin stories are about love and knowledge. Of course, love is never innocent, often disturbing, given to betrayal, occasionally aggressive, no stranger to domination, and regularly not reciprocated in the ways the lovers desire. In the practices I inherited, the love and knowledge of nature—embedded inextricably in the histories of colonialism, racism, and sexism—are no Edenic legacies. A white South African does not accidentally grace my text. Finally, love is relentlessly particular, specific, contingent, historically various, and resistant to anyone having the last word. Colonialism, racism, and sexism are not the last words even on a Westerner's love of nature; but they are an enduring part of the lexicon.

The major ethical and epistemological issue for me, in trying to understand what kinds of undertakings the biological and anthropological sciences are, is that knowledge is *always* an engaged material practice and *never* a disembodied set of ideas. Knowledge is embedded in projects; knowledge is always *for*, in many senses, some things and not others, and knowers are always formed by their projects, just as they shape what they can know. Such shapings never occur in some unearthly realm; they are always about the material and meaningful interactions of located humans and nonhumans—machines, organisms, people, land, institutions, money, and many other things. Because scientific knowledge is not "transcendent," it can make solid claims about material beings that are neither reducible to opinion nor exempt from interpretation. Those solid claims and material beings are *irreducibly* engaged in cultural practice and practical culture; i.e., in the traffic in meanings and bodies, or acts of love, with which all things begin. Semiosis is about the physiology of meaning-making; science studies is about the behavioral ecology and optimal foraging strategies of scientists and their subjects; and primatology seems to me to be about the historically dynamic, material-semiotic webs where important kinds of knowledge are at stake.

In this introduction I want to offer a slightly confessional account of my own noninnocent act of love in writing about primatology and primatologists over the last two decades. I care about primatology for many reasons, not the least of which is the pleasure of *knowledge* about these animals, who include us, *Homo sapiens*, but who exceed us in their varied ways of life. The other primates who are different from us are at least as interesting and consequential as those beasts taken to be like us. But mainly, it is a personal fact that I *identify* as a member of a species and a zoological order. My view of myself is shaped by bioscientific accounts, and that is a source of intense interest and pleasure. My life has been shaped at its heart and soul by material-semiotic practices through which I know and relate to myself and others as organisms. To identify, internally and subjectively, as a member of a zoological species and order is an odd thing to do, historically speaking. I am intensely interested in how such a practice came to be possible for many millions of different kinds of people over a couple hundred years. So, my act of love with primatology is more like sisterly incest than alien surveillance of another family's doings.

I am in love with words themselves, as thick, living, physical objects that do unexpected things. My paragraphs are peppered with words like "semiosis" because I am in love with the barnacles that crust such seedy, generative, seemingly merely "technical" terms. Words are weeds—pioneers, opportunists, and survivors. Words are irreducibly "tropes" or figures. For many commonly used words, we forget the figural, metaphoric qualities; these words are silent or dead, metaphorically speaking. But the tropic quality of any word can erupt to enliven things for even the most literal minded. In Greek, *trópos* means a turning; and the verb *trepein* means to swerve, not to get directly somewhere. Words trip us, make us swerve, turn us around; we have no other options. Semiosis is the *process* of meaning-making in the discipline called semiotics. Primatologists, beginning with C.R. Carpenter, have drawn richly from the human science of semiotics, and I have a playful and serious relationship with the ways communications sciences, linguistics, information sciences, and their motley offspring have infused primatology since the 1930s.

Science and science studies depend constitutively upon troping. Unless we swerve, we cannot communicate; there is no direct route to the relationship we call knowledge, scientific or otherwise. *Technically,* we cannot know, say, or write exactly what we mean. We *cannot* mean literally; that negative gift is a condition of being an animal and doing science. No alternative exists to going through the medium of thinking and communicating, no alternative to swerving. Mathematical symbolisms and experimental protocols do not escape from the troping quality of any communicative medium. Facts are tropic; otherwise they would not matter. Material-semiotic is one word for me. I also know that there is a fine line between an exuberant love affair with words and a pornographic fascination with jargon. Tropes are tools, and, female or not, endowed with only the minor instrument of the *mentula mulieribus,*[1] I am a practicing member of *Homo faber.*

Embedded in narrative practices, stories are thick, physical entities. If storytelling is intrinsic to the practice of the life sciences, that is no insult or dismissal. Stories are not "merely" anything. Rather, narrative practice is a compelling part of the semiosis of making primatology. Some sciences reduce narrative to the barest minimum, but primate studies have never had the questionable privilege of an antiseptic narrative sterilization. Many other practices make up primatology, but not to attend lovingly to stories seems worse than abstemious to me; it seems a kind of epistemological contraception. "For thus all things must begin, with an act of love."

There is a doubled quality to what I love in primatology, or in any science. *First,* I am physically hypersensitive to the historically specific, materially-semiotically dense practices that constitute science made, as well as science in the making (Latour 1987). Science is practice and culture (Pickering 1992) at every level of the onion. There is no core, only layers. It is "elephants all the way down," in my purloined origin story about science. Everything is supported, but there is no final foundation, only the infinite series of carrying all there is. Or, in another tropic effort to say what I mean, nothing insoluble precipitates out of the solution of science as cultural practice (Hess 1997). Remaining in "solution" is the permanent condition of knowledge-making.

Linked to my meaning of a solution, science and science studies are both generative mixings of natures and cultures, where the lively and heterogeneous actors blast the categories used to contain them. Conventionally in Western discussions, nature is both outside of culture and posited as a resource for culture's transforming power. Culture is tropically layered onto nature in a quasi-geological sedimentation. Simultaneously, culture is figured as the force that transforms natural resource into social product. Nature is needed for foundations; culture—in its manifestation as science, for example—is indispensable for direction and progress. Elaborate linked binaries are like stem cells in the marrow of this conventional discourse, in which primatology initially differentiated. Animal-human, body-mind, individual-society, resource-product, nature-culture; the monotonous litany is interrupted by science studies, which both foregrounds the tropic quality of these tools to think with and also suggests the possibilities of other tropes, other tools. As the British social anthropologist Marilyn Strathern insists, "it matters what ideas one uses to think other ideas (with)" (Strathern 1992, 10). This approach to primatology strengthens the perception of diverse, complex, nonanthropomorphic actors.

Second, my primate loves demand an appreciation of the solidity and nonoptional, if also open and revisable, quality of scientific projects. How else could I continue to argue that teaching Christian creationism in biology classes in the public schools is serious child abuse? The articulations that constitute knowledge are fragile, precious, historical achievements. If no god intervenes to grant certainty to the representations we call knowledge, no devil reduces knowledge to illusion either. The strength of scientific articulations is a practical matter, involving the development of analytical tools, narrative possibilities, representational technologies, patterns of training, institutional ecologies, structures of power, money, and, not least, the ability to craft diverse connections with nonhumans of many kinds. Primatology is made of such things.

My first job after graduate school in Yale's Department of Biology was in a general science department in a large state university from 1970 to 1974. My task was to teach biology and the history of science to "non-science majors," a wonderful ontological category, to make them better citizens. I was part of a team of young faculty led by a senior teacher, who had designed a course to fill an undergraduate science requirement for hundreds of students each year. In the middle of the Pacific Ocean, home of the Pacific Strategic Command, so critical to the Vietnam War with its electronic battlefield and chemical herbicides, this University of Hawaii biology course aimed to persuade students that natural science, not politics or religion, offered hope, promising secular progress not infected by ideology. I and the other younger members of the course staff could not teach the subject that way. Our post-Enlightenment epistemological confidence was messier than that.

For us, science and history had a more contradictory and interesting texture than the allegory of purity and prophylactic separation we were supposed to teach. Many of my graduate-school biology faculty and fellow graduate students were activists against the war partly because we knew how intimately science, including biology, was woven into that conflict—and into every aspect of our lives and beliefs. Without ever giving up our commitments to biology as knowledge, many of us left that period of activism and teaching committed to understanding the historical specificity and conditions of solidity of what counts as nature, for whom, where, and at what cost. It was the epistemological, semiotic, technical, political, and material connection—not the separation—of science and cultural-historical specificity that riveted our attention. Biology was not interesting because it transcended historical practice in some positivist epistemological liftoff from earth, but because natural science was part of the lively action on the ground.

I still use biology, animated by heterodox organisms burrowing into the nooks and crannies of the New World Order's digestive systems, to persuade readers and students about ways of life that I believe might be more sustainable and just. I have no intention of stopping and no expectation that this rich resource will or should be abandoned by others. Biology *is* a political discourse, which we should engage at every level—technically, semiotically, morally, economically, institutionally. Besides all that, biology is a source of intense intellectual, emotional, and physical pleasure. Nothing like that should be given up lightly—or approached only in a scolding or celebratory mode. From the establishment of biology as part of the curriculum in urban high schools early in the twentieth century to the training of environmental managers and molecular geneticists in the 1990s, the teaching of biology in the United States has been part of civics. From complex systems to flexible bodies, imploded natural-cultural worlds are modeled and produced. As biologist Scott Gilbert argues, in the post–Cold War United States, biology is the functional equivalent of the once required course in Western Civilization as the obligatory passage point through which educated citizens must pass on their way to careers from law to finance to medicine, as well as to a liberal arts degree (Gilbert 1997, 48–52). By the late 1990s, biology has become the foundation for the educated citizen in complexly global and local worlds.

What I want to understand in science and science studies—for which I use primatology metonymically as part for whole—is the *simultaneity* of both the facts and explanatory theoretical power and also the relentlessly tropic, historically contingent, and practical materiality of science. The world we know is made, in committed projects, to be in the shapes our knowledge shows us; humans are not the only actors in such projects; and the world thrown up by knowledge-making projects *might* (still) be otherwise, but *is* not otherwise. Science is revisable from what its practitioners think of as "outside" as well as from "inside." What counts "semiotically" as inside and outside is the result of ongoing work *inflected by* and *constitutive of* power of all sorts.

One instance of the simultaneity of power-laden historical contingency and material facticity that characterizes scientific objects of knowledge has riveted my attention since that first job in the militarized tourist fields of Hawaii. I came to know what a command-control-communications-intelligence (C^3I) system was. As a late twentieth-century American organism, I was such a system, both literally and tropically. Teaching biology as civics to non-science majors was a revelation. I began to get it that discourse is practice, and participation in the materialized world, including one's own naturalcultural (one word) body, is not a choice. Practitioners of immunology, genetics, social theory, insurance analysis, cognitive science, military discourse, and behavioral and evolutionary sciences all invoked the same eminently material, theoretically potent stories to do real work in the world, epistemologically and ontologically. That is, I learned that I was a cyborg, in cultural-natural fact. Like other beings that both scientists and laypeople were coming to know, I too, in the fabric of my flesh and soul, was a hybrid of information-based organic and machinic systems.

The term "cyborg" was coined by Manfred Clynes and Nathan Kline to refer to the enhanced man—a cyb(ernetic) org(anism)—who could survive in extraterrestrial environments. They imagined the cyborgian man-machine hybrid would be needed in the next great technohumanist challenge—space flight. A designer of physiological instrumentation and electronic data-processing systems, Clynes was the chief research scientist in the Dynamic Simulation Laboratory at Rockland State Hospital in New York. Director of research at Rockland State, Kline was a clinical psychiatrist. Enraptured with cybernetics, Clynes and Kline (1960, 27) thought of cyborgs as "self-regulating man-machine systems." Their first cyborg was the "standard" laboratory rat implanted with an osmotic pump designed to inject chemicals continuously to modify and regulate homeostatic states.

Brain-implanted lab animals showed up regularly in psychiatric research projects in the United States. Following up ideas he had in 1938 for behavioral experiments using primates with brain ablations, C. R. Carpenter, despite misgivings recorded in his research notes, took advantage of the convergence of Cold War and psychiatric research agendas for a brief study of experimentally brain-damaged, free-living gibbons on Hall's Island in Bermuda in 1971. The gibbons had been operated on before Carpenter entered the story; he addressed questions of "anti-social behavior" among the one adult and five juvenile gibbons. Carpenter was brought into the work by José Delgado, then at Yale's psychiatry department, who collaborated with the Rockland State researchers on behavioral-control technologies and psychopharmacology in the context, stated in their research applications, of "stress" and "alienation" in U.S. cities. Like Delgado, Carpenter lectured widely in the 1960s on the relations of war, aggression, stress, and territoriality; he believed the other primates could teach "modern man" important lessons about these subjects (Haraway 1989, 108–10). Evident everywhere by the late twentieth century, the implosion of informatics and biologics has a cyborg pedigree.

This account of myself and other organisms as communications systems is a representation; but it is more than that. This kind of representation shapes lived worlds, even as the account is shaped by all the naturally/culturally situated human and nonhuman collaborators that have to be articulated to do such representations. In the late twentieth century, in a globally distributed pattern affecting billions of people, we really do know and relate to the biological world, in material-semiotic-practical fact, as energetic, economic, and informational processes. Similar formulations can and do show up interchangeably in economics textbooks, immunology journals, evolutionary discourses, family policy documents, and military strategy conferences. What is going on? How does the "mangle of practice" (Pickering 1995) that is science, including primatology, produce the zoological order to which I am committed? How do commitment, anger, hope, pleasure, and work all come together in the practice of love we call science? And science studies?

PRIMATE REVISIONS

Near a drawer with a chimp skull implanted with a box-like telemetric monitoring device dating from the Cold War's space race, the bones of Gombe's old Flo—the first chimpanzee to receive an obituary in the *New York Times*—rest in the laboratory of my colleague at the University of California at Santa Cruz, Adrienne Zihlman. By 1980, baboons at Amboseli were understood to be dual-career mothers (Altmann 1980), juggling the demands of making a living and raising kids. At the same time, many of these monkeys' trans-specific human sisters in the U.S.

represented their lives in similar terms. In the 1930s, howler monkeys in Panama and gibbons in Siam lived in different sorts of societies; but both types seemed to be socially managed by a mechanism called a "socio-nomic sex ratio" (Carpenter 1964). In the 1980s, a sign language–using, middle-aged, middle-class gorilla living in California while awaiting per-mission to move to Hawaii sought IVF in a desperate effort to conceive a child (Haraway 1989, 143–46).

If primates in past decades were sometimes represented by monoma-niacal models of species-typical behavior, impoverished minds, sexual rigidity, and ecological stereotypes, by the 1990s human and nonhu-man primates alike appear to be flexible strategists, with multifactorial cost-benefit analyses guiding the order's behavioral and evolutionary investments. Diversity is everything.[2] If communications dominated professional and popular scientific discourse in the 1970s and 1980s, diversity and flexibility name the high-stakes game at the turn of the mil-lennium. Indeed, in their position paper for the conference on "Chang-ing Images of Primate Societies," the organizers argued—with a touch of irony—that, even in a fiercely dangerous world, "[b]y 1995, baboons everywhere have more options than ever before" (Strum and Fedigan 1996, 45). The biological world these days acts like a flexible accumu-lation system. Appropriately, Strum and Fedigan emphasized multifac-torial flexibility: "[P]rimates are smart actors. Individuals, regardless of sex or age, are strategists in an intricate evolutionary game. Their op-tions, choices and successes depend on a variety of factors including environment, demography, age, sex, development, personality, biology, and historical accident" (1996, 48).

It is too easy reading the last two paragraphs to mumble about bias, cultural relativism, ideology, popularization, storytelling as opposed to proper conceptual models and testable hypotheses, and all the other things that are supposed to act as obstacles to the hard-won prize of real scientific knowledge. I don't believe it. In particular, I don't believe bias is a very interesting idea for thinking about primate studies. Bias exists, and goddess knows primate studies (as well as feminist theory and science studies) provide truckloads of examples. It is edifying for an historian of science to watch a notion—say, the "man-the-hunter" hypothesis or the competitive sexual access model of macaque social organization—move from state-of-the-art theory to surpassed science to pseudoscience and sometimes back into fashion again among folks with doctoral science credentials, no math anxiety, and solid fieldwork in the right subfields. But "bias" tells the scientist or historian little about how a field practice, story, or theory travels and the work that gets done with the miscreant tools' aid. Scrubbing away bias is like cleaning one's toilet—it's got to be done, but more has to be said about how life gets lived in different sorts of houses.[3]

Rather, I want the intersecting worlds of primate lives and scientific practice—in which "[b]y 1995, baboons everywhere have more options than ever before"—to signal a different kind of approach to the historical contingency, tropic and narrative thickness, situated interactions of subjects and objects (like simians and people), and explanatory power. I see primatology (like feminist theory and science studies) to be a zone of implosion, where the technical, mythic, organic, cultural, textual, oneiric (dream-like), political, economic, and formal lines of force converge and tangle, bending and warping both our attention and the objects that enter the gravity well. Imploded zones interest me because that is where knowledge-making projects are emerging, at stake, and alive. It is possible to discuss mythic and textual axes separately from technical and organic ones, but the (often hidden) work it takes to keep the lines separate is stupendous and counterproductive. Getting important things done in the world—like building a creditable scientific account of primate lives—requires forcefully converging threads. My mode of attention causes me to mix things up that sometimes others have high stakes in keeping separate, and I might often be wrong-headed. But my way of working will also, sometimes, usefully avoid reductive notions of what is "inside" and "outside" scientific primatology, what is popular and professional, and what is "cultural" or "political" and what is "scientific" about our notions of primates.

I will enter a zone of implosion by concentrating on a persistent question in primate studies: What is the correct social unit of analysis for behavioral and evolutionary understanding of one species, the common chimpanzee, *Pan troglodytes*? Most of my account will draw from and revise bits of two chapters in *Primate Visions*, "A Pilot Plant for Human Engineering: Robert Yerkes and the Yale Laboratories of Primate Biology, 1924–1942" and "Apes in Eden, Apes in Space: Mothering as a Scientist for National Geographic" (Haraway 1989). I am also instructed by *The Chimpanzees of the Mahale Mountains* (Nishida 1990) and a spate of reports in 1997. I will highlight analytical moves taught me by science studies scholars, feminist and antiracist writers and activists, and primate scientists. Sometimes the same people inhabit all of those categories, and the boundaries among them are permeable. The point is to learn how to navigate in a gravity well.

Primate Visions was often reviewed as if it were about gender and science. I read the book to be about race, gender, nature, generation, simian doings, and primate sciences, as well as about many other things as they co-constitute each other, not as they are retrospectively narrated as already formed variables. Neither gender nor science—or race, field, and nation—preexist the heterogeneous encounters we call practice. "Gender" does not refer to preconstituted classes of males and females. Rather, "gender" (or "race," "national culture," etc.) is an asymmetrical, power-saturated, symbolic, material, and social relationship that is constituted and sustained—or not—in heterogeneous naturalcultural practice, such as primate studies. Doing science studies, my eye is as much on "gender-in-the-making" or "race-in-the-making," as on "science-in-the-making." Category names like "gender" or "science" are crude indicators for a mixed traffic.

In this light, "Apes in Eden, Apes in Space" cannot make sense outside of its dramatic setting in the theaters of Cold War; nation-making; oil multinationals sponsoring natural history television specials in the age of ecology; changing field practices in primate studies; expatriate practices situated in decolonizing white settler colonies; relations between foreign scientists and primate habitat-nation populations, field staff, and officials; publication conventions; "first-world" feminism; racialized gender narratives in both of the mythic-material spaces called "Africa" and "the West"; the histories of academic disciplines, institutions, and cohorts in several nations (U.S., U.K., Japan, the Netherlands, Tanzania); and pedagogical and popular magazines, film, and TV.

From the point of view of many practicing scientists, perhaps the part of "Apes in Eden, Apes in Space" that seems to be about what they do— as opposed to what affects scientists from the outside—occupies two subsections, "Intermission at Gombe: History of a Research Site" and "Crafting Data." There, at last, we hear about such things as changes in data collection and analysis, from field diaries to computerized databases and efforts to make variously collected data comparable; field site development; theoretical alternatives with varying kinds of empirical support; and career patterns and contending cohorts of primate scientists.[4] The wording of the subtitle "Intermission at Gombe" was an impolite troping device meant to urge the reader to swerve in order to pay attention to the traffic between professional and popular practices, to the unpredictable direction of arrows of influence, and to the *constitutive*, and not merely *contextual*, doings of many communities of practice in shaping our knowledge of other primates.[5] My analysis is full of promiscuously presented material that perhaps ought to be hygienically sorted into science, on the one hand, and contexts for and influences on science on the other. No such luck, or so my mind, shaped on Hawaiian beaches and mountains as much as in labs at Woods Hole and Yale, insists.

"Apes in Eden, Apes in Space" opens with a section on the fetish that ruled popular and technical practice in post–World War II transnational contexts: "communication." A serious joke, "fetish" is a trope that stands in for a missing organ (guess which) and represents a disavowal of the dangerous, castrated condition of the matrix of our origin, or "mother." In my story about post-World War II science, including primatology, "communication" stands in for a disavowed "history." "Communication" was everywhere in 1960s and 1970s scientific and popular discourse, not least in the intercourse that humans engaged in with simians. In most representational practices in professional and popular primate studies, "nature" and "science" substitute for the trauma-inducing traffic of naturecultures called history, which I think is what science is really all about.

I pay attention to "communication" as a fetish in a web of practices: (1) *National Geographic* television advertising for the Goodall, Fossey, and Galdikas specials ("Understanding Is Everything," one Gulf Oil ad proclaimed); (2) Goodall's ethograms; (3) streams of data pouring into the space race's computers from the bodies of orbiting captured chimpanzee children acting as "surrogates for man"; and (4) representations of the Ameslan-using gorilla, Koko, who showed the (dehistoricized) universal signs of being "man" in her naming pets, referring to herself, knowing what is naughty and nice, and taking her picture in a mirror with a Polaroid camera. In the face of such thick histories and dense collections of human and nonhuman actors, I am fascinated by the technologies that accomplished magical things like making "Jane Goodall" appear to be "alone in nature"—especially in 1960, the year fifteen primate-habitat nations in Africa achieved independence. With their own social, technical, and rhetorical practices, Japanese primatologists were not much taken by the device of representing themselves to be "alone in nature."

The chapter continues its investigation of ways of "reading out history" by exploring scientific/cultural productions from the 1960s through the 1980s. Beginning with Frans de Waal's and Dian Fossey's hybrid technical/popular books, the chapter turns its attention to a synergistic triple code—gender, race, and science—needed to read *National Geographic*'s accounts of monkeys and apes. That task required paying attention to the details of how and why U.S. and U.K. "white" women filled the narrative function they did in those stories, how "black" women and men got the kinds of scientific credentials they did in the same years in the United States, and how writers for *National Geographic,* like Shirley Strum, struggled with modest success to control visual and prose narration of their scientific work. Missing throughout these accounts was the contemporary ape fieldwork being done near Gombe in the Mahale Mountains by the Japanese.

Just before the Intermission, "Apes in Eden, Apes in Space" heads for the movies. The bill of fare includes both *King Kong* and his ongoing mutants and the cascade of sober pedagogical films on primate behavior. These edifying celluloid records began with Julian Huxley and Solly Zuckerman's 1938 *Monkey into Man*—which predictably linked family, race, and technology in a functionalist and evolutionary great chain of being—and then moved to C. R. Carpenter's "positivistic" films about free-ranging primate species, made in the 1940s from his prewar field footage, and lastly settled on Sherwood Washburn and Irven DeVore's (1966) "objective" baboon behavior and society films. The visual and verbal rhetoric of the films produced the epistemological and aesthetic *effect* of objective vision. *How* such important effects get produced commands rapt attention among science studies scholars, who investigate the relation of the filmic effect of direct, objective observation to the messy doings of human and nonhuman primates. It would be hard to overestimate the influence of Carpenter's and DeVore's films, which for generations of novice viewers of nonhuman primates warranted belief in species-typical behavior and grouping patterns.

By the time we get to Gombe as a research site, the reader of "Apes in Eden, Apes in Space" is saturated with the messy cross-traffic in the midst of which knowledge of primates is crafted. I mean materially solid knowledge, not biased opinion or ideological illusion. Like any good science, primate studies produces revisable and complexly progressive knowledge that travels beyond its sites of emergence. Scientific practice never yields knowledge that precipitates out of the solution of situated histories and material-semiotic apparatuses. If communication was the fetish that "read out history," then my tool for learning to inhabit natureculture will be "memory"—i.e., a trained practice of retelling scientific accounts to situate them as thickly as I can.

As promised, my focus will be on one kind of contested object, namely the unit of chimpanzee social life. I will trace a few threads in a complex fabric; but perhaps enough can be said to show what I mean by a material-semiotic object of knowledge located in apparatuses of knowledge production, for which the concepts of bias, ideology, and cultural relativism are weak tools. Not a neutral observer, I, like the primate scientists, am less "biased" than "engaged."

I can't start my story in East Africa, at Gombe and in the Mahale Mountains. Instead, I have to go to caged pairs of adult chimpanzees engaged in a test of motivation for taking food treats at Robert Yerkes' Florida breeding station that was part of the Laboratories of Comparative Psychobiology at Yale University in the late 1930s (Yerkes 1939). Yerkes firmly believed that "the family" was the organic unit of primate social life, and that "dominance" organized "cooperation" and "integration." None of these words between quotation marks was transparent; all were "boundary objects," grounded in action, which traveled among many communities of practice with just enough continuity in reference to sustain projects and debates (Star and Griesemer 1989). Believing that chimpanzees approximated the state of monogamy in nature, Yerkes caged his animals in male-female pairs when he could. Chimpanzees were a model for man; their natural family life, occurring just on the other side of the border from "culture," was a mirror and testing ground for theories and policies. In Yerkes' framework of functionalist associationism, the family was its members, which could be analyzed into constituent organic drives functionally integrated by the nervous system. The family economy, like the mental one, involved division of labor, (re)productive efficiency, and unity resulting from an integrating hierarchical principle of higher functional adaptation in an evolutionary (but not Darwinian) chain of being.

Yerkes was committed to the intelligent interaction of apes and people in the cooperative enterprise called the laboratory. "Dominance" did not mean exploitative domination, but rather assured natural positioning in organic hierarchies that maximized group efficiency and harmony. That was true among animals and between animals and people. Yerkes' secular, New England, Protestant love for his science and for the animals he studied was intimately intertwined with his beliefs in both himself and the chimps as servants of science for a better world (Yerkes 1943, 11).

Organic drives, such as the "hunger for social status" (Yerkes 1943, 46), shaped role differentiation. Drives varied in strength and effectiveness of expression, so it was important to measure them, just as it was important to measure cognitive capacities in the plethora of mental tests that Yerkes excelled at designing for both human and nonhuman primates. Neither males nor females were inherently dominant; position in a hierarchy was a question of relative strength of organic drives. Status motivation was conditioned by sexual hungers and opportunities. The food-chute test for caged pairs of "mates" measured the interaction of drives for dominance and sex, as observers registered who grabbed bits of banana against stages of sexual swelling of the female and the personality of the animals. "Personality" was "the product of integration of all the psychobiological traits and capacities of the organism" (Yerkes 1939, 130). Individuality mattered, but functional integration of organic systems was a higher level of organization. If females were seen to trade sex for tasty favors, that was simply a view made possible by the research apparatus.[6]

For Yerkes, dominance was a physiological, psychological, and social principle linked to the processes of competition and cooperation, both of which were central to his overall project, which he called human engineering. The chimpanzee lab was a pilot plant for human engineering.

Yerkes' work in the Personnel Research Federation, the Committee for Research in Problems of Sex, the Yale Laboratories of Psychobiology, the Boston Psychopathic Hospital, the Surgeon General's Office in the Army in World War I, the Rockefeller Foundation, and many other locations was geared toward fulfilling his vocation of shaping man for more efficient organic modern social life. Such scientific projects were intrinsically part of building democracy in the contest with "authoritarianism," especially fascism. Racial hierarchies, sex-role relations, and democratic cooperation were part of the great evolutionary, non-Darwinian organic scheme that chimpanzees were asked to clarify. Classifying individuals, ape or human, in accordance with their organic capacities—whether through intelligence tests or scales of motivation—was a fundamental scientific practice. Cage design, building architecture, experimental protocols, and data collection practices in Yerkes' laboratory only make sense within these frameworks.

So, for Yerkes, the monogamous heterosexual pair was the natural chimpanzee social unit, and the food-chute test yielded important data about role differentiation in the family. Concepts of bias and ideology get us almost nowhere in understanding this woeful situation. Yerkes was practicing good science, and he got good data, by the standards of his community of practice (which, it must be said, was a bit short on statisticians). That does not mean he was right about chimpanzees or immune from criticism in 1939 or now; but he was not doing science "influenced" by subjective and cultural "biases." Rather, his science as naturalcultural practice *built* an apparatus of knowledge production that crafted the world in a particular semiotic-material way. Strip the "biases" and not much is left of the scientific apparatus. "The family," "personality," and "intelligence" were solid material-semiotic entities that Yerkes' apparatus helped put together in the world. Their real materiality was an effect of their constructedness (by humans and nonhumans). In humans as well as animals, sexual "role differentiation" (the word "gender" would not have made sense to any of the communities of practice in 1939) was as much a product as a preexisting variable in the Laboratories of Comparative Psychobiology. The *projects* and *commitments* were what Yerkes engaged in as a scientist. The whole messy web of articulations was Yerkes' science. It deserved critical engagement over the practical material-semiotic work that produces knowledge, not ideology critique or celebratory hagiography in the history of primatology.

Jane Goodall's (1967) early descriptions of chimpanzee social organization at Gombe identified only one stable social grouping, the mother and her dependent offspring. Otherwise, chimpanzees were described to associate fluidly and mostly peacefully in nomadic bands, without defended social or territorial boundaries among bands or parties. The first phase of research, 1960 to 1966, when Goodall got her doctorate and more observers began arriving at Gombe, seemed to reveal a primate utopia—mother-centered, but with outstanding male personalities engaged in status competitions, which did not seem to be the organizing axis of chimpanzee society. The material-semiotic unit of the mother-infant pair remained deep in Goodall's naturalcultural practice (Goodall 1984). The unit was crucial to many scientific constructions of objects of knowledge, including Robert Hinde's. Goodall's later collaboration with David Hamburg of Stanford's psychiatry department, in the context of work on "stress" in modern society, is another part of a wide-ranging process of constructing a natural-technical unit of observation in the field.

Symbolically, in Goodall's writing the chimpanzee mother and infant, especially Flo and her newborn, constituted a perfect model after which she portrayed her own relationship with her infant son, Grub. Her personal motivations are unknowable, but the textual narrative of the personal emphasizes the congruence of her own mothering, the utopian model, and the scientific inquiry. The forest's peaceable, open chimpanzee society, full of strong personalities, was a counterpoint to the dominance-organized and closed baboon unit on the dangerous dry savanna. Narrative mattered. Culturally, politically, and technically, the early Gombe accounts participated in contemporary European and Euro-American concerns. The accounts offered a peaceable kingdom, one part of the dual code of a culture obsessed with psychological explanations and therapies for all kinds of historical conflict and pain. Male aggression concerned Goodall, but it did not define what counted as chimpanzee society. These matters were part of the practices that shaped research at Gombe, not some suspect "outside" to the real action "inside" science.

In Japanese accounts of their chimpanzee study population in the Mahale Mountains, observed from 1965, the concept of a "unit-group"— a multimale, bisexual group of 20–100 animals—was emphasized (Itani and Suzuki 1967; Nishida 1990). The group was described as fluid, breaking into different subgroups with exchange of members among neighboring unit-groups. Resulting from their search to identify the social unit as the first task of a proper study, the Japanese emphasis on a unit-group was consistent with their general methods. For the Japanese, the rational starting point of an explanation was not the autonomous individual; they did not begin by seeking to explain the slightly scandalous (to a Westerner) fact that many animals live in groups whose members, beyond the mother-infant primal One, seem mostly to like being with each other.

After the Japanese reports, Gombe workers began to describe chimpanzee groupings in terms of the concept of a "community." The community concept at Gombe was constructed from observations of male associations and interactions (Bygott 1979; Wrangham 1979a, 1979b). Females were assigned to communities as a function of the frequency of their interaction with males, whose own interactions were the independent variables. Chimpanzee males engaged in more overt violent and affiliative behaviors with each other than the females did, and the patterns established the core and boundaries of a social unit. Bygott described females as living in the male community, more or less as valuables within the shared male ranges (Bygott 1979, 407). This meaning of a bisexual community was not what Japanese workers in the Mahale Mountains meant by a bisexual unit group. The focus among the European and American men who followed male chimpanzees at Gombe seemed to be the problem of "human" aggression, as that essentialized attribute of "human nature" was materially-semiotically constructed in psychological, evolutionary, and mental health practices, including primate studies. Goodall shared this framework with the students, and the issue was basic to Hamburg's interest in the chimps. The chimpanzee community was the ahistorical natural-technical object for examining "male" violence and cooperation. These kinds of studies were tools for constituting what it meant to be male in Western scientific societies.

In a situation that would later be seen as a logical scandal, female behavior was not at the center of early sociobiological formulations of natural selection and inclusive fitness, as they began to seep into the increasingly Darwinian Gombe accounts. The gender-stereotypic (and gender-constituting) interest of male observers in chimpanzee male behavior of certain types, leading to a natural-technical object of knowledge called a community defined in terms of male associations, was initially unchallenged by the emerging "new" explanatory frameworks.

Meanwhile, at Gombe women quite different from Goodall were producing accounts of female lives, like Anne Pusey's study of female transfers between the male-defined communities. She noted the similarity of her picture to Japanese descriptions of female movements. The absence of data on female-female interactions and female behavioral ecology began to be remarked in the literature, and graduate students planned field studies to explore the topics (Pusey 1979, 479; Smuts, interview, 18 March 1982). Primate workers began to understand that sociobiological explanatory strategies destabilized the centrality of male behavior for defining social organization. Female reproductive strategies came to look critical, unknown, and complicated, rather than like dependent (or silent and unformulated) variables in a male drama.

Human female observers at Gombe pressed their arguments with their male associates in the field and in informal transnational networks. In general, since the non-Tanzanian men were not then taking many data on females, they were not in a position to see the new possibilities first. I think the Western women generally had higher motivation to reconsider what it meant to be female. Several of the women in my interviews in the mid-1980s reported personal and cultural affirmation and legitimation for focusing scientifically on females from the atmosphere of feminism in their own societies. The men I interviewed also reported a growing sense of legitimation in the 1970s for taking females more seriously, coming from the emerging sociobiological framework, from the data and arguments of women scientific peers, from the prominence of feminist ideas in their culture, and from their experience of friendships with women influenced by feminism. It is not possible in principle to build a causal argument from these reports, even if unanimous, but the construction of scientific knowledge is implausible without these dimensions, where "inside" and "outside" are unstable rhetorical emphases. Implosion is more evident than separation.

In that context, especially in light of the scientific-personal friendship with Barbara Smuts in the key period of rethinking, I read Richard Wrangham's use of sociobiological resources to formulate his papers on chimpanzee behavioral ecology. In interviews, Smuts and Wrangham both recalled a rich brew of conversation about females, selection theory, Robert Trivers's ideas about females as limiting resources for males, and missing data on female behavioral ecology. Published during this period of intense interaction, Wrangham's papers developed the theoretical perspective of behavioral ecology to redraw ape society. His explanations centered female foraging and social strategies as independent variables, in relation to which male patterns would have to be explained. Simultaneously, similar ideas were developed for evolutionary theory of vertebrate society generally (Wrangham 1979a, 1979b, interview on 13 August 1982; Wrangham and Smuts 1980).

Sociobiological theory really must be "female centered" in ways not true for previous paradigms, where the "mother-infant" unit substituted for females. The "mother-infant" unit had not been theorized as a rational autonomous individual; its material-semiotic functions were different, located in the space called "personal" or "private" in Western narrative practice. The sociobiological kind of female-centering remains firmly within Western economic and liberal theoretical frames and succeeds in reconstructing what it means to be female by a complex elimination of this older special female sphere. In sociobiological narrative, the female becomes the calculating, maximizing machine that males had long been. In locally relevant gender symbolism, the "private" collapses into the "public" (Keller 1992, 148). The female is no longer assigned to male-defined "community" when she is restructured ontologically as a fully "rational" creature. The female ceases to be a dependent variable when males and females both are defined as liberal man, i.e., "rational" calculators. The practical effect of constructing this "female male" was to legitimate data-collection practices that made both men and women watch females more and differently. The picture that emerged of female lives has been full of rich contradictions for the logical model of stripped-down individualism that legitimated the investigation.

It is impossible to account for these developments without appealing to personal friendship and conflict, webs of people planning books and conferences, disciplinary developments in several fields (including practices of narration, theoretical modeling, and hypothesis testing with quantitative data), the history of economics and political theory, and recent feminism among particular national, racial, and class groups. The concept of situatedness, not bias, is crucial.

Female-centered behavioral ecology, however, is not the "good" ending to a story that began with Yerkes' caged mates bumping each other aside for food while modeling heterosexual family life for rapt scientists with data sheets. The boundary object called the unit of chimp society remains in the hot trading zones of scientific practice, where data systems, personal and cohort friendship and enmity, theoretical narratives, national and institutional inheritances, local chimp doings, gender-in-the-making, and more are the machine tools for crafting scientific knowledge.

The chimpanzees of Gombe structure my program for "Morphing in the Order." And so, appropriately, on the front cover of *Science* magazine for August 8, 1997, a touching portrait of old Flo's adult daughter Fifi (now thirty-eight years old with seven surviving children of her own) and baby grandson Fred highlights updated accounts for my primate revisions. Several threads come together. The lineage of Gombe workers reproduces itself, even as the scientists focus on the differential reproductive success of the chimpanzees. A graduate student at the University of Minnesota (Jennifer Williams) publishes with her senior mentor Anne Pusey, from the generation of sociobiologically influenced researchers that followed and in many ways challenged Jane Goodall, who is the third author of the 1997 *Science* article (Pusey, Williams, and Goodall 1997).

The central achievement of the publication is a statistically significant demonstration that differential female reproductive success—measured as infant survival, rate of maturation of daughters, and the rate of annual production of babies—can be correlated to female dominance rank. At least at Gombe, for these fifteen adult female chimpanzees, the correlation holds and is pregnant with testable behavioral ecological questions gestating in the pages of *Science*. That is, the correlation holds if the highest ranking female is excluded because she remained sterile for the twenty-eight years the scientists could account for. Appropriately, Flo's lineage shines with reproductive achievement at Gombe. The achievement of the scientists rested on thirty-five years of collective work by Tanzanian and foreign observers embedded in diverse institutional, cultural, and individual matrices. The central artifact that allowed the important new knot to be tied in the web of collective knowledge was the record of the exchange of pant-grunts between chimp females from 1970 to 1992. It is on just such homely stuff that the credibility of Darwinian understandings of life depends. That labor-intensive examination was only possible because of the initial and subsequent systems of record-keeping at Gombe and the transcription of those data into a computer-based data retrieval system beginning in the late 1970s at Stanford. Those data systems are materializations of mostly invisible conflictual and collaborative work to produce "good enough" categories, practices of collection, and mobility and comparability of records.

Examination of the noises made infrequently by female chimps to each other over twenty-two years made sense because of prior narrative and theoretical transformations. In particular, the drama of evolution had to feature the idea that *females* evolved—that they differ *from each other* in ways consequential for natural selection, that is, for differential reproductive success of individuals (or some other bounded unit in the story). Females had to be "strategists" in the great games of productivity and efficiency. Females had to be inventive in Darwinian terms. The ability to state such a thing explicitly, in testable formulations, took the same kind of conflictual and collaborative work by scientists and nonscientists as did sustaining a field site with its transnational and multimedia data tendrils (Brody 1996). In Yerkes' world, such a thing was unthinkable, literally.

Such a thing is also unthinkable in naturalcultural worlds that do not *think* action in terms of bounded possessive individuals. Again, I recall Strathern's admonition that it matters what ideas are used to think other ideas with. The people she worked with in Papua New Guinea do property, reproduction, gender, and dominance differently (Strathern 1988, 1994). If these Melanesians did primate studies *with their own categories for thinking person, action and interaction,* the Japanese, Europeans, Americans, and other primate science producers would have to reimagine and retheorize the history of life in order to do good science. Tropes matter, literally.

Comparing three commentaries on the Pusey, Williams, and Goodall (1997) paper collects up remaining lines of force imploding in the gravity well of Gombe. The first is by the authors themselves, who speculate about the consequences for genetic diversity in this endangered species if the kind of reproductive skew at Gombe over the last thirty years prevails in other populations. The authors' speculation highlights the consequences of habitat destruction and fragmentation that intensify genetic depletion for endangered species worldwide. In "Perspectives" in the same issue of *Science*, besides commenting on the importance of support for the idea that female chimps vary in fitness, Wrangham (1997) translates the rare, low-key expressions of dominance among females (the pant-grunts) into the idea of "covert rivalry," which he analogizes to "cuckoldry" by chimp females in a study in the Ivory Coast; these females got pregnant at high rates from copulations with males outside their "community." Females seem mighty secretive in Wrangham's story. Speculating about costs and benefits for these extra-group matings, he did not suggest that genetic diversity, rather than his inference of "choosing genes," might be the payoff for females. He emphasizes that "until this year no one suspected that female chimpanzees were so active in pursuit of their reproductive interests, yet they are probably doing still more than we appreciate" (1997, 775).

Despite his provocative tropes, Wrangham abstemiously cautions against analogies with humans, but my third commentator is not so severe. The prize-winning *New York Times* science writer Natalie Angier could never be accused of deemphasizing sex, competition, and violence in her riveting accounts of life's ways for the Science Times. True to form, Angier finds grounds for reading the cunning and power of Federal Reserve Chair Alan Greenspan in the doings of chimp supermom Fifi. Her message, however, is close to Wrangham's and Pusey et al.'s, and her language is no more ripe than Wrangham's. "Beneath the females' apparently distracted exteriors skulked true political animals" (Angier 1997, B11). Pusey et al. and the commentators agree in the speculation that the females' subtle, consequential dominance might be exercised through conferring "better access to food, both by enabling a female to acquire and maintain a core area of high quality and by affording her priority of access to food in overlap areas"—a testable socioecological idea favored in 1990s evolutionary biology (Pusey, Williams, and Goodall 1997, 830).

Several things imploded in the 1997 Gombe report. First, the bidirectional traffic between professional and popular science remains thick. The way of troping individuals and action by an Angier shapes a Pusey or Wrangham just as much as the reverse, and all are shaped in the cauldron of naturalcultural life in the New World Order, Inc., where flexible accumulation and diversity management are the high-stakes transnational games. Second, science is crafted in this world not as ideology, but as materialized action in thick histories of work, where tools for thinking and doing are relentlessly tropic. That means the ways of doing science are contingent on levels that make many people, scientists and not, nervous. Next, in the 1990s the power and agency of human and nonhuman females are still produced as front-page news in science journals and the daily press, at least in the United States. Finally, studying primate science shows the precious achievement of such knowledge even while emphasizing the situated character of the achievement.

I want to give the last word in this unfinishable essay to workers with experience watching chimpanzees in the Mahale Mountains. In his conference paper for this volume, Takasaki Hiroyuki[7] noted that members of the Kyoto school of primate studies do not generally find Wrangham's approaches to describing or explaining chimpanzee social grouping very fruitful (e.g., Hasegawa 1990). Part of the explanation Takasaki offered is a Japanese "culture-language complex" that, from the angle of common Western perspectives, reverses the relations of part and whole, individual and society, and other organizing polarities for explanations in biosocial sciences. Happy as I was with his account of difference, I had trouble sleeping on the soft bed of cultural relativism. Like the seed under the mattress in the fairy tale about the princess and the pea, there was something unsettling to the soothing surface of contrasts between East and West.

I found the pea in Hiraiwa-Hasegawa Mariko's (1990) paper on "Maternal Investment before Weaning" in *The Chimpanzees of the Mahale Mountains*. There, in a study rigorously focused on the mother-infant pair, but not in Goodall's 1960s social-functionalist frameworks, Hiraiwa-Hasegawa fluently deployed sociobiological explanatory narratives and associated quantitative methods to examine

> aspects of chimpanzee maternal care before weaning: nursing, infant transport, and grooming. The first two aspects were selected because they apparently inflict costs on the mother. The third is a typical primate social activity on which a considerable amount of time is spent during the day.... Because an individual's time for social behavior is limited, the time a mother allocates for grooming her infant is regarded as a form of maternal investment. (1990, 257–58)

The languages of cost-benefit investment strategies flow freely in this paper. Her acknowledgments hint at the transnational webs of primate studies, where interactions among the bisexual and multinational groupings of scientists weaned at Gombe and Mahale are a microcosm of the disciplinary, institutional, narrative, personal, and other trading zones in primate studies.[8]

This seems a good place to close my own far-from-innocent account. Remembering that in the world of human and nonhuman primates, "all things must begin with an act of love," I hope that my shaggy-dog story of intercourse among science studies, primate studies, and feminist studies can participate a little bit in making it true someday that "[primates] everywhere have more options than ever before." For that hope to be realized, the old naturalcultural issues of survival, justice, diversity, agency, and knowledge in science and politics are as sharply relevant as ever in the primate order. If, as Strum and Fedigan put it, the lines between science and advocacy and between basic and applied science are increasingly blurred for field biologists—and, I would add, for science studies and feminist scholars—perhaps primatology is, after all, mission science.

NOTES

1. *Mentula mulieribus*, the "little mind of women," is an Early Modern term for the clitoris.
2. For "diversity" as an object-in-the-making, see Wilson 1992, Shiva 1993, and World Resources Institute et al. 1993. On the biological world in terms of flexible strategies and the traffic between political and biological economies, see Martin 1992, 1994, Harvey 1989, 147–97, and Haraway 1991, 203–30. This traffic dates from circulations of the concept of division of labor among political economists and biologists from the late 1700s (Limoges 1994).
3. Who cleans up after whom can tell much about how the world is built, including the scientific world, even in its squeaky-clean theoretical game rooms. Studying phenomena from the angle of those who do the cleaning up—from the position of those who must live in relation to standards that they cannot fit (Star 1991)—can be the most powerful scientific (and moral) approach. What happens in 2000 to the humans and nonhumans who cannot be flexible strategists?
4. Early doctoral dissertations by Gombe researchers, with graduate and/or undergraduate degrees mainly from Cambridge and Stanford, showed men largely writing about males, and women about the females and kids (Haraway 1989, 174, 404). The significance is neither self-evident nor the pattern necessarily typical. Tanzanian male field staff, without Ph.D.s, have also shaped and been shaped by primate studies (Goodall 1986, 597–608).
5. Useful for primatology, Clarke and Montini (1993) use social arenas analysis to show how communities of practice constitute and contest for the abortifacient RU486.
6. One female ape dissented from Yerkes' scoring practices (Herschberger 1948, 7, 11).
7. I keep the Japanese convention for ordering names as a reminder of the ways Japanese practices have to be translated into Western formats to be known by Westerners; the reverse is not true.
8. Nishida was Hiraiwa-Hasegawa's advisor; Kelly Stewart, Sandy Harcourt, and Timothy Clutton-Brock got thanks; English-language translation and Japanese science-funding systems were noted.

REFERENCES

Altmann, Jeanne. 1980. *Baboon Mothers and Infants.* Cambridge: Harvard University Press.
Angier, Natalie. 1997. "In the Society of Female Chimps, Subtle Signs of Vital Status." *New York Times,* August 21, B11.
Brody, Jane E. 1996. "Gombe Chimps Archived on Video and CD-ROM." *New York Times,* February 20, B5, 10.
Bygott, J. D. 1979. "Agonistic Behavior, Dominance, and Social Structure in Wild Chimpanzees of the Gombe National Park." In *The Great Apes,* edited by D. Hamburg and E. McCown. Menlo Park, CA: Benjamin/Cummings, 405–28.
Carpenter, C. Ray. 1964. *Naturalistic Behavior of Nonhuman Primates.* University Park: Pennsylvania State University Press.
Clarke, Adele, and Theresa Montini. 1993. "The Many Faces of RU486." *Science, Technology, and Human Values* 18(1): 42–78.
Clynes, Manfred E., and Nathan S. Kline. 1960. "Cyborgs and Space." *Astronautics* (Sept. 26–27): 75–76.
Gilbert, Scott. 1997. "Bodies of Knowledge." In *Changing Life,* edited by P. Taylor, S. Halfon, and P. Edwards. Minneapolis: University of Minnesota Press, 36–52.
Goodall, Jane. 1967. "Mother-Offspring Relationships in Chimpanzees." In *Primate Ethology,* edited by D. Morris. London: Weindenfeld & Nicolson, 287–346.
———. 1984. "The Nature of the Mother-Child Bond and the Influence of the Family on the Social Development of Free-Living Chimpanzees." In N. Kobayshi and T. B. Brazelton, eds. *The Growing Child in Family and Society.* Tokyo: University of Tokyo Press, pp. 47–66.
———. 1986. *The Chimpanzees of Gombe.* Cambridge: Harvard University Press.
Haraway, Donna J. 1989. *Primate Visions.* New York: Routledge.
———. 1991. *Simians, Cyborgs and Women.* New York: Routledge.
Harvey, David. 1989. *The Condition of Postmodernity: An Enquiry into the Origins of Cultural Change.* Oxford: Basil Blackwell.
Herschberger, Ruth. 1948. *Adam's Rib.* New York: Harper & Row.
Hess, David J. 1997. *Science Studies.* New York: New York University Press.
Hiraiwa-Hasegawa, Mariko. 1990. "Maternal Investment before Weaning." In *The Chimpanzees of the Mahale Mountains,* edited by T. Nishida. Tokyo: University of Tokyo Press, 257–66.
Itani, Junichiro, and Suzuki Akira. 1967. "The Social Unit of Chimpanzees." *Primates* 8: 355–82.
Keller, Evelyn Fox. 1992. *Secrets of Life, Secrets of Death.* New York: Routledge.
Latour, Bruno. 1987. *Science in Action.* Cambridge: Harvard University Press.
Limoges, Camille. 1994. "Milne-Edwards, Darwin, Durkheim and the Division of Labour." In *The Natural and the Social Sciences,* edited by I. B. Cohen. Dordrecht, The Netherlands: Kluwer.
Martin, Emily. 1992. "The End of the Body?" *American Ethnologist* 19(1): 121–40.
———. 1994. *Flexible Bodies: Tracking Immunity in American Culture from the Days of Polio to the Days of AIDS.* Boston: Beacon Press.
Nishida, Toshisada, ed. 1990. *The Chimpanzees of the Mahale Mountains.* Tokyo: University of Tokyo Press.
Pickering, Andrew, ed. 1992. *Science as Practice and Culture.* Chicago: University of Chicago Press.
———. 1995. *The Mangle of Practice.* Chicago: University of Chicago Press.
Pusey, Anne E. 1979. "Intercommunity Transfer of Chimpanzees in Gombe National Park." In *The Great Apes,* edited by D. Hamburg and E. McCown. Menlo Park CA: Benjamin/Cummings, 465–80.

Pusey, Anne E., Jen Williams, and Jane Goodall. 1997. "The Influence of Dominance Rank on the Reproductive Success of Female Chimpanzees." *Science* (August 8): 828–31.

Shiva, Vandana. 1993. *Monocultures of the Mind: Perspectives on Biodiversity and Biotechnology.* London: Zed Books.

Star, Susan Leigh. 1991. "Power, Technology and the Phenomenology of Conventions." In *A Sociology of Monsters,* edited by J. Law. Oxford: Basil Blackwell.

Star, Susan Leigh, and James R. Griesemer. 1989. "Institutional Ecology, 'Translations,' and Boundary Objects: Amateurs and Professionals in Berkeley's Museum of Vertebrate Zoology, 1907–39." *Social Studies of Science* 19:387–420.

Strathern, Marilyn. 1988. *The Gender of the Gift.* Berkeley: University of California Press.

———. 1992. *Reproducing the Future: Anthropology, Kinship, and the New Reproductive Technologies.* New York: Routledge.

———. 1994. "The New Modernities." Paper for the Conference of the European Society for Oceanists, Basel, December.

Strum, Shirley C., and Linda Marie Fedigan. 1996. "Theory, Method, and Gender: What Changed Our Views of Primate Society?" Paper for the Wenner Gren conference, "Changing Images of Primate Society."

Takasaki, Hiroyuki. 1996. "Traditions of the Kyoto School of Field Primatology in Japan." Paper for the Wenner Gren conference, "Changing Images of Primate Society."

Wilson, Edward O. 1992. *The Diversity of Life.* Cambridge: Harvard University Press.

World Resources Institute et al. 1993. *Biodiversity Prospecting* (a contribution to the WRI/IUCN/UNEP global biodiversity strategy, May 1993).

Wrangham, Richard. 1979a. "On the Evolution of Ape Social Systems." *Social Science Information* 18(3): 335–69.

———. 1979b. "Sex Differences in Chimpanzee Dispersion." In *The Great Apes,* edited by D. Hamburg and E. McCown. Menlo Park, CA: Benjamin/Cummings, 481–89.

———. 1997. "Subtle, Secret Female Chimpanzees." *Science* 277 (8 August): 774–75.

Wrangham, Richard, and Barbara B. Smuts. 1980. "Sex Differences in the Behavioral Ecology of Chimpanzees in Gombe National Park. *Journal of Reproduction and Fertility* 28, suppl: 13–31.

Yerkes, Robert M. 1939. "Social Dominance and Sexual Status in the Chimpanzee." *Quarterly Review of Biology* 14:115–36.

———. 1943. *Chimpanzees: A Laboratory Colony.* New Haven: Yale University Press.

7

MODEST_WITNESS@SECOND_MILLENNIUM

A man whose narratives could be credited as mirrors of reality was a modest man: his reports ought to make that modesty visible.
   —Steven Shapin and Simon Schaffer, *Leviathan and the Air-Pump*

MODEST WITNESS

The modest witness is the sender and receiver of messages in my e-mail address. So let us investigate how this subject position is woven into the nets traced here. The modest witness is a figure in the narrative net of this book, which works to *refigure* the subjects, objects, and communicative commerce of technoscience into different kinds of knots.[1] I am consumed by the project of materialized refiguration; I think that is what's happening in the worldly projects of technoscience and feminism. A figure collects up the people; a figure embodies shared meanings in stories that inhabit their audiences. I take the term *modest witness* from the important book by Steven Shapin and Simon Schaffer (1985), *Leviathan and the Air-Pump: Hobbes, Boyle, and the Experimental Life.* In order for the modesty, referred to in the epigraph above, to be visible, the man—the witness whose accounts mirror reality—must be invisible, that is, an inhabitant of the potent "unmarked category," which is constructed by the extraordinary conventions of self-invisibility. In Sharon Traweek's wonderfully suggestive terms, such a man must inhabit the space perceived by its inhabitants to be the "culture of no culture"[2] (1988). This is the culture within which contingent facts—the real case about the world—can be established with all the authority, but none of the considerable problems, of transcendental truth. This self-invisibility is the specifically modern, European, masculine, scientific form of the virtue of modesty. This is the form of modesty that pays off its practitioners in the coin of epistemological and social power. This kind of modesty is one of the founding virtues of what we call modernity. This is the virtue that guarantees that the modest witness is the legitimate and authorized ventriloquist for the object world, adding nothing from his mere opinions, from his biasing embodiment. And so he is endowed with the remarkable power to establish the facts. He bears witness: he is objective; he guarantees the clarity and purity of objects. His subjectivity is his objectivity. His narratives have a magical power—they lose all trace of their history as stories, as products of partisan projects, as contestable representations, or as constructed documents in their potent capacity to define the facts.[3] The narratives become clear mirrors, fully magical mirrors, without once appealing to the transcendental or the magical. In what follows, I would like to queer the elaborately constructed and defended confidence of this civic man of reason in order to enable a more corporeal, inflected, and optically dense, if less elegant, kind of modest witness to matters of fact to emerge in the worlds of technoscience.

Robert Boyle (1627–1691) is memorialized in the narratives of the Scientific Revolution and of the Royal Society of London for Improving Natural Knowledge as the father of chemistry and, even more important, father of the experimental way of life. In a series of crucial developments in the 1650s and 1660s in post–civil war Restoration England, Boyle played a key role in forging the three constitutive technologies for such a new life form: "a *material technology* embedded in the construction and operation of the air-pump; a *literary technology* by means of which the phenomena produced by the pump were made known to those who were not direct witnesses; and a *social technology* that incorporated the conventions experimental philosophers should use in dealing with each other and considering knowledge-claims" (Shapin and Schaffer 1985:25).[4] Experimental philosophy—science—could only spread as its materialized practices spread. This was a question not of ideas but of the apparatus of production of what could count as knowledge.

At the center of this story is an instrument, the air-pump. Embedded in the social and literary technologies of proper witnessing, and sustained by the subterranean labor of its building, maintenance, and operation, the air-pump acquired the stunning power to establish matters of fact independent of the endless contentions of politics and religion. Such contingent matters of fact, such "situated knowledges," were constructed to have the earth-shaking capacity to ground social order *objectively*, literally. This separation of expert knowledge from mere opinion as the legitimating knowledge for ways of life, without appeal to transcendent authority or to abstract certainty of any kind, is a founding gesture of what we call modernity. It is the founding gesture of the separation of the technical and the political. Much more than the existence or nonexistence of a vacuum was at stake in Boyle's demonstrations of the air-pump. As Shapin and Schaffer put it, "The matter of fact can serve as the foundation of knowledge and secure assent insofar as it is not regarded as man-made. Each of Boyle's three technologies worked to achieve the appearance of matters of fact as *given* items. That is to say, each technology functioned as an *objectifying resource*" (1985:77). The three technologies, metonymically integrated into the air-pump itself, the neutral instrument, factored out human agency from the product. The experimental philosopher could say, "It is not I who say this; it is the machine" (77). "It was to be nature, not man, that enforced assent" (79). The world of subjects and objects was in place, and scientists were on the side of the objects. Acting as objects' transparent spokesmen, the scientists had the most powerful allies. As men whose only visible trait was their limpid modesty, they inhabited the culture of no culture. Everybody else was left in the domain of culture and of society.

But there were conditions for being able to establish such facts credibly. To multiply its strength, witnessing should be public and collective. A public act must take place in a site that can be semiotically accepted as public, not private. But "public space" for the experimental way of life had to be rigorously defined; not everyone could come in, and not everyone could testify credibly. What counted as private and as public was very much in dispute in Boyle's society. His opponents, especially Thomas Hobbes (1588–1679), repudiated the experimental way of life precisely because its knowledge was dependent on a practice of *witnessing* by a special community, like that of clerics and lawyers. Hobbes saw the experimentalists as part of private, or even secret, and not civil, public space. Boyle's "open laboratory" and its offspring evolved as a most peculiar "public space," with elaborate constraints on who legitimately occupied it. "What in fact resulted was, so to speak, a public space with restricted access" (Shapin and Schaffer 1985:336).

Indeed, it is even possible today, in special circumstances, to be working in a top-secret defense lab, communicating only to those with similar security clearances, and to be *epistemologically* in public, doing leading-edge science, nicely cordoned off from the venereal infections of politics. Since Boyle's time, only those who could disappear "modestly" could really witness with authority rather than gawk curiously. The laboratory was to be open, to be a theater of persuasion, and at the same time it was constructed to be one of the "culture of no culture's" most highly regulated spaces. Managing the public/private distinction has been critical to the credibility of the experimental way of life. This novel way of life *required* a special, bounded community. Restructuring that space—materially and epistemologically—is very much at the heart of late-twentieth-century reconsiderations of what will count as the best science.

Also, displaying the labor expended on stabilizing a matter of fact comprised its status. The men who worked the bellows in Boyle's home laboratory were *his* men; they sold their labor power to him; they were not independent. "As a free-acting gentleman, [Boyle] was the author of their work. He spoke for them and transformed their labor into his truth" (Shapin 1994:406). Unmasking this kind of credible, unified authorship of the labor required to produce a fact showed the possibility of a rival account of the matter of fact itself—a point not lost on Boyle's famous opponent, Thomas Hobbes. Furthermore, those actually physically present at a demonstration could never be as numerous as those virtually present by means of the presentation of the demonstration through the literary device of the written report. Thus, the rhetoric of the modest witness, the "naked way of writing," unadorned, factual, compelling, was crafted. Only through such naked writing could the facts shine through, unclouded by the flourishes of any human author. Both the facts and the witnesses inhabit the privileged zones of "objective" reality through a powerful writing technology. And, finally, only through the routinization and institutionalization of all three technologies for establishing matters of fact could the "transposition onto nature of experimental knowledge" be stably effected (Shapin and Schaffer 1985:79).

All of these criteria for credibility intersect with the question of modesty. Transparency is a peculiar sort of modesty. The philosopher of science Elizabeth Potter, of Mills College, gave me the key to this story in her paper "Making Gender/Making Science: Gender Ideology and Boyle's Experimental Philosophy"[5] (2001). Shapin and Schaffer attended to the submerging, literally, as represented by engravings of the regions under the room with the visible air-pump, of the labor of the crucial artisans who built and tended the pump—and without whom nothing happened—but they were silent on the structuring and meaning of the specific civil engineering of the modest witness. They took his masculine gender for granted without much comment. Like the stubbornly reproduced lacunae in the writing of many otherwise innovative science studies scholars, the gap in their analysis seems to depend on the unexamined assumption that gender is a preformed, functionalist category, merely a question of preconstituted "generic" men and women, beings resulting from either biological or social sexual difference and playing out roles, but otherwise of no interest.

In a later book, Shapin (1994) does look closely at the exclusion of women, as well as of other categories of nonindependent persons, from the preserves of gentlemanly truth-telling that characterized the relations of civility and science in seventeenth-century England. As "covered" persons, subsumed under their husbands or fathers, women could not have the necessary kind of honor at stake. As Shapin noted, the "covered" status of women was patently social, not "biological," and understood to be such, irrespective of whatever beliefs a seventeenth-century man or woman might also hold about natural differences between the sexes.[6] Shapin saw no reason to posit that gender was at stake, or remade, by any of the processes that came together as the experimental way of life. The preexisting dependent status of women simply precluded their epistemological, and for the most part their physical, presence in the most important scenes of action in that period in the history of science. The issue was not whether women were intelligent or not. Boyle, for example, regarded his aristocratic sisters as his equal in intellectually demanding religious discussions. The issue was whether women had the independent status to be modest witnesses, and they did not. Technicians, who were physically present, were also epistemologically invisible persons in the experimental way of life; women were invisible in both physical and epistemological senses.

Shapin's questions are different from mine. He notes exclusions, but his focus is on other matters. In contrast, my focus in this chapter is to ask if gender, with all its tangled knots with other systems of stratified relationships, was at stake in key reconfigurations of knowledge and practice that constituted modern science. If Shapin perhaps erred in seeing only conservation, my excesses will be in the other direction.

There are several ways to contest Shapin's judgment that gender was merely conserved, and not redone, or at least hardened in consequential ways, in the seventeenth-century meeting of science and civility. In this regard, historians emphasize the critical role of the defeat of the hermetic tradition in the establishment of scientific mechanistic orthodoxy and the correlated devaluation of much that was gendered feminine (which did not necessarily have to do with real women) in science. The virulence of the witch hunts in Europe in the sixteenth and seventeenth centuries, and the involvement of men who saw themselves as rationalist founders of the new philosophy, testifies to the crisis in gender in that molten period in both knowledge and religion.[7] David Noble (1992:205–43) points out that the "disorderly" public activities of women in the period of religious and political turmoil before the Restoration, as well as women's association with the alchemical tradition, made wise gentlemen scramble to dissociate themselves from all things feminine, including oxymoronic independent women, after mid-century, if not before.

Shapin (1994:xxii) is openly sympathetic to efforts to foreground the voices and agencies of the excluded and silenced in history, but he is emphatic about the legitimacy of doing the history of what he only half jokingly calls "Dead White European Males" where their activities and ways of knowing are what mattered—and not just to themselves. I agree completely with Shapin's insistence on focusing on men, of whatever categories, when it is their doings that matter. Masculine authority, including the seventeenth-century gentlemanly culture of honor and truth, has been widely taken as legitimate by both men and women, across many kinds of social differentiation. It would not serve feminism to obscure this problem. I do not think Shapin or Shapin and Schaffer should have written their books about women; and besides, Shapin (1994) has a great deal that is interesting to say about the agencies of, among others, Boyle's aristocratic and pious sister in religious and domestic realms. Without focusing on "Dead White European Males" it would be impossible to understand gender at all, in science or elsewhere. However, what I think Shapin does not interrogate in his formulations was whether and how precisely the world of scientific gentlemen was *instrumental* in both sustaining old and in crafting new "gendered" ways of life. Insofar as the experimental way of life built the exclusion of actual women, as well as of cultural practices and symbols deemed feminine, into what could count as the truth in science, the air-pump was a technology of gender at the heart of scientific knowledge. It was the general absence, not the occasional presence, of women of whatever class or lineage/color—and the historically specific ways that the semiotics and psychodynamics of sexual difference worked—that gendered the experimental way of life in a particular way.

My question is, How did all this matter to what could count as knowledge in the rich tradition we know as science? Gender is always a relationship, not a preformed category of beings or a possession that one can have. Gender does not pertain more to women than to men. Gender is the relation between variously constituted categories of men and women (and variously arrayed tropes), differentiated by nation, generation, class, lineage, color, and much else. Shapin and Schaffer assembled all the elements to say something about how gender was one of the products of the air-pump; but the blind spot of seeing gender as women instead of as a relationship got in the way of the analysis. Perhaps Shapin in his later book is right that nothing very interesting happened to gender in the meeting of civility and science in the experimental way of life, with its practices of truth-telling. But I suspect that the way he asked his questions about excluded categories precluded having much to say about the two questions that vex me: (1) In what ways in the experimental way of life was gender in-the-making? (2) Did that matter or not, and how or how not, to what could count as reliable knowledge in science during and after the seventeenth century? How did gender-in-the-making become part of negotiating the continually vexed boundary between the "inside" and the "outside" of science? How did gender-in-the-making relate to establishing what counted as objective and subjective, political and technical, abstract and concrete, credible and ridiculous?

The effect of the missing analysis is to treat race and gender, at best, as a question of empirical, preformed beings who are present or absent at the scene of action but are not generically constituted in the practices choreographed in the new theaters of persuasion. This is a strange analytical aberration, to say the least, in a community of scholars who play games of epistemological chicken trying to beat each other in the game of showing how all the entities in technoscience are constituted *in* the action of knowledge production, not *before* the action starts.[8] The aberration matters, for, as David Noble argues in his synthesis on the effect of Western Christian clerical culture on the culture and practice of science, "any genuine concern about the implications of such a culturally distorted science-based civilization, or about the role of women within it, demands an explanation. For the male identity of science is no mere artifact of sexist history; throughout most of its evolution, the culture of science has not simply excluded women, it has been defined in defiance of women and their absence. . . . How did so strange a scientific culture emerge, one that proclaimed so boldly the power of the species while at the same time shrinking in horror from half the species?" (1992:xiv).

Elizabeth Potter, however, has a keen eye for how men became man in the practice of modest witnessing. Men-in-the-making, not men, or women, already made, is her concern. Gender was *at stake* in the experimental way of life, she argues, not predetermined. To develop this suspicion, she turns to the early-seventeenth-century English debates on the proliferation of genders in the practice of sexual cross-dressing. In the context of anxieties over gender manifested by early modern writers, she asks how Robert Boyle—urbane, celibate, and civil—avoided the fate of being labeled a *haec vir*, a feminine man, in his insistence on the virtue of modesty? How did the masculine practice of modesty, by appropriately civil (gentle)men, enhance agency, epistemologically and socially, while modesty enforced on (or embraced by) women of the same social class simply removed them from the scene of action? How did some men become transparent, self-invisible, legitimate witnesses to matters of fact, while most men and all women were made simply invisible, removed from the scene of action, either below stage working the bellows that evacuated the pump or offstage entirely? Women lost their security clearances very early in the stories of leading-edge science.

Women were, of course, literally offstage in early modern English drama, and the presence of men acting women's roles was the occasion for more than a little exploring and resetting of sexual and gender boundaries in the foundational settings of English drama in the sixteenth and seventeenth centuries. As the African American literary scholar Margo Hendricks (1992, 1994 and 1996) tells us, Englishness was also at stake in this period, for example, in Shakespeare's *Midsummer Night's Dream*.[9] And, she notes, the story of Englishness was part of the story of modern gendered racial formations, rooted still in lineage, civility, and nation, rather than in color and physiognomy. But the discourses of "race" that were cooked in this cauldron, which melted nations and bodies together in discourses on lineage, were more than a little useful throughout the following centuries for demarcating the differentially sexualized bodies of "colored" peoples around the world, locally and globally, from the always unstably consolidated subject positions of self-invisible, civil inquirers.[10] Gender and race never existed separately and never were about preformed subjects endowed with funny genitals and curious colors. Race and gender are about entwined, barely analytically separable, highly protean, *relational* categories. Racial, class, sexual, and gender formations (not essences) were, from the start, dangerous and rickety machines for guarding the chief fictions and powers of European civil manhood. To be unmanly is to be uncivil, to be dark is to be unruly: Those metaphors have mattered enormously in the constitution of what may count as knowledge.

Let us attend more closely to Potter's story. Medieval secular masculine virtue—noble manly valor—required patently heroic words and deeds. The modest man was a problematic figure for early modern Europeans, who still thought of nobility in terms of warlike battles of weapons and words.[11] Potter argues that in his literary and social technologies, Boyle helped to construct the new man and woman appropriate to the experimental way of life and its production of matters of fact. "The new man of science had to be a chaste, modest, heterosexual man who desires yet eschews a sexually dangerous yet chaste and modest woman" (2001).[12] Female modesty was of the body; the new masculine virtue had to be of the mind. This modesty was to be the key to the gentleman-scientist's trustworthiness; he reported on the world, not on himself. Unadorned "masculine style" became English national style, a mark of the growing hegemony of the rising English nation. An unmarried man in Puritan England, which valued marriage highly, Boyle pursued his discourse on modesty in the context of the vexed *hic mulier/haec vir* (masculine woman/feminine man) controversies of the late sixteenth and early seventeenth centuries. In that anxious discourse, when gender characteristics were transferred from one sex to another, writers worried that third and fourth sexual kinds were created, proliferating outside all bounds of God and Nature. Boyle could not risk his modest witness's being *a haec vir*. God forbid the experimental way of life have queer foundations.

Two additional taproots for the masculinity of Boyle's brand of modesty exist: the King Arthur narratives and the clerical monastic Christian tradition. Bonnie Wheeler (1992) argues that the first reference to the Arthur figure in the sixth century referred to him as a *vir modestus*, and the qualifier followed Arthur through his many literary incarnations. This tradition was probably culturally available to Boyle and his peers looking for effective new models of masculine reason. *Modestus* and *modestia* referred to measure, moderation, solicitude, studied equilibrium, and reticence in command. This constellation moves counter to the dominant strand of Western heroism, which emphasizes self-glorification by the warrior hero. The *vir modestus* was a man characterized by high status and disciplined ethical restraint. *Modestia* linked high class, effective power, and masculine gender. Wheeler finds in the King Arthur figure "one alternative norm of empowered masculinity for post-heroic culture" (1992:1).

David Noble emphasizes the reappropriation of clerical discourse in a Royal Society sanctioned by crown and church. "As an exclusively male retreat, the Royal Society represented the continuation of the clerical culture, now reinforced by what may be called a scientific asceticism" (Noble 1992:231). The kind of gendered self-renunciation practiced in this masculine domain was precisely the kind that enhanced epistemological-spiritual potency. Despite the importance of marriage in the Protestant Reformation's attack on the Catholic church, even celibacy in the experimental way of life was praised by lay Puritans of the early Restoration, and especially by Robert Boyle, who served as a model of the new scientist. Potter quotes Boyle's praise of male chastity in the context of man's right to a priesthood rooted in reason and knowledge of the natural world. As Potter puts it, female chastity served male chastity, which allowed men to serve God undistractedly through experimental science. For Boyle, "the laboratory has become the place of worship; the scientist, the priest; the experiment, a religious rite" (Potter 2001).

Within the conventions of modest truth-telling, women might watch a demonstration; they could not witness it. The definitive demonstrations of the working of the air-pump had to take place in proper civil public space, even if that meant holding a serious demonstration late at night to exclude women of his class, as Boyle did. For example, reading Boyle's *New Experiments Physico-Mechanical Touching the Spring of the*

Air, which describes experiments with the air-pump, Potter recounts a demonstration attended by high-born women at which small birds were suffocated by the evacuation of the chamber in which the animals were held. The ladies interrupted the experiments by demanding that air be let in to rescue a struggling bird. Boyle reports that to avoid such difficulties, the men later assembled at night to conduct the procedure and attest to the results. Potter notes that women's names were never listed among those attesting the veracity of experimental reports, whether they were present or not. Several historians describe the tumult caused in 1667 at the Royal Society when Margaret Cavendish (1623–1673), Duchess of Newcastle, generous patron of Cambridge University, and a substantive writer on natural philosophy who intended to be taken seriously, requested permission to visit a working session of the all-male society.[13] Not wanting to offend an important personage, "the leaders of the society ultimately acceded to her request, arranging for her to visit several scientific demonstrations by, among others, Hooke and Boyle" (Noble 1992:231). There was no return visit, and the first women admitted to the Royal Society, after lawyers' advice made it clear that continued exclusion of women would be illegal, entered in 1945, almost 300 years after Cavendish's unwelcome appearance.[14]

Enhancing their agency through their masculine virtue exercised in carefully regulated "public" spaces, modest men were to be self-invisible, transparent, so that their reports would not be polluted by the body. Only in that way could they give credibility to their descriptions of other bodies and minimize critical attention to their own. This is a crucial epistemological move in the grounding of several centuries of race, sex, and class discourses as objective scientific reports.[15]

All of these highly usable discourses feed into the conventions of masculine scientific modesty, whose gendering came to be more and more invisible (transparent) as its masculinity seemed more and more simply the nature of any non-dependent, disinterested truth-telling. The new science redeemed Boyle's celibate, sacred-secular, and nonmartial man from any gender confusion or multiplicity and made him a modest witness as the type specimen of modern heroic, masculine action— of the mind. Depleted of epistemological agency, modest women were to be invisible to others in the experimental way of life. The kind of visibility—the body—that women retained glides into being perceived as "subjective," that is, reporting only on the self, biased, opaque, not objective. Gentlemen's epistmological agency involved a special kind of transparency. Colored, sexed, and laboring persons still have to do a lot of work to become similarly transparent to count as objective, modest witnesses to the world rather than to their "bias" or "special interest." To be the object of vision, rather than the "modest," self-invisible source of vision, is to be evacuated of agency.[16]

The self-invisibility and transparency of Boyle's version of the modest witness—that is, the "independence" based on power and on the invisibility of others who actually sustain one's life and knowledge—are precisely the focus of late-twentieth-century feminist and multiculultural critique of the limited, biased forms of "objectivity" in technoscientific practice, insofar as it produces itself as "the culture of no culture." Antiracist feminist science studies revisit what it meant, and means, to be "covered" by the modest witnessing of others who, because of their special virtue, are themselves transparent. "In the beginning," the exclusion of women and laboring men was instrumental to managing a critical boundary between watching and witnessing, between who is a scientist and who is not, and between popular culture and scientific fact. I am not arguing that the doings of Boyle and the Royal Society are the whole story in crafting modern experimental and theoretical science; that would be ridiculous. Also, I am at least as invested in the continuing need for stabilizing contingent matters of fact to ground serious claims on each other as any child of the Scientific Revolution could be. I am using the story of Boyle and the experimental way of life as a figure for technoscience; the story stands for more than itself. My claim is double: (1) There have been practical inheritances, which have undergone many reconfigurations but which remain potent; and (2) the stories of the Scientific Revolution set up a narrative about "objectivity" that continues to get in the way of a more adequate, self-critical technoscience committed to situated knowledges. The important practice of credible witnessing is still at stake.

A further central issue requires compressed comment: the structure of heroic action in science. Several scholars have commented on the proliferation of violent, misogynist imagery in many of the chief documents of the Scientific Revolution.[17] The modest man had at least a tropic taste for the rape of nature. Science made was nature undone, to embroider on Bruno Latour's (1987) metaphors in his important *Science in Action*. Nature's coy resistance was part of the story, and getting nature to reveal her secrets was the prize for manly valor—all, of course, merely valor of the mind. At the very least, the encounter of the modest witness with the world was a great trial of strength. In disrupting many conventional accounts of scientific objectivity, Latour and others have masterfully unveiled the self-invisible modest man. At the least, that is a nice twist on the usual direction of discursive unveiling and heterosexual epistemological erotics.[18] In *Science, the Very Idea!* Steve Woolgar (1988) keeps the light relentlessly on this modest being, the "hardest case" or "hardened self" that covertly guarantees the truth of a representation, which ceases magically to have the status of a representation and emerges simply as the fact of the matter. That crucial emergence depends on many kinds of transparency in the grand narratives of the experimental way of life. Latour and others eschew Woolgar's relentless insistence on reflexivity, which seems not to be able to get beyond self-vision as the cure for self-invisibility. The disease and the cure seem to be practically the same thing, if what you are after is another kind of world and worldliness. Diffraction, the production of difference patterns, might be a more useful metaphor for the needed work than reflexivity.

Latour is generally less interested than his colleague in forcing the Wizard of Oz to see himself as the linchpin in the technology of scientific representation. Latour wants to follow the action in science-in-the-making. Perversely, however, the structure of heroic action is only intensified in this project—both in the narrative of science and in the discourse of the science studies scholar. For the Latour of *Science in Action*, technoscience itself is war, the demiurge that makes and unmakes worlds.[19] Privileging the younger face as science-in-the-making, Latour adopts as the figure of his argument the double-faced Roman god, Janus, who, seeing both ways, presides over the beginnings of things. Janus is the doorkeeper of the gate of heaven, and the gates to his temple in the Roman Forum were always open in time of war and closed in times of peace. War is the great creator and destroyer of worlds, the womb for the masculine birth of time. The action in science-in-the-making is all trials and feats of strength, amassing of allies, forging of worlds in the strength and numbers of forced allies. All action is agonistic; the creative abstraction is both breathtaking and numbingly conventional. Trials of strength decide whether a representation holds or not. Period. To compete, one must either have a counterlaboratory capable of winning in these high-stakes trials of force or give up dreams of making worlds. Victories and performances are the action sketched in this seminal book. "The list of trials becomes a thing; it is literally reified" (Latour 1987:92).

This powerful tropic system is like quicksand. *Science in Action* works by relentless, recursive mimesis. The story told is told by the same story. The object studied and the method of study mime each other. The analyst and the analysand all do the same thing, and the reader is sucked into the game. It is the only game imagined. The goal of the book is "penetrating science from the outside, following controversies and accompanying scientists up to the end, being slowly led out of science in the making" (15). The reader is taught how to resist both the scientist's and the false science studies scholar's recruiting pitches. The prize is not getting stuck in the maze but exiting the space of technoscience a victor, with the strongest story. No wonder Steven Shapin began his review of this book with the gladiator's salute: "Ave, Bruno, morituri te salutant" (1988:533).

So, from the point of view of some of the best work in mainstream science studies of the late 1980s, "nature" is multiply the feat of the hero, more than it ever was for Boyle. First, nature is a materialized fantasy, a projection whose solidity is guaranteed by the self-invisible representor. Unmasking this figure, s/he who would not be hoodwinked by the claims of philosophical realism and the ideologies of disembodied scientific objectivity fears to "go back" to nature, which was never anything but a projection in the first place. The projection nonetheless tropically works as a dangerous female threatening manly knowers. Then, another kind of nature is the result of trials of strength, also the fruit of the hero's action. Finally, the scholar too must work as a warrior, testing the strength of foes and forging bonds among allies, human and nonhuman, just as the scientist-hero does. The self-contained quality of all this is stunning. It is the self-contained power of the culture of no culture itself, where all the world is in the sacred image of the Same. This narrative structure is at the heart of the potent modern story of European autochthony.[20]

What accounts for this intensified commitment to virile modesty? I have two suggestions. First, failing to draw from the understandings of semiotics, visual culture, and narrative practice coming specifically from feminist, post-colonial, and multicultural oppositional theory, many science studies scholars insufficiently examine their basic narratives and tropes. In particular, the "self-birthing of man," "war as his reproductive organ," and "the optics of self-origination" narratives that are so deep in Western philosophy and science have been left in place, though so much else has been fruitfully scrutinized. Second, many science studies scholars, like Latour, in their energizing refusal to appeal to society to explain nature, or vice versa, have mistaken other narratives of action about scientific knowledge production as functionalist accounts appealing in the tired old way to preformed categories of the social, such as gender, race, and class. Either critical scholars in antiracist, feminist cultural studies of science and technology have not been clear enough about racial formation, gender-in-the-making, the forging of class, and the discursive production of sexuality *through the constitutive practices of technoscience production themselves,* or the science studies scholars aren't reading or listening—or both. For the oppositional critical theorists, both the facts and the witnesses are constituted in the encounters that are technoscientific practice. Both the subjects and objects of technoscience are forged and branded in the crucible of specific, located practices, some of which are global in their location. In the intensity of the fire, the subjects and objects regularly melt into each other. It is past time to end the failure of mainstream and oppositional science studies scholars to engage each other's work. Immodestly, I think the failure to engage has not been symmetrical.

Let me close this meditation on figures who can give credible testimony to matters of fact by asking how to queer the modest witness this time around so that s/he is constituted in the furnace of technoscientific practice as a self-aware, accountable, anti-racist FemaleMan, one of the proliferating, uncivil, late-twentieth-century children of the early modern *haec vir* and *hic mulier*. Like Latour, the feminist philosopher of science Sandra Harding is concerned with strength, but of a different order and in a different story. Harding (1992) develops an argument for what she calls "strong objectivity" to replace the flaccid standards for establishing matters of fact instaurated by the literary, social, and material technologies inherited from Boyle. Scrutiny of what constitutes "independence" is fundamental. "A stronger, more adequate notion of objectivity would require methods for systematically examining all of the social values shaping a particular research process, not just those that happen to differ between members of a scientific community. Social communities, not either individuals, or 'no one at all,' should be conceptualized as the 'knowers' of scientific knowledge claims. Culture-wide beliefs that are not critically examined within scientific processes end up functioning as evidence for or against hypotheses" (Harding 1993:18).

Harding maintains that democracy-enhancing projects and questions are most likely to meet the strongest criteria for reliable scientific knowledge-production, with built-in critical reflexivity. That is a hope in the face of, at best, ambiguous evidence. It is a hope that needs to be made into a fact by practical work. Such labor would reconstitute the relationships we call gender, race, nation, species, and class in unpredictable ways. Such reformed semiotic, technical, and social practice might be called, after Deborah Heath's term for promising changes in standards for building knowledge in a molecular biology she studies ethnographically, "modest interventions" (1997).

So, agreeing that science is the result of located practices at all levels, Harding concurs with Woolgar that reflexivity is a virtue the modest witness needs to cultivate. But her sense of reflexivity is closer to my sense of diffraction and to Heath's modest interventions than it is to Woolgar's rigorous resistance to making strong knowledge claims. The point is to make a difference in the world, to cast our lot for some ways of life and not others. To do that, one must be in the action, be finite and dirty, not transcendent and clean. Knowledge-making technologies, including crafting subject positions and ways of inhabiting such positions, must be made relentlessly visible and open to critical intervention. Like Latour,

Harding is committed to science-in-the-making. Unlike the Latour of *Science in Action*, she does not mistake the constituted and constitutive practices that generate and reproduce systems of stratified inequality—and that issue in the protean, historically specific, marked bodies of race, sex, and class—for preformed, functionalist categories. I do not share her occasional terminology of macrosociology and her all-too-self-evident identification of the social. But I think her basic argument is fundamental to a different kind of strong program in science studies, one that really does not flinch from an ambitious project of symmetry that is committed as much to knowing about the people and positions from which knowledge can come and to which it is targeted as to dissecting the status of knowledge made.

Critical reflexivity, or strong objectivity, does not dodge the world-making practices of forging knowledges with different chances of life and death built into them. All that critical reflexivity, diffraction, situated knowledges, modest interventions, or strong objectivity "dodge" is the double-faced, self-identical god of transcendent cultures of no culture, on the one hand, and of subjects and objects exempt from the permanent finitude of engaged interpretation, on the other. No layer of the onion of practice that is technoscience is outside the reach of technologies of critical interpretation and critical inquiry about positioning and location; that is the condition of articulation, embodiment, and mortality. The technical and the political are like the abstract and the concrete, the foreground and the background, the text and the context, the subject and the object. As Katie King (1993) reminds us, following Gregory Bateson, these are questions of pattern, not of ontological difference. The terms pass into each other; they are shifting sedimentations of the one fundamental thing about the world—relationality. Oddly, embedded relationality is the prophylaxis for both relativism and transcendence. Nothing comes without its world, so trying to know those worlds is crucial. From the point of view of the culture of no culture, where the wall between the political and the technical is maintained at all costs, and interpretation is assigned to one side and facts to the other, such worlds can never be investigated. Strong objectivity insists that both the objects and the subjects of knowledge-making practices must be located. Location is not a listing of adjectives or assigning of labels such as race, sex, and class. Location is not the concrete to the abstract of decontextualization. Location is the always partial, always finite, always fraught play of foreground and background, text and context, that constitutes critical inquiry. Above all, location is not self-evident or transparent.

Location is also partial in the sense of being *for* some worlds and not others. There is no way around this polluting criterion for strong objectivity. Sociologist and ethnographer Susan Leigh Star (1991) explores taking sides in a way that is perhaps more readily heard by science studies scholars than Harding's more conventional philosophical vocabulary. Star is interested in taking sides with some people or other actors in the enrollments and alliance formations that constitute so much of technoscientific action. Her points of departure are feminist and symbolic interactionist modes of inquiry that privilege the kind of witness possible from the point of view of those who suffer the trauma of not fitting into the standard. Not to fit the standard is another kind of oxymoronically opaque transparency or invisibility: Star would like to see if this kind is conducive to crafting a better modest witness. Not fitting a standard is not the same thing as existing in a world without that standard. Instructed by the kinds of multiplicity that result from exposure to violence, from being outside a powerful norm, rather than from positions of independence and power, Star is compelled by the starting point of the monster, of what is exiled from the clean and light self. And so she suspects that the "voices of those suffering from the abuses of technological power are among the most powerful analytically" (Star 1991:30).

Star's own annoying but persistent allergy to onions, and the revealing difficulty of convincing service people in restaurants that such a condition is real, is her narrative wedge into the question of standardization. In order to address questions about power in science and technology, Star looks at how standards produce invisible work for some while clearing the way for others, and at how consolidated identities for some produce marginalized locations for others. She adopts what she calls a kind of "cyborg" point of view: Her "cyborg" is the "relationship between standardized technologies and local experience," where one falls "between the categories, yet in relationship to them" (39).

Star thinks "that it is both more analytically interesting and more politically just to begin with the question *cui bono*, than to begin with a celebration of the fact of human/non-human mingling" (43). She does not question the fact of the implosion of categorical opposites; she is interested in who lives and dies in the force fields generated. "Public" stability for some is "private" suffering for others; self-invisibility for some comes at the cost of public invisibility for others. They are "covered" by what is conventionally made to be the case about the world. I think that such coverings reveal the grammatical structure of "gender," "race," "class," and similar clumsy categorical attempts to name how the world is experienced by the nonstandard, who nonetheless are crucial to the technologies of standardization and others' ease of fitting.

In Star's account, we are all members of many communities of practice. Multiplicity is in play with questions of standardization, and no one is standard or ill-fitted in all communities of practice. Some kinds of

Fig. 7.1. Millennial Children, Lynn Randolph, oil on canvas, 58″ × 72″, 1992.

Stalked by hyenas and mocked by a dancing clown-devil with a leering face mask for a stomach, two embracing girls kneel on the flaming ground outside the burning city of Houston on the banks of an oil-polluted bayou. Facing the viewer, these millennial children ask if there can still be a future on this earth. Vultures perch on the limbs of a blasted tree, its roots miming the bird feet of the cavorting demon, whose stomach is a portrait of George Bush. Smoking towers of a nuclear power plant loom in the background and a Stealth bomber dives toward the ground out of a lightning-scorched sky. Reds, blacks, and slashing yellows dominate the large canvas, relieved by the sepia flesh and pastel dresses of the children and the greens of the not-yet-burned bushes. The girls are whole, firm, and flanked by diminutive guardian angels. Sober in their regard the children are not destroyed, but they are menaced by the apocalypse that engulfs the world. They are in the dangerous borderlands between reality and nightmare, between the comprehensive futurelessness that is only a dire possibility and the blasted futures of hundreds of millions of children that are a fierce reality now. These are the children whose witness calls the viewer to account for both the stories and the actualities of the millennium.

standardization matter more than others, but all forms work by producing those that do not fit as well as those who do. Inquiry about technoscience from the point of view of Star's monsters does not necessarily focus on those who do not fit, but rather on the contingent material-semiotic articulations that bring such ill-fitting positions into being and sustain them. Star's monsters also ask rather uncivilly how much it costs, and who pays, for some to be modest witnesses in a regime of knowledge-production while others get to watch. And monsters in one setting set the norm in others; innocence and transparency are not available to feminist modest witnesses.

Double vision is crucial to inquiring into the relations of power and standards that are at the heart of the subject- and object-making processes of technoscience. Where to begin and where to be based are the fundamental questions in a world in which "power is about *whose* metaphor brings worlds together" (Star 1991:52). Metaphors are tools and tropes. The point is to learn to remember that we might have been otherwise, and might yet be, as a matter of embodied fact. Being allergic to onions is a niggling tropic irritant to the scholarly temptation to forget one's own complicity in apparatuses of exclusion that are constitutive to what may count as knowledge. Fever, nausea, and a rash can foster a keen appreciation of located knowledges.

So I close this evocation of the figure of the modest witness in the narrative of science with the hope that the technologies for establishing what may count as the case about the world may be rebuilt to bring the technical and the political back into realignment so that questions about possible livable worlds lie visibly at the heart of our best science.

SECOND MILLENNIUM

They did not know for sure, but they suspected that the dances were beyond nasty because the music was getting worse and worse with each passing season the Lord waited to make Himself known.
—Toni Morrison, *Jazz*

I have not written a narrative Leviathan. Did you really want another one?
—Sharon Traweek, "Border Crossings"

From a millennarian perspective, things are always getting worse. Evidence of decay is exhilarating and mobilizing. Oddly, belief in advancing disaster is actually part of a trust in salvation, whether deliverance is expected by sacred or profane revelations, through revolution, dramatic scientific breakthroughs, or religious rapture. For example, for radical science activists like me, the capitalist commodification of the dance of life is always advancing ominously; there is always evidence of nastier and nastier technoscience dominations. An emergency is always at hand, calling for the need for transformative politics. For my twins, the true believers in the church of science, a cure for the trouble at hand is always promised. That promise justifies the sacred status of scientists, even, or especially, outside their domains of practical expertise. Indeed, the *promise* of technoscience is, arguably, its principal social weight. Dazzling promise has always been the underside of the deceptively sober pose of scientific rationality and modern progress within the culture of no culture. Whether unlimited clean energy through the peaceful atom, artificial intelligence surpassing the merely human, an impenetrable shield from the enemy within or without, or the prevention of aging ever materializes is vastly less important than always living in the time zone of amazing promises. In relation to such dreams, the impossibility of ordinary materialization is intrinsic to the potency of the promise. Disaster feeds radiant hope and bottomless despair, and I, for one, am satiated. We pay dearly for living within the chronotope of ultimate threats and promises.

Literally, *chronotope* means topical time, or a *topos* through which temporality is organized. A topic is a commonplace, a rhetorical site. Like both place and space, time is never "literal," just there; *chronos* always intertwines with *topos*, a point richly theorized by Bakhtin (1981) in his concept of the chronotope as a figure that organizes temporality. Time and space organize each other in variable relationships that show any claim to totality, be it the New World Order, Inc., the Second Millennium, or the modern world, to be an ideological gambit linked to struggles to impose bodily / spatial / temporal organization. Bakhtin's concept requires us to enter the contingency, thickness, inequality, incommensurability, and dynamism of cultural systems of reference through which people enroll each other in their realities. Bristling with ultimate threats and promises, drenched with the tones of the apocalyptic and the comic, the gene and the computer both work as chronotopes throughout *Modest_Witness@Second_Millennium.*

So, replete with such costs, the Second Millennium is this book's space-time machine; it is the machine that circulates the figures of the modest witness, the FemaleMan, and OncoMouse in a common story. The air-pump is itself a chronotope closely related to my mechanical-millennial address. Both machines have to do with a narrative space-time frame associated with millennarian hopes for new foundations. The air-pump was an actor in the drama of the Scientific Revolution. The device's potent agency in civil matters and its capacity to bear witness exceeded that of most of the humans who attended its performances and looked after its functioning. Those humans to whom could be attributed a power of agency approaching that of the air-pump and its progeny over the next centuries had to disguise themselves as its ventriloquists. Their subjectivity had to become their objectivity, guaranteed by their close kinship with their machines. Inhabiting the culture of no culture, these modest witnesses were transparent spokesmen, pure mediums transmitting the objective word made flesh as facts. These humans were self-invisible witnesses to matters of fact, the new world's guarantors of objectivity. The narrative frames of the Scientific Revolution were a kind of time machine that situated subjects and objects into dramatic pasts, presents, and futures.

If belief in the stable separation of subjects and objects in the experimental way of life was one of the defining stigmata of modernity, the implosion of subjects and objects in the entities populating the world at the end of the Second Millennium—and the broad recognition of this implosion in both technical and popular cultures—are stigmata of another historical configuration. Many have called this configuration "postmodern." Suggesting instead the notion of the "metamodern" for the current moment, Paul Rabinow (1992) rejects the "postmodern" label for two main reasons: (1) Foucault's three axes of the modern episteme—life, labor, and language—are all still very much in play in current knowledge-power configurations; and (2) the collapse of metanarratives that is supposed to be diagnostic of postmodernism is nowhere in evidence in either technoscience or transnational capitalism. Rabinow is correct about both of these important points, but for my taste he does not pay enough attention to the implosion of subjects and objects, culture and nature, in the warp fields of current biotechnology and communications and computer sciences as well as in other leading domains of technoscience. This implosion issuing in a wonderful bestiary of cyborgs is different from the *cordon sanitaire* erected between subjects and objects by Boyle and reinforced by Kant. It is not just that objects, and nature, have been shown to be full of labor, an insight insisted on most powerfully in the last century by Marx, even if many current science studies scholars have forgotten his priority here. More pregnantly, in the wombs of technoscience, as well as of postfetal science studies, chimeras of humans and nonhumans, machines and organisms, subjects and objects, are the obligatory passage points, the embodiments and articulations, through which travelers must pass to get much of anywhere in the world. The chip, gene, bomb, fetus, seed, brain, ecosystem, and database are the wormholes that dump contemporary travelers out into contemporary worlds. These chimeras are not close cousins of the air-pump, although the air-pump is one of their distant ancestors.

Instead, entities like the chip, gene, bomb, fetus, seed, brain, ecosystem, and database are more like OncoMouse™. And those who attest to matters of fact are less like Boyle's modest man than they are like the FemaleMan© . We will meet both of these genetically strange, inflected, proprietary beings soon, as they are made to encounter each other and discover their kinship. Bruno Latour (1993) suggested the useful notion of the *amodern* for the netherlands in which the really interesting chimeras of humans and nonhumans gestate. But, for my taste, he still sees too much continuity in the late twentieth century with Boyle's practice. I think something is going on in the world vastly different from the constitutional arrangements that established the separations of nature and society proper to "modernity," as early modern Europeans and their offspring understood that historical configuration; and recent technoscience is at the heart of the difference. Instead of naming this difference—postmodern, metamodern, amodern, late modern, hyper-modern, or just plain generic Wonder Bread modern—I give the reader an e-mail address, if not a password, to situate things in the net.

But, obviously, I did not name my e-mail address innocently. I am appealing to the disreputable history of Christian realism and its practices of figuration; and I am appealing to the love/hate relation with apocalyptic disaster-and-salvation stories maintained by people who have inherited the practices of Christian realism, not all of whom are Christian, to say the least. Like people allergic to onions eating at McDonald's, we are forced to live, at least in part, in the material-semiotic system of measure connoted by the Second Millennium, whether or not we fit that story. Following Eric Auerbach's arguments in *Mimesis* (1953), I consider figures to be potent, embodied—incarnated, if you will—fictions that collect up the people in a story that tends to fulfillment, to an ending that redeems and restores meaning in a salvation history. After the wounding, after the disaster, comes the fulfillment, at least for the elect; God's scapegoat has promised as much. I think contemporary technoscience in the United States is deeply engaged in producing such stories, slightly modified to fit the conventions of secular realism.

In that sense the "human genome" in current biotechnical narratives regularly functions as a figure in a salvation drama that promises the fulfillment and restoration of human nature. As a symptomatic example, consider a short list of titles of articles, books, and television programs in the popular and official science press about the Human Genome Project to map and sequence all of the genes on the 46 human chromosomes: "Falling Asleep over the Book of Life," "Genetic Ark," "Gene Screening: A Chance to Map our Body's Future," "Genesis, the Sequel," "James Watson and the Search for the Holy Grail," "A Guide to Being Human," "Thumbprints in Our Clay," "In the Beginning Was the Genome," "A Worm at the Heart of the Genome Project," "Genetics and Theology: A Complementarity?" "Huge Undertaking—Goal: Ourselves," "The Genome Initiative: How to Spell 'Human,' " "Blueprint for a Human," *The Code of Codes, Gene Dreams, Generation Games, Mapping the Code, Genome*, and, finally, on the BBC and NOVA television,

"Decoding the Book of Life." Genes are a bit like the Eucharist of biotechnology. Perhaps that insight will make me feel more reverent about genetically engineered food.

Instrinsic to placing my modest witnesses in a conventional millennarian machine is the evocation of the impending time of tribulations. There is no shortage of such narratives of disasters in the technical and popular cultures of technoscience. The time machine of the Second Millennium churns out expectations of nuclear catastrophe, global economic collapse, planetary pandemics, ecosystem destruction, the end of nurturing families, private ownership of the commons of the human genome, and many other kinds of silent springs. Of course, just as within any other belief system, all these things look eminently real, eminently possible, perhaps even inevitable, once we inhabit the chronotope that tells the

◄——————————————————————————————

Fig. 7.2. The Laboratory, or The Passion of OncoMouse, Lynn Randolph, oil on masonite, 10″ × 7″, 1994.

Modest_Witness@Second_Millennium FemaleMan©_Meets_OncoMouse™ was revised, literally, under the portrait of The Laboratory, or The Passion of OncoMouse. Set in the simultaneously globally distributed and parochial timescape of the end of the Second Christian Millennium, this is a book about the figurations, tools, tropes, and articulations of technoscience as I have lived it in the United States in the 1990s. The biotechnical, biomedical laboratory animal is one of the key figures inhabiting my book, world, and body. Figures cohabit with subjects and objects inside stories. Figures take up and transform selves. Lynn Randolph painted her transspecific human-mouse hybrid in response to the first draft of "Mice into Wormholes." That paper examines sticky threads extruding from the natural-technical body of the world's first patented animal—OncoMouse™, a breast cancer research model produced by genetic engineering. As a model, the transgenic mouse is both a trope and a tool that reconfigures biological knowledge, laboratory practice, property law, economic fortunes, and collective and personal hopes and fears. In Randolph's rendering, the white, female, breast-endowed, transspecific, cyborg creature is crowned with thorns. She is a Christ figure, and her story is that of the passion. She is a figure in the sacred-secular dramas of technoscientific salvation history, with all of the disavowed links to Christian narrative that pervade U.S. scientific discourse. The laboratory animal is sacrificed: her suffering promises to relieve our own; she is a scapegoat and a surrogate. She is the object of transnational technoscientific surveillance and scrutiny, the center of a multicolored optical drama. Her passion transpires in a box that mimes the chamber of the air-pump in Robert Boyle's house in seventeenth-century England. Small animals expired in that experimental chamber to show to credible witnesses the workings of the vacuum air-pump so that contingent matters of fact might ground less deadly social orders. This mouse is a figure in secularized Christian salvation history and in the linked narratives of the Scientific Revolution and the New World Order—with their promises of progress; cures; profit; and, if not of eternal life, then at least of life itself. Randolph's OncoMouse invites reflection on the terms and mechanism of these noninnocent genetic stories. Her figure invites those who inhabit this book to take up and reconfigure technoscientific tools and tropes in order to practice the grammar of a mutated experimental way of life that does not issue in the New World Order, Inc.

story of the world that way. I am not arguing that such threats aren't threatening. I am simply trying to locate the potency of such "facts" about the contemporary world, which is so enmeshed in technoscience, with its threats and its promises. There is no way to rationality—to actually existing worlds—outside stories, not for our species, anyway. This book, like all of my writing, is anxious much more than it is optimistic. I am not arguing for complacency when I list the narrative setup of threats and promises, only for taking seriously that no one exists in a culture of no culture, including the critics and prophets as well as the technicians. We might profitably learn to doubt our fears and certainties of disasters as much as our dreams of progress. We might learn to live without the bracing discourses of salvation history. We exist in a sea of powerful stories: They are the condition of finite rationality and personal and collective life histories. There is no way out of stories; but no matter what the One-Eyed Father says, there are many possible structures, not to mention contents, of narration. Changing the stories, in both material and semiotic senses, is a modest intervention worth making. Getting out of the Second Millennium to another e-mail address is very much what I want for all mutated modest witnesses.

NOTES

1. Commerce is a variant of conversation, communication, intercourse, passage. As any good economist will tell you, commerce is a procreative act.
2. Traweek was studying the legitimate sons of Robert Boyle; her physicists' detector devices are the mechanical descendants of his air-pump as well. Humans and nonhumans have progeny in the odd all-masculine reproductive practices of technoscience. "I have presented an account of how high energy physicists construct their world and represent it to themselves as free of their own agency, a description, as thick as I could make it, of an extreme culture of objectivity: a culture of no culture, which longs passionately for a world without loose ends, without temperament, gender, nationalism, or other sources of disorder—for a world outside human space and time" (Traweek 1988:162).
3. Of course, what counts as a warrant for disinterestedness, or lack of bias, changes historically. Shapin (1994:409–17) stresses the difference between the face-to-face, gentlemanly standards for assessing truth telling in seventeenth-century England and the anonymous, institutionally and professionally warranted practices of science in the twentieth century. Inside concrete laboratories, however, Shapin suggests that members of the community based on face-to-face interactions continue to assess credibility in ways Robert Boyle would have understood. Part of the problem scientists face today is legitimation of their criteria in the eyes of "outsiders." One of my goals in this book is to trouble what counts as insiders and outsiders in setting standards of credibility and objectivity. "Disinterested" cannot be allowed to mean "dislocated"; i.e., unaccountable for, or unconscious of, complex layers of one's personal collective historical situatedness in the apparatuses for the production of knowledge. Nor can "politically committed" be allowed to mean "biased." It is a delicate distinction, but one fundamental to hopes for democratic and credible science. Etzkowitz and Webster (1995)

discuss how the "norms of science," and so of what counts as objective, have changed during the twentieth century in the United States. For example, in molecular biology university-based investigators formerly doing tax- and foundation-supported "pure science," which semiotically warranted their credibility and disinterestedness, as the grants economy eroded became much more closely tied to corporations, where intellectual property and science implode. Perhaps some of the anxiety about objectivity in the "science wars"—in which science studies scholars, feminist theorists, and the like are seen as threatening broad-based belief in scientific credibility and objectivity through their irresponsible "perspectivalism" and "relativism"—should really be traced to transformed standards of disinterestedness among scientists themselves. See especially the attacks by Gross and Levitt (1994).

4. Shapin (1994) writes almost exclusively about the social technology for warranting credibility. He analyzes the transfer of the code of gentlemanly honor, based on the *independence* of the gentleman, that man of means who owes no one anything but the truth, from established social regions to a new set of practices—experimental science. The most original contribution of Shapin and Schaffer (1985) is their analysis of the weave of all three technologies, and especially of the heart of the experimental life form—the sociotechnical apparatus that built and sustained the air-pump, which I take to be metonymic for the technoscientific instrument in general.

5. Potter (2001). In writing this chapter, I worked from an earlier manuscript version of Potter's paper in which she discussed the *hic mulier/haec vir* controversy from the 1570s through 1620 in the context of gender anxieties evident in English Renaissance writers, and extending to Boyle and other post-Restoration authors. Therefore, I do not give page numbers. Potter relied on Woodbridge (1984).

6. On that topic, see Schiebinger (1989) and Laqueur (1990). "Biological" sexual difference is my own anachronistic adjective in this sentence.

7. See Merchant and Easlea (1980).

8. See the series of essays and counteressays that begins with Collins and Yearley's (1992:301–26) "Epistemological Chicken." Bruno Latour, Steve Woolgar, and Michel Callon were the other combatants, some better humored than others. The stakes were what got to count as the really real.

9. Hendricks (1996 and 1994). *A Midsummer Night's Dream* was composed about 1600.

10. Exploring how "race" was constructed in early modern England, Boose (1994) cautions against hearing twentieth-century meanings of color in sixteenth- and seventeenth-century writing. Boose argues that the almost unrepresentable narrative of love and sexual union between a dark African woman and an English man, tied to European patriarchal questions about lineage and the fidelity of transmission of the image of the father; was an important node in the production of modern race discourse. Inflected also by discourse on Jews and on the Irish, English constitutions of race were changing across the seventeenth century, not unlinked to the fact that by mid-century, "England would be competing with the Dutch for the dubious distinction of being the world's largest slave trader" (1994:40). These issues are vastly understudied in accounting for the shapes taken by early modern science.

11. The ambiguities and tensions between the two chief aristocratic and gentlemanly qualities, civility and heroic virtue, should be examined in the context of the experimental way of life in this period. Shapin (1994) assembles compelling evidence about the nature and importance of civility for establishing truth-telling.

12. Because the published page numbers will differ, I omit page references to both Potter's manuscript and forthcoming paper.

13. Schiebinger (1989:25–26); Noble (1992:230–31); Potter (2001).

14. See Rose (1994:115–35) for the story of women in England's Royal Society.

15. "From this perspective the proper subject of gender and science thus becomes the analysis of the web of forces that supports the historic conjunction of science and masculinity, and the equally historic disjunction between science and femininity. It is, in a word, the conjoint making of 'men,' 'women,' and 'science' " (Keller 1990:74). If "gender" here means "kind," and thus includes *constitutively* the complex lineages of racial, sexual, class, and national formations in the production of differentiated men, women, and science, I could not agree more.

16. Recall the trope of the eye of God in Linnaeus's vision of the second Adam as the authorized namer of the new plants and animals revealed by eighteenth-century explorations. Nature can be seen and warranted; it is not the witness to itself. This narrative epistemological point is part of the apparatus for the repeated placing of "white" women and people of "color" in nature. Only as objects can they enter science; their only subjectivity in science is called bias and special interest unless they become honorary honorable men. This is an ethnospecific story of representation, requiring surrogacy and ventriloquism as part of its technology. The self-acting agent who is the modest witness is also "agent" in another sense—as the delegate for the thing represented, as its spokesperson and representative. Agency, optics, and recording technologies are old bedfellows.

17. Merchant (1980); Easlea (1980); Keller (1985); Jordanova (1989); Noble (1992); Schiebinger (1989).

18. The veil is the chief epistemological element in Orientalist systems of representation, including much of technoscience. The point of the veil is to promise that something is behind it. The veil guarantees the worth of the quest more than what is found. The metaphoric system of discovery that is so crucial to the discourse about science depends on there being things hidden to be discovered. How can one have breakthroughs if there is no resistance, no trial of the hero's resolve and virtue? The explorer is a hero, another aspect of epistemological manly valor in technoscience narratives. See Yegenoglu (1993). Feminist narratologists have spent a lot of time on these issues. Science studies scholars should spend a little more time with feminist and postcolonial narratology and film theory.

19. Remember that the author is a fiction, a position, and an ascribed function. And writing is dynamic; positions change. There are other Latours, in and out of print, who offer a much richer tropic tool kit than that in *Science in Action*. In particular, in writing and speaking in the mid-1990s, Latour, as well as Woolgar and several other scholars, evidence serious, nondefensive interest in feminist science studies, including the criticism of their own rhetorical and research strategies in the 1980s. I focus on *Science in Action* in this chapter because that book was taken up so widely in science studies. But see Woolgar (1995); Latour (1996).

REFERENCES

Auerbach, Erich. 1953. *Mimesis: The Representation of Reality in Western Literature.* Princeton: Princeton University Press.

Bakhtin, Mikhail. 1981. *The Dialogical Imagination.* Edited by M. Holquist. Austin: University of Texas Press.

Boose, Lynda E. 1994. " 'The Getting of a Lawful Race': Racial Discourse in Early Modern England and the Unrepresentable Black Woman." In *Women, "Race," and Writing in the Early Modern Period,* edited by M. Hendricks and P. Parker. New York: Routledge, 35–54.

Collins, H. M., and Steven Yearley. 1992. "Epistemological Chicken." In *Science as Practice and Culture,* edited by A. Pickering. Chicago: University of Chicago Press, 302–26

CRICT (Center for Research into Innovation, Culture, and Technology). 1995. Workshop on the Subject(s) of Technology: Feminism, Constructivism, and Identity.

Easlea, Brian. 1980. *Witch-Hunting, Magic, and the New Philosophy.* Brighton: Harvester Press.

Etzkowitz, Henry, and Andrew Webster. 1995. "Science as Intellectual Property." In *Handbook of Science and Technology Studies,* edited by S. Jasanoff, G. E. Markle, J. C. Petersen, and T. Pinch. Thousand Oaks, CA: Sage, 480–505.

Gross, Paul R., and Norman Levitt. 1994. *Higher Superstition: The Academic Left and Its Quarrels with Science.* Baltimore: Johns Hopkins University Press.

Harding, Sandra. 1992. *Whose Science? Whose Knowledge? Thinking from Women's Lives.* Ithaca: Cornell University Press.

———. 1993. *The "Racial" Economy of Science: Toward a Democratic Future.* Bloomington: Indiana University Press.

Heath, Deborah. 1997. "Bodies, Antibodies, and Modest Interventions." In *Cyborgs and Citadels: Anthropological Interventions in Emerging Sciences and Technologies,* edited by G. Downey and J. Dumit. Santa Fe: SAR Press, 67–82.

Hendricks, Margo. 1992. "Managing the Barbarians: The Tragedy of Dido, Queen of Carthage." *Renaissance Drama,* New Series 23:165–88.

———. 1994. "Civility, Barbarism, and Aphra Behn's *The Widow Ranter.*" In *Women, "Race," and Writing in the Early Modern Period,* edited by M. Hendricks and P. Parker. New York: Routledge, 225–39.

———. 1996. "Obscured by Dreams: Race, Empire, and Shakespeare's *A Midsummer Night's Dream.*" *Shakespeare Quarterly* 47(1): 37–60.

Jordanova, Ludmilla. 1989. *Sexual Visions: Images of Gender in Science and Medicine between the Eighteenth and Twentieth Centuries.* Madison: University of Wisconsin Press.

Keller, Evelyn Fox. 1985. *Reflections on Gender and Science.* New Haven: Yale University Press.

———. 1990. "Gender and Science: 1990." *The Great Ideas Today.* Chicago: Encyclopaedia Britannica, Inc.

King, Katie. 1993. "Feminism and Writing Technologies." Paper read at conference on Located Knowledges, April 8–10, at UCLA.

Laqueur, Thomas. 1990. *Making Sex: Body and Gender from the Greeks to Freud.* Cambridge: Harvard University Press.

Latour, Bruno. 1987. *Science in Action: How to Follow Scientists and Engineers through Society.* Cambridge, MA: Harvard University Press.

———. 1993. *We Have Never Been Modern.* Translated by Catherine Porter. Cambridge: Harvard University Press.

———. 1996. *Petit réflexion sur le culte moderne des dieux faitiches.* Le Plessis-Robinson: Synthélabo.

Merchant, Carolyn. 1980. *The Death of Nature: Women, Ecology, and the Scientific Revolution.* New York: Harper & Row.

Morrison, Toni. 1992. *Jazz.* New York: Knopf.

Noble, David F. 1992. *A World Without Women: The Christian Clerical Culture of Western Science.* New York: Oxford University Press.

Potter, Elizabeth. 2001. *Gender and Boyle's Law of Gases.* Bloomington: Indiana University Press.

Rabinow, Paul. 1992. "Metamodern Milieux." Paper read at American Anthropological Association Meetings, December 4, at San Francisco.

Rose, Hilary. 1994. *Love, Power, and Knowledge: Towards a Feminist Tranformation of the Sciences.* Bloomington: Indiana University Press.

Schiebinger, Londa. 1989. *The Mind Has No Sex? Women in the Origins of Modern Science.* Cambridge: Harvard University Press.

Shapin, Steven. 1988. "Following Scientists Around." *Social Studies of Science* 18(3): 533–50.

Shapin, Steven. 1994. *A Social History of Truth: Civility and Science in Seventeenth-Century England*. Chciago: University of Chicago Press.

Shapin, Steven, and Simon Schaffer. 1985. *Leviathan and the Air-Pump: Hobbes, Boyle and the Experimental Life*. Princeton: Princeton University Press.

Star, Susan Leigh. 1991. "Power, Technology and the Phenomenology of Conventions." In *A Sociology of Monsters*, edited by J. Law. Oxford: Basil Blackwell.

Traweek, Sharon. 1988. *Beam Times and Life Times: The World of High Energy Physicists*. Cambridge, MA: Harvard University Press.

———. 1992. "Border Crossings: Narrative Strategies in Science Studies and among Physicists in Tsukuba Science City, Japan." In *Science as Practice and Culture*, edited by A. Pickering. Chicago: University of Chicago Press, 429–65.

Wheeler, Bonnie. 1992. "The Masculinity of King Arthur: From Gildas to the Nuclear Age." *Quondam et Futuras: A Journal of Arthurian Interrelations* 2(4): 1–26.

Woodbridge, Linda. 1984. *Women and the English Renaissance: Literature and the Nature of Womankind, 1540–1620*. Urbana: University of Illinois Press.

Woolgar, Steve. 1988. *Science, the Very Idea!* London: Tavistock.

———, ed. 1995. *Feminist and Constructivist Perspectives on New Technology*. Special Issue, *Science, Technology, and Human Values* 20 (Summer:3).

Yegenoglu, Meyda. 1993. "Veiled Fantasies: Towards a Feminist Reading of Orientalism." Ph.D. diss., Sociology Board, University of California at Santa Cruz.

8

RACE: UNIVERSAL DONORS IN A VAMPIRE CULTURE. IT'S ALL IN THE FAMILY: BIOLOGICAL KINSHIP CATEGORIES IN THE TWENTIETH-CENTURY UNITED STATES

RACE

The starting point for my story is the racial discourse in place at the end of the nineteenth century in Europe and the United States. As the historian George Stocking put it, "'blood' was for many a solvent in which all problems were dissolved and processes commingled." "Race" meant the "accumulated cultural differences carried somehow in the blood" (Stocking 1993:6). The emphasis was on "somehow," for blood proved a very expansible and inclusive fluid. Four major discursive streams poured into the cauldron in which racial discourse simmered well into the early decades of the twentieth century, including the ethnological, Lamarckian, polygenist, and evolutionist traditions. For each approach, the essential idea was the linkages of lineage and kinship. No great distinction could be maintained between linguistic, national, familial, and physical resonances implied by the terms *kinship* and *race*. Blood ties were the proteinaceous threads extruded by the physical and historical passage of substance from one generation to the next, forming the great nested, organic collectives of the human family. In that process, where race was, sex was also. And where race and sex were, worries about hygiene, decadence, health, and organic efficiency occupied the best of minds of the age, or at least the best published.

These same minds were uniformly concerned about the problems of progress and hierarchy. Organic rank and stage of culture from primitive to civilized were at the heart of evolutionary biology, medicine, and anthropology. The existence of progress, efficiency, and hierarchy were not in question scientifically, only their proper representation in natural-social dramas, where race was the narrative colloid or matrix left when blood congealed. The plenum of universal organic evolution, reaching from ape to modern European with all the races and sexes properly arrayed between, was filled with the bodies and measuring instruments proper to the life sciences. Craniometry and the examination of sexual/reproductive materials both focused on the chief organs of mental and generative life, which were the keys to organic social efficiency. Brains were also sexual tissues, and reproductive organs were also mental structures. Furthermore, the face revealed what the brain and the gonad ordained; diagnostic photography showed as much. The evolution of language, the progress of technology, the perfection of the body, and the advance of social forms seemed to be aspects of the same fundamental human science. That science was constitutively physiological and hierarchical, organismic and wholist, progressivist and developmental.

To be sure, in the early twentieth century Franz Boas and social-cultural anthropology broadly were laying the foundations of a different epistemological order for thinking about race. But, encompassing immigration policy, mental-health assessments, military conscription, labor patterns, nature conservation, museum design, school and university curricula, penal practices, field studies of both wild and laboratory animals, literary evaluation, the music industry, religious doctrine, and much more, race—and its venereal infections and ties to sexual hygiene—was real, fundamental, and bloody. If the skeptic of poststructuralist analysis still needs to be convinced by an example of the inextricable weave of historically specific discursive, scientific, and physical reality, race is the place to look. The discursive has never been lived with any greater vitality than in the always undead corpus of race and sex. For many in the first decades of the twentieth century, race mixing was a venereal disease of the social body, producing doomed progeny whose reproductive issue was as tainted as that of lesbians, sodomites, Jews, overeducated women, prostitutes, criminals, masturbators, or alcoholics. These were the subjects, literal and literary, of the commodious discourse of eugenics, where intraracial hygiene and interracial taxonomy were two faces of the same coin.[1]

Even radicals and liberals, to name them anachronistically, who fought the reproductive narrative and social equations named in the preceding paragraph, accepted race as a meaningful object of scientific knowledge. They had little choice. These writers and activists worked to reshape race into a different picture of collective human health (Stepan and Gilman 1993).[2] Scientific racial discourse—in the sense that did not insist on the separation of the physical and the cultural and spoke in the idiom of organic health, efficiency, and familial solidarity—accommodated writers from great American liberators such as W.E.B. Du Bois and Charlotte Perkins Gilman to middle-of-the-road, Progressive Era, unabashed racists such as Madison Grant.[3] Du Bois is particularly interesting because he most consistently rejected "biologism" in his approach to race and racism, but the broad discourse that assimilated race feeling to family feeling and invited discussion on the childhood and maturity of collective human groups called races was inescapable (Du Bois 1989:8). Although he retracted such language a decade or so later, in 1897 Du Bois wrote that the history of the world is the history of races: "What is race? It is a vast family ... generally of common blood and language, always of common history" (Du Bois 1971:19; see also Appiah 1985; 1990:16n3; Stepan and Gilman 1993:192n7).

George Stocking's thumbnail portrait of the Social Marking System developed by the U.S. sociologist Franklin H. Giddings around 1900 to 1910 collects up the ways that race and nation, passing through kinship of many ontological kinds and degrees of closeness, were held together on a continuum of social-biological differences. "The essential element of the race concept was the idea of kinship. . . . 'Race' and 'nation' were simply the terms applied to different levels of a single pyramid" (Stocking 1993:7–8). Giddings attempted to provide a quantitative notation to distinguish degrees of kinship, arrayed across eight different kinds of relatedness. Types such as the Hamitic, the Semitic, the Celtic, and so on filled the taxonomic slots. The specifics of Gidding's classification are less important here than their illustration of the exuberance of racial taxonomizing in the United States. In these taxonomies, which are, after all, little machines for clarifying and separating categories, the entity that always eluded the classifier was simple: race itself. The pure Type, which animated dreams, sciences, and terrors, kept slipping through, and endlessly multiplying, all the typological taxonomies. The rational classifying activity masked a wrenching and denied history. As racial anxieties ran riot through the sober prose of categorical bioscience, the taxonomies could neither pinpoint nor contain their terrible discursive product.

To complete my brief caricature of race as an object of bioscientific knowledge in the period before World War II, I will turn to a family portrait that innocently embodies the essence of my argument. The portrait slips down the developmental chain of being to racialized urban humanity's ultimate other and intimate kin, the gorilla in nature[4] [Figure 8.1. Gorilla Group in the American Museum of Natural History].

Fig. 8.1. Gorilla Group in African Hall. Animals by Carl E. Akeley. Background by Willima Leigh. Neg. #314824. Courtesy of the American Museum of Natural History. Photograph by Wurts Brothers.

Figure 8.1 shows a taxidermic reconstruction of a gorilla group, with a striking silverback male beating his chest, a mother at one side eating calmly, and a toddler. A young blackback male is in the diorama but out of the photograph. The primal ape in the jungle is the doppelgänger and mirror to civilized white manhood in the city. Culture meets nature through the looking glass at the interface of the Age of Mammals and the Age of Man. Preserved in changeless afterlife, this vibrant gorilla family is more undead than it is alive. The members of this (super)natural gorilla family were hunted, assembled, and animated by the art of taxidermy to become the perfect type of their species. Dramatic stories about people, animals, tools, journeys, diseases, and money inhere in each precious corpse, from the chest-beating male called the Giant of Karisimbi to the ape-child speared as it screamed in terror on the steep volcanic mountainside. The blood was drained; face masks taken from the corpses; the skins stripped and preserved, shipped across continents, and stretched over special light mannequins. Lit from within and surrounded by the panoramic views made possible by Hollywood set painting and the new cameras of the 1920s, the perfect natural group—the whole organic family in nature—emerged in a lush Eden crafted out of detailed reconstructions of leaves, insects, and soils. In these ways, the gorilla was reborn out of the accidents of biological life, a first birth, into epiphanic perfection, a second birth, in a diorama in the Akeley African Hall in the American Museum of Natural History in New York City.

Behind the dioramic re-creation of nature lies an elaborate world of practice. The social and technical apparatus of the colonial African scientific safari and the race-, class-, and gender-stratified labor systems of urban museum construction organized hundreds of people over three continents and two decades to make this natural scene possible. To emerge intact, reconstructed nature required all the resources of advanced guns, patented cameras, transoceanic travel, food preservation, railroads, colonial bureaucratic authority, large capital accumulations, philanthropic institutions, and much more. The technological production of a culturally specific nature could hardly be more literal. The intense realism of the diorama was an epistemological, technological, political, and personal-experiential achievement. Natural order was simply there, indisputable, luminous. Kinship was secure in the purity of the achieved vision.

Walt Disney Studios and *National Geographic* might do better in the decades to come, but they needed the magic of motion pictures. The achievement of the prewar natural history diorama relied more on a sculptural sensibility that was also manifest in the elegant bronzes, placed just outside the African Hall, of "primitive natural man," the East African Nandi lion-hunters. Their perfection was sought by the same scientist-artist, Carl Akeley, who designed the dioramas for the American Museum. Organicism and typology ruled unchallenged in these practices, in which the earth's great racial dramas, constructed in a white, imperial, naturalist, and progressive frame, were displayed as pedagogy, hygiene, and entertainment for an urban public.

After the successful scientific hunt for the perfect specimen, the superior nobility of hunting with the camera was urged in a conservationist doctrine that downplayed further hunting with the gun. To strengthen the conservationist argument, white women and children came on the final hunt for the museum's gorillas to prove that the great violent drama of manhood in confrontation across species could give way to a gentler tale. In part because of the efforts of the members of this collecting expedition in 1921–1922 and of the officers of the American Museum, the area where the Giant of Karisimbi died became a Belgian national park, the Parc Albert in the Belgian Congo, where nature, including "primitive" people as fauna in the timeless scene, was to be preserved for science, adventure, uplift, and moral restoration as proof against civilization's decadence. No wonder universal nature has been a less than appealing entity for those who were not its creators and its beneficiaries. Undoing this inherited dilemma has never been more urgent if people and other organisms are to survive much longer.

The hunt for the Giant of Karisimbi took place in 1921, the same year that the American Museum of Natural History hosted the Second International Congress of Eugenics. Collected proceedings from the congress were titled "Eugenics in Family, Race, and State." The Committee on Immigration of the Eugenics Congress sent its exhibit on immigration to Washington, D.C., as part of its lobbying for racial quotas. In 1924 the U.S. National Origins Act restricted immigration by a logic that linked race and nation. For officials of the American Museum, nature preservation, germ plasm protection, and display work were all of a piece. Exhibition, conservation, and eugenics were part of a harmonious whole. Race was at the center of that natural configuration, and racial discourse, in all of its proliferating diversity and appalling sameness, reached deep into the family of the nation.

POPULATION

The community of race, nation, nature, language, and culture transmitted by blood and kinship never disappeared from popular racialism in the United States, but this bonding has not been meaningfully sustained by the biological sciences for half a century. Rather than dwell on the scientific and political processes that led to the biosciences' reversal on the reality and importance of race to evolutionary, genetic, physiological, therapeutic, and reproductive explanations in the middle decades of the twentieth century, I will leap to the other side of the divide, to where the Wizard of Oz has changed the set in the theater of nature. The major difference is that an entity called the population is now critical to most of the dramatic action.

A population, a relatively permeable group within a species, differed by one or more genes from other such groups. Changes of gene frequencies within populations were fundamental evolutionary processes, and gene flow between populations structured the traffic that bound the species together. Genes and genotypes were subject to Darwinian natural selection in the context of the functioning phenotypes of whole organisms within populations. Occasionally still a convenient notion, race was generally a misleading term for a population. The frequency of interesting genes, such as those coding for immunological markers on blood cells or for different oxygen-carrying hemoglobins, might well differ more for individuals within a population than between populations. Or they might not; the question was an empirical one and demanded an explanation that included consideration of random drift, adaptational complexes, and the history of gene exchange. The populations' history of random genetic mutation and gene flow, subjected to natural selection resulting in adaptation, constituted the history of the species. Populations were not types arranged hierarchically but dynamic assemblages that had to function in changing environments. Measurements had to be of structures important to adaptational complexes related to current function. For example, craniometry producing brain-volume values on a putative hierarchical chain of being gave way to measurements of structures critical to dynamic action in life, such as facial regions critical to chewing and subject to physical and functional stresses during the development of the organism. Highly variable and permeable natural populations seemed to be the right kind of scientific object of knowledge, and the racial type seemed to be a residue from a bad nightmare.

The construction of the category of the population occurred over several decades. Leading parts were taken by naturalists studying geographical variation and speciation; geneticists learning that mutations were inherited in discrete Mendelian fashion; population geneticists constructing mathematical models showing how mutation, migration, isolation, and other factors could affect the frequency of genes within populations; and experimentalists demonstrating that natural selection could operate on continuous variations to alter the characteristics of a population. The synthesis of these lines of research—which was effected by the Russian-trained immigrant U.S. geneticist Theodosius Dobzhansky; the English scion of the scientific Huxley clan, Julian Huxley; the polymath German-trained immigrant U.S. systematist Ernst Mayr; and the U.S. paleontologist George Gaylord Simpson, among others, from the late 1930s to the late 1940s—changed the face of dominant evolutionary theory. The result was called the modern synthesis or the neo-Darwinian evolutionary theory.[5] Several of the men who put the modern synthesis together were also popular writers, published by the major university presses, who developed an antiracist, liberal, biological humanism that held sway until the 1970s.[6] This was a scientific humanism that emphasized flexibility, progress, cooperation, and universalism.

This was also precisely the humanism enlisted by M. F. Ashley Montagu, former student of Franz Boas and organizer of the United Nations Educational, Scientific, and Cultural Organization's (UNESCO) statements on race in 1950 and 1951 (UNESCO 1952). Perched on the cusp between the Allied victory over the Axis powers, the ideological contest for defining human nature waged by "socialism" and "capitalism" in the Cold War, and the struggles for third world decolonization that sharpened after World War II, the U.N.–sponsored documents were intended to break the bioscientific tie of race, blood, and culture that had fed the genocidal policies of fascism and still threatened doctrines of human unity in the emerging international scene. Since biologists had to bear so much of the responsibility for having constructed race as a scientific object of knowledge in the first place, it seemed essential to marshal the authority of the architects of the new synthesis to undo the category and relegate it to the slag heap of pseudo-science. It would not have done for the UNESCO statement to have been authored by social scientists. The crafting of the UNESCO race statements provides a unique case study for the discursive reconstitution of a critical epistemological and technical object for policy and research, where science and politics, in the oppositional sense of those two slippery terms, form the tightest possible weave.

The concept of the population was in the foreground as the authors argued that plasticity was the most prominent species trait of *Homo sapiens*. While the strong statement that the range of mental talent is the same in all human groups did not survive controversy over the 1950 version, the negative argument that science provides no evidence of inherited racial inequality of intelligence remained. The contentious 1950 statement that universal brotherhood (*sic*) is supported by a specieswide, inborn trait of a drive toward cooperation also did not live through the rewriting in 1951. Nonetheless, the latter document—signed by 96 internationally prominent scientific experts before it was released—remained uncompromising on the key ideas of plasticity, educability, the invalidity of the race-and-culture tie, and the importance of populationist evolutionary biology.[7] To cast group differences typologically was to do bad science— with all the penalties in jobs, institutional power, funding, and prestige that flow from such labeling. Needless to say, biological racialism did not disappear overnight, but a palace coup had indeed taken place in the citadel of science.[8]

Walking out of UNESCO House in Paris, the new universal man turned up fossilized in East Africa almost immediately. In honor of this timely geological appearance, the *Harvard Lampoon* dubbed Olduvai Gorge, made famous by the paleo-anthropological investigations of the Leakey family, the "Oh Boy! Oh Boy! Gorge" for its stunning hominid fossils and the associated accounts of the dawn of human history and of the species-defining characteristics of human nature. Deeply indebted to the modern synthesis, the New Physical Anthropology developed from the 1950s to become a major actor in identifying those adaptational complexes that made "us" human and in installing them in both pedagogical and research practice. Public and intradisciplinary antiracist lectures, new undergraduate and graduate curricula in physical anthropology sustained by the expanding institutional prosperity of the postwar era in the United States, field studies of natural primate populations, and major programs of research on African hominid fossils were all part of the program of the new physical anthropology. Its objects of attention were not typologically constructed taxonomies but systems of action that left their residue in the enduring hard structures in fossil beds or under the skin of still living animals. Adaptational behavior is what these biological anthropologists cared about, whether they were looking at pelvic bones, crania, living monkeys and apes, or modern hunter-gatherers. In the new framework, people who were typical "primitives" to the earlier expeditions of the American Museum of Natural History were fully modern humans, exhibiting clearly the fundamental adaptational complexes that continue to characterize all populations of the species. Indeed, lacking the stresses of too much first world abundance, the former "primitives," like modern hunter-gatherers, became especially revealing "universal" human beings.

The most important adaptational complex for my purposes in this chapter is the species-defining sharing way of life, rooted in hunting and the heterosexual nuclear family. Man the Hunter, not the urban brother of the Giant of Karisimbi or the Nandi lion spearmen, embodied the ties of technology, language, and kinship in the postwar universal human family. Parent to technology and semiology—to the natural sciences and the human sciences—in the same adaptational behavior, Man the Hunter crafted the first beautiful and functional objects and spoke the first critical words. Hunting in this account was not about competition and aggression but about a new subsistence strategy possible for striding, bipedal protohumans with epic hand-eye coordination. Acquiring big brains and painful births in the process, these beings developed cooperation, language, technology, and a lust for travel, all in the context of sharing the spoils with mates, children, and each other. Males were certainly the active motor of human evolution in the hunting hypothesis of the 1950s and 1960s, but the logic was not too much strained in the 1970s by foregrounding Woman the Gatherer and a few useful family reforms, such as female orgasms and mate choice favoring males who made themselves useful with the kids.[9] Still, baby slings, carrying bags for roots and nuts, daily adult gossip, and talking to children could hardly compete for originary drama with elegant projectiles, adventurous travel, political oratory, and male bonding in the face of danger.[10]

Two powerful photographic documents of the universal human family conclude my meditation on the hopeful, but fatally flawed, biological humanism of the mid-twentieth century: the late-1970s painting called *Fossil Footprint Makers of Laetoli* by the anatomical illustrator Jay Matternes, and the New York Museum of Modern Art's publication from its 1955 epic photographic exhibit called *The Family of Man*. Both documents stage the relations of nature and culture mediated by the heterosexual, reproductive, nuclear family as the figure of human unity and diversity. Both renderings of the human story are starkly under the visible sign of the threat of nuclear destruction, and both suggest a saga of unity, danger, and resilience that permeated accounts of science, progress, and technology in the post–World War II era.

Accompanying an international museum exhibit of hominid fossils in the 1980s, Matternes's painting shows the hominid First Family walking across the African savanna under the cloud of an erupting volcano, the sign of destruction by fire.[11] These transitional figures between apes and modern humans recall the gorilla family in the American Museum of Natural History. But for earthlings in the last chilling years of the Cold War, the thick cloud of dust spewing into the sky to obscure the sun in Matternes's reconstruction could not help but evoke the looming threat of nuclear winter. Expulsion from Eden had particular narrative resonances in nuclear culture. In the era of nuclear superpowers facing off in fraternal rivalry, threats came in centralized apocalyptic packages. In the New World Order of the post–Cold War era, nuclear threats, like all else, have a more dispersed and networked structure of opportunity and danger—for example, criminal smuggling of plutonium from the former Soviet Union and the apocalypse-lite of plutonium poisoning of urban water supplies or dirty minibombs backing up political disputes. Matternes's painting is a reconstruction of the life events that might have been responsible for the 3.7-million-year-old footprints found in the volcanic ash at Laetoli, near the Olduvai Gorge, by Mary Leakey and others in the late 1970s. The space-faring descendants of the First Family put their footprints in moon dust in 1969 in Neil Armstrong's "one small step for mankind," just as the *Australopithecus afarensis* trekkers, at the dawn of hominization, made their way through the volcanic dust of the human travel narrative.

The great myths of birth and death, beginnings and endings, are everywhere in this painting. The reconstructed hominids are members of a highly publicized ancestor-candidate species that has been at the center of scientific debates about what counts as human. Perhaps the best-known fossil in this media and scientific fray has been the 3.5-million-year-old skeleton of a diminutive female named Lucy by her Adamic founders, after the Beatles' "Lucy in the Sky with Diamonds." The African plain in the painting, scene of the passage of Lucy's relatives, is both rich with the signs of abundant animal life and thickly encrusted with the smothering ash that must drive all the animals, including these early hominids, in search of food. The three family members vividly dramatize the central adaptive complexes that made "us" human. The elements for the universal sharing way of life are unmistakable. The male strides ahead, carrying a serviceable tool, although not quite the future's elegant projectiles that were critical to the hunting hypothesis as well as to Stanley Kubrick's *2001: A Space Odyssey*. *A. afarensis* would have to wait for somewhat larger heads before they improved their aesthetic sense. The antiracist universals of the evolutionary drama scripted according to the humanist doctrines of the modern synthesis left in place the durable essentials of the sexual division of labor, male-headed heterosexual families, and child-laden females—here pictures *without* the baby-carrying sling that many anthropologists argue was likely to have been among the first human tools. In Matternes's Adamic imagination, the child-carrying female follows behind, looking to the side, while the male leads, looking into the future. The germ of human sociality was the couple and their offspring, not a mixed foraging group, a group of related females with their kids, two males with one carrying a kid, or any other of the many possibilities for those first small steps for mankind left in the dust at Laetoli.[12]

If it is the numbing and hegemonic sameness of the universal way of life that I resist in the new physical anthropology, including many of its feminist versions, and in Matternes's painting, then perhaps an earlier document, the popular coffee-table book of Edward Steichen's photographic exhibit called *The Family of Man*, can settle my dyspeptic attack of political correctness. If I detect the unself-conscious ethnocentricity of those who crafted the natural-technical object of knowledge called the First Family and the universal hominizing way of life, perhaps the global scope of the 1955 document will allow a more capacious field for imagining human unity and difference. Yet, once I have learned to see the Sacred Image of the Same and the Edenic travelogue of so much Western historical narrative, I have a hard time letting go of this perhaps monomaniacal critical vision, which might be worse than the objects it complains about. My own perverse skill at reading the sameness of my own inherited cultural stories into everything is one of the symptoms that drives this chapter. Still, I believe that this capacity of reproducing the Same, in culpable innocence of its historical, power-charged specificity, characterizes not just me but people formed like me, who are liberal, scientific, and progressive—just like those officials of the American Museum of Natural History who sent their eugenic immigration exhibit to Washington in 1921. I am worried that too little has changed in hegemonic bioscientific discourse on nature, race, unity, and difference, even in the face of seeming major change. So let me pursue my suspicion that the Sacred Image of the Same is not just my problem but is also one of the tics that reproduces sexually charged racist imaginations even in the practices most consciously dedicated to antiracism.

In this mood, I am not surprised that Steichen's 1955 photo album does not settle my dyspepsia. My queasiness is not just with the title and its conventional familial trope for binding together humanity, with all the resonances that metaphor evokes of kinship, lineage, and blood ties. There is much to love in *The Family of Man*, including its vivid photos of working, playing, and fighting. Old age, infirmity, and poverty are no barriers to liveliness here. Even the staging of everybody and everything into one grandly decontextualized narrative, which culminates in the United Nations and the hopes for peace in nuclear times after the ravages of depression, fascism, and war, can almost be forgiven. After all, *The Family of Man* is a lot less sanitized than most 1990s versions of multiculturalism. Despite decades of critical visual theory, I am susceptible, even now, to the images of this book. That helps, because it is a rule for me not to turn a dissolving eye onto straw problems, not to "deconstruct" that to which I am not also emotionally, epistemologically, and politically vulnerable.

The Family of Man is ruled throughout its organic tissues by a version of unity that repeats the cyclopean story that collects up the people into the reproductive heterosexual nuclear family, the potent germ plasm for the Sacred Image of the Same. The opening photos show culturally varied young men and women in courtship; then marriage; then all sorts of women in pregnancy and labor; then birth (mediated by a male scientific-medical doctor), nursing, babyhood, and parenting by both genders. The photo album then opens out into culturally and nationally varied scenes of work on the land and in factories. Food, music, education, religion, technology, tragedy and mercy, aging and death, anger and joy, hunger and suffering all find their place. The icons of nuclear war and of other wars, as well as images of racism and fascism, cast a deep shadow. The pall is lifted by the images of democracy (voting) and internationalism (the United Nations), which locate hope for this family story solidly in the signifiers of the "free world." The last pages of the exhibit are full of multihued children, seeds of the future. The last photo (before the unfortunate ocean wave on the inside back cover) is of a little boy and little girl moving away from the viewer, walking hand in hand in a sylvan nature toward the sunny light of a possible future. This book about human universals is vehemently antiracist and simultaneously deeply enmeshed in an ethnospecific, teleological story that continues to make the human collective bleed, or at least to hunger for other stories of what it means to be members of a species and a community. What's not collected in a reproductive family story does not finally count as human. For all the photo narrative's emphasis on difference, this is the grammar of indifference, of the multiplication of sameness.

The desire for a child, for a future, in that potent image permeating *The Family of Man* is at least as fierce as the yearning sustaining the New Reproductive Technologies of the 1980s and 1990s. The genetic imagination never dimmed under the sign of the population. Genetic desire would be no less when the genome became the signifier of human collectivity.

GENOME

If universal humanity was plastic under the sign of the population at midcentury, then human nature is best described as virtual in present, end-of-the-millennium regimes of biological knowledge and power. Specifically, human nature is embodied, literally, in an odd thing called a genetic database, held in a few international locations such as the three large public databases for genetic map and sequence data: the U.S. GenBank©, the European Molecular Biological Laboratory, and the DNA Data Bank of Japan. The Genome Data Base at Johns Hopkins University is a massive central repository of all gene-mapping information. In the world of gene sequencing, intellectual property rights vie with human rights for the attention of lawyers and scientists alike. Criminal as well as corporate lawyers have a stake in the material and metaphoric representation of the genome. Funding and policy strongly support rapid public access to genome databases in the interests of research and development. For example, in 1993 the French researcher Daniel Cohen, of the Centre d'Etude du Polymorphisme Humaine in Paris, made his first complete draft map of the human genome available through the Internet. GenInfo, developed by the U.S. National Center for Biotechnology Information of the National Library of Medicine, is a kind of meta-database containing both protein and nucleic acid sequence data "to which other databases can add, refer, annotate, interpret, and extrapolate" (Corteau 1991:202).[13] In part because of the tremendous physical computing power and human expertise that resulted from nuclear weapons research, informatics development in the U.S. Human Genome Project began under the auspices of GenBank© at the U.S. National Laboratories at Los Alamos, New Mexico. It was there also that the expertise and machines existed that built the matrix for the flourishing of artificial life research at the nearby Santa Fe Institute.

A database is an information structure. Computer programs are the habitats of information structures, and an organism's genome is a kind of nature park among databases. Just as racial hygiene and eugenics were committed to science and progress, and populationist doctrines of human universals were unambiguously on the side of development and the future, the genome is allied with all that is up-to-the-minute. Yet, something peculiar happened to the stable, family loving, Mendelian gene when it passed into a database, where it has more in common with LANDSAT photographs, Geographical Information Systems, international seed banks, and the World Bank than with T. H. Morgan's fruitflies at Columbia University in the 1910s or UNESCO's populations in the 1950s. Banking and mapping seem to be the name of the genetic game at an accelerating pace since the 1970s, in the corporatization of biology to make it fit for the New World Order, Inc.[14] If the modern synthesis, ideologically speaking, tended to make everyone his brother's keeper, then, in its versions of kin selection and inclusive fitness-maximization strategies, the sociobiological synthesis runs to making everyone his or her sibling's banker.[15]

Biotechnology in the service of corporate profit is a revolutionary force for remaking the inhabitants of planet Earth, from viruses and bacteria right up the now repudiated chain of being to *Homo sapiens* and beyond. Biological research globally is progressively practiced under the direct auspices of corporations, from the multinational pharmaceutical and agribusiness giants to venture-capital companies that fascinate the writers for the business sections of daily newspapers. Molecular biology and molecular genetics have become nearly synonymous with biotechnology as engineering and redesign disciplines. Beings like Man the Hunter and Woman the Gatherer reappear for their roles on the stage of nature enterprised up as Man™ and Woman™—copyrighted; registered for commerce; and, above all, highly flexible.[16] In a world where the artifactual and the natural have imploded, nature itself, both ideologically and materially, has been patently reconstructed. Structural adjustment demands no less of bacteria and trees as well as of people, businesses, and nations.

The genome is the totality of genetic "information" in an organism, or, more commonly, the totality of genetic information in all the chromosomes in the nucleus of a cell. Conventionally, the genome refers only to the nucleic acid that "codes" for something and not to the dynamic, multipart structures and processes that constitute functional, reproducing cells and organisms. Thus, not even the proteins critical to nuclear chromosomal organization or DNA structures such as mitochondrial chromosomes outside the nucleus are part of the genome, much less the whole living cell. Embodied information with a complex time structure is reduced to a linear code in an archive outside time. This reduction gives rise to the curious, ubiquitous, mixed metaphor of "mapping the code," applied to projects to represent all the information in the genome. DNA in this view is a master molecule, the code of codes, the foundation of unity and diversity. Much of the history of genetics since the 1950s is the history of the consolidation and elaboration of the equation of "gene = information" in the context of master-molecule metaphors. I consider this representational practice for thinking about genetics to constitute a kind of artificial life research itself, where the paradigmatic habitat for life—the program—bears no necessary relationship to messy, thick organisms.

The convergence of genomics and informatics, in technique and personnel as well as in basic theory and shared tropes, is immensely consequential for bioscientific constructions of human nature. The technical ability to manipulate genetic information, in particular to pass it from one kind of organism to another in a regulated manner in the lab, or to synthesize and insert new genes, has grown exponentially since the first successful genetic engineering experiments of the early 1970s. In principle, there is no naturally occurring genome that cannot be experimentally redesigned. This is a very different matter compared to the genetic traffic among populations of a species studied within the midcentury evolutionary synthesis, much less compared to the genetic, natural racial types that inhabited the biological world earlier in the century. Genetic engineering is not eugenics, just as the genome does not give the same kind of account of a species as does organic racial discourse.[17]

From the point of view of the 1990s, the genome is an information structure that can exist in various physical media. The medium might be the DNA sequences organized into natural chromosomes in the whole organism. Or the medium might be various built physical structures, such as yeast artificial chromosomes (YACs) or bacterial plasmids, designed to hold and transfer cloned genes or other interesting stretches of nucleic acid. The entire genome of an organism might be held in a "library" of such artifactual biochemical information structures. The medium of the database might also be the computer programs that manage the structure, error checking, storage, retrieval, and distribution of genetic information for the various international genome projects that are under way for *Homo sapiens* and for other model species critical to genetic, developmental, and immunological research. Those species include mice, dogs, bacteria, yeast, nematodes, rice, and a few more creatures indispensable for international technoscientific research.

The U.S. Human Genome Project officially began in 1988 under the management of the Department of Energy and the National Institutes of Health. As a whole, the global Human Genome Project is a multinational, long-term, competitive and cooperative, multibillion-dollar (yen, franc, mark, etc.) effort to represent exhaustively—in genetic, physical, and DNA sequence maps—the totality of information in the species genome.[18] The data are all entered into computerized databases, from which information is made available around the world on terms still very much being worked out. Computerized database design is at the leading edge of genomics research. Design decisions about these huge databases shape what can be easily compared to what else, and so determine the kinds of uses that can be made of the original data. Such decisions structure the kinds of ideas of the species that can be sustained. National science bodies, tax- and foundation-supported universities, international organizations, private corporations, communities, indigenous peoples, and many configurations of political and scientific activists all play a part in the saga.

Questions about agency—who is an actor—abound in the world of the genome, as in the worlds of technoscience in general. For example, in the discourse of genome informatics, data are exchanged among "agents" and sent to "users" of databases. These entities could as easily be computers or programs as people (Erickson 1992).[19] It does not solve the trouble to say that people are the end users. That turns out to be a contingent, technical, design decision—or a way of representing ongoing flows of information—more than an ontological necessity. People are in the information loop, but their status is a bit iffy in the artificial life world. Compared to the biological humanism of the modern synthesis, technohumanism has had to make a few timely ideological adjustments. Genomics is neither taxidermy nor the reconstruction practices of the new physical anthropology, and the emerging techniques of animation occupy the minds of more than the *Jurassic Park* special-effects programmers at Industrial Light and Magic.

Issues of *agency* permeate practices of *representation* in many senses of both terms: Who, exactly, in the human genome project represents whom? A prior question has to be a little different, however. Who, or what, is the human that is to be exhaustively represented? Molecular geneticists are consumed with interest in the variability of DNA sequences. Their databases are built to house information about both stable and variable regions of genes or proteins. Indeed, for actors from drug designers to forensic criminologists, the uniqueness of each *individual's* genome is part of the technical allure of the human genome projects' spinoffs. More fundamentally, however, the genome projects produce entities of a different ontological kind than flesh-and blood organisms, "natural races," or any other sort of "normal" organic being. At the risk of repeating myself, the human genome projects produce ontologically specific things called databases as objects of knowledge and practice. The human to be represented, then, has a particular kind of totality, or species being, as well as a specific kind of individuality. At whatever level of individuality or collectivity, from a single gene region extracted from one sample through the whole species genome, this human is itself an information structure whose program might be written in nucleic acids or in the artificial intelligence programming language called Lisp®.

Therefore, variability has its own syntax in genome discourse as well. There is no illusion in the 1990s about single "wild-type" genes and various mutant deviants.[20] That was the terminology of Mendelian genetics of the early twentieth century, when the languages of the normal and the deviant were much more sanitary. Racial hygiene and its typological syntax are not supported by genome discourse, or by artificial life discourses in general. Genetic investment strategies, in the sense of both evolutionary theory and business practice, *are* supported. The populationist thinking of the modern synthesis blasted an entire toolkit of resources for believing in norms and types. Flexibility, with its specific grammars of human unity and diversity, is the name of the game at the end of this millennium. However, for all of their commitment to variability, most molecular geneticists are not trained in evolutionary population biology, or even in population genetics. This disciplinary fact has given rise to a most interesting project and ensuing controversy for the purposes of this chapter. Let us pick up questions of agency and representation, as well as unity and difference, through the Human Genome Diversity Project (HGDP).

If the human genome databases are exhaustively to represent the species—and to provide information to users who demand that kind of knowledge, in dreams of totality as well as in practical projects—the repositories must contain physical and electronic data about the specific molecular constitution and frequency of genes on a truly global scale. Population geneticists were critical both of molecular biologists' sampling protocols for human genetic material and of their woeful statistical grasp of the structure, distribution, history, and variability of human populations. The population geneticists were also worried that many human populations around the world were becoming extinct—either literally or through interbreeding and swamping of their diversity in larger adjoining populations—with the consequent loss of genetic information forever impoverishing the databases of the species. What it means to be human would have irredeemable informational gaps. There would be a biodiversity information loss in the lifeworld of the genome. Like the vanishing of a rainforest fungus or fern before pharmaceutical companies could survey the species for promising drugs, the vanishing of human gene pools is a blow to technoscience. Prompt and thorough genetic collection and banking procedures as well as preservation of the source of variation, if possible, are the solution.

I am being a bit mordant in my reading of purposes in this account, for the organizers of the Human Genome Diversity Project were largely liberal biological humanists of the old stamp. Also, I remain sympathetic to the desire to produce a human species database that draws from as large a concept of humanity as possible. I want there to be a way to reconfigure this desire and its attendant humanism. However, it was precisely the doctrines of difference, representation, and agency of "universal" humanism that got the project and its well-meaning organizers into well-deserved trouble.[21]

Beginning about 1991, the organizers of the Human Genome Diversity Project proposed to amend the evolutionary population thinking, or lack of thinking, of the mainline Human Genome Project by collecting hair-root, white blood-cell, and cheek-tissue samples, to be held in the American Type Culture Collection, from over 700 groups of indigenous peoples on six continents. Over five years, the cost would be about $23 to $35 million (compared to more than $3 billion for the Human Genome Project as a whole). Unfortunately, unself-conscious, modernist perspectives distorted the definition of the categories of people from whom samples were to be sought, leading to a vision of dynamic human groups as timeless "isolates of historic interest." Also, other potentially genetically distinct ethnic communities did not appear on the sampling list.

The planning of the project did not involve members of the communities to be studied in any formative way in the science. The people to be sampled might give or withhold permission, to be more or less carefully sought and thoroughly explained, but they were not regarded as partners in knowledge production who might have ends and meanings of their own in such an undertaking. Their versions of the human story, complexly articulated with the genetic science of the visitors, did not shape the research agenda. Permission is not the same thing as collaboration, and the latter could lead to fundamental changes in who and what would count as science and as scientists. All the trappings of universal science notwithstanding, amending a database is a pretty culturally specific thing to want to do. Just why should other people, much less folks called "isolates of historic interest," help out with that project? That is not a rhetorical question, and there can be very strong answers coming from counterintuitive as well as obvious viewpoints for any actor. The question is a fundamental one about the rhetoric of persuasion and the practical processes through which people—including scientists and everybody else—get reconstituted as subjects and objects in encounters. How should the many discourses in play within and between people like the Guaymi of Panama and the Population Geneticists of California be articulated with each other in a power-sensitive way? This is an ethical question, but it is much more than that. It is a question about what may count as modern knowledge and who will count as producers of that knowledge.[22]

Not surprisingly, it turned out that indigenous people were more interested in representing themselves than in being represented in the human story. The encounter was most certainly not between "traditional" and "modern" peoples but between contemporaneous people (and peoples) with richly interlocking and diverging discourses, each with its own agendas and histories. Functioning as boundary objects, "genes" and "genomes" circulated among many of the languages in play.[23] Members of communities to be sampled, as well as other spokespeople, had several concerns. Some were adamant that genes or other products derived from indigenous material not be patented and used for commercial profit. Others were worried that the genetic information about tribal and marginalized peoples could be misused in genocidal ways by national governments. Some argued that medical and social priorities of the communities could be addressed by the money that would go to funding the genetic sampling and the HGDP did not give benefits back to the people. Some were quite willing to have indigenous genetic material contribute to a medically useful world knowledge fund, but only under United Nations or similar auspices that would prevent exploitation and profit-making. Ethics committee members of the HGDP tried to assure skeptics that the project had no commercial interests and that the HGDP would try to make sure that any commercial benefits that did result from the sampled material flowed back to the communities. But overall, the general issue was the question of the agency of people who did not consider themselves a biodiversity resource. Diversity was about both their *object* status and their *subject* status.

In May 1993, at a nongovernmental conference meeting parallel to the UN Human Rights Conference in Vienna, the Rural Advancement Foundation International (RAFI) and indigenous peoples urged the HGDP to "halt current collection efforts, convene a meeting with Indigenous peoples to address ethical and scientific issues, incorporate Indigenous organizations in every aspect of the HGDP and grant them veto power, and place the HGDP under direct United Nations control, with decision making delegated to a management committee dominated by Indigenous people" (RAFI 1993:13). Leaders of the HGDP tried to address the objections, but by fall of 1993 they had not set up mechanisms acceptable to the critics to include indigenous peoples in project organizing. The World Council of Indigenous Peoples monitored the project skeptically. It is important to me to note, however, that the HGDP was a minority effort in the Human Genome Project (HGP) and not at the center of the prestigious action. To get the research done at all in the face of the nonpopulationist molecular genetic orthodoxy that guided ordinary practice in the HGP would have been no small trick. It has proved easier to slow down or stop the HGDP, a kind of oppositional effort, than to question the powerful HGP itself. That makes the trouble with "difference" built into this potentially positive scientific project all the more disturbing—and important.

Inescapably, independently of the HGDP but fatally glued onto it, the all-too-predictable scandal happened. Like all pathologies, the scandal revealed the structure of what passes for normal in bioscientific regimes of knowledge and power. The Guaymi people carry a unique virus and its antibodies that might be important in leukemia research. Blood taken in 1990 from a 26-year-old Guaymi woman with leukemia, with her "informed oral consent," in the language of the U.S. Center for Disease Control in Atlanta, was used to produce an "immortalized" cell line deposited at the American Type Culture Collection. The U.S. Secretary of Commerce proceeded to file a patent claim on the cell line. Pat Moony of the Rural Advancement Foundation International found out about the claim in August 1993 and informed Isidoro Acosta, the president of the Guaymi General Congress. Considering the patent claim to be straight-forward biopiracy, Acosta and another Guaymi representative went to Geneva to raise the issue with the Biological Diversity Convention, which had been adopted at the 1993 Earth Summit in Brazil.[24] That convention had been intended to deal with plant and animal material, but the Guaymi made strategic use of its language to address technoscientifically defined human biodiversity. The Guaymi also went to the GATT secretariat to argue against the patentability of material of human origin in the intellectual property provisions of the new GATT treaty then being drafted.

In late 1993, the U.S. Secretary of Commerce withdrew the patent application, although by early 1994 the cell culture had not been returned, as demanded, to the Guaymi. The property and sovereignty battles are far from being resolved; they are at the heart of bioscientific regimes of knowledge and power worldwide. Scientific and commercial stakes are high. The stakes are also the ongoing configuration of subjects and objects, of agency and representation, inside of and by means of these disputes about biopower. The stakes are about what will count as human unity and diversity. The human family is at stake in its databases. I am instructed by the encounter of discourses, where genes are the circulating boundary objects. The Guaymi and the U.S. actors engaged each other in biogenetic terms, and they struggled for shaping those terms in the process. Perhaps the Guaymi did not initiate biotechnological and genetic engineering discourses, including their business and legal branches, but the indigenous Panamanians are far from passive objects in these material and linguistic fields. They are actors who are reconfiguring these powerful discourses, along with others they bring to the encounter. In the process, the Guaymi are changing themselves, the international scientists, and other policy elites.

The organizers of the HGDP continued to try to reorganize the research plan to satisfy both funding agencies and people to be sampled, and in late 1994 the project's International Executive Committee released a document that aimed to establish trust with indigenous peoples' organizations (Kahn 1994). The revised plan promised local control over the survey and protection of the research subjects' patent rights as well as an independent committee established by UNESCO to advise project organizers on ongoing ethical and other controversial matters. A key provision is that in order to develop scientific priorities and ethical guidelines based on local conditions and cultures, the research be done as much as possible in the countries or regions where the sampled populations live. But localism will not solve key problems. International biodiversity property issues will not go away, and the cosmopolitan nature, as well as local cultural dimensions, of science provide both the attraction and the danger in the HGDP. Issues of cultural meaning, as well as technical and financial matters, are at stake in the global-local dialectic of technoscience, and people categorized as "indigenous" might well be more cosmopolitan than those labeled "Westerners" in key respects. Global/local does not translate as western/elsewhere or modern/traditional.[25] The biotechnology involved in the HGDP is of interest to prospective host countries, and several groups have also expressed interest in possible medical benefits as well as in participating in a project that contributes to defining humanity transnationally.

Europeans were among the first indigenous peoples to proceed with HGDP research. In 1994, the European Union provided $1.2 million to set up 25 labs from Barcelona to Budapest to study questions about European genetic diversity and paleoanthropological history. Of course, the "races" of Europe were also central to the scientific constructions of human unity and diversity in the nineteenth century, and people elsewhere in the world have not always been so convinced this is the way to think about the matter. But regional committees to pursue the HGDP have been set up in North America, South and Central America, and Africa as well as Europe, while India, China, and Japan had declined by late 1994 (Kahn 1994:722). Organized Native Americans in the United States predictably have been divided. The Euchees and Apaches of Oklahoma decided to participate in the HGDP, in part because of their interest in research on the genetics of diabetes, a major health problem of Native

Americans. At the same time, in the summer of 1994 a broad coalition of consumer, indigenous, environmental and nongovernmental organizations working on development issued a statement calling on all participants "to work with parallel movements led by indigenous nations to eliminate federal funding to the Human Genome Diversity Project" (Bereano 1994). The major reason was the potential for commercialization, especially in the form of patents on human genes and proteins, without benefit to the sampled populations whose body parts would become museum specimens in an updated form. The Europeans have also shown considerable resistance to the patent fever that grips biotechnology in North America, and the European Parliament legislated that publicly funded research should not give rise to privately held patents (Bereano 1995).

A troubling leitmotiv in the Guaymi cell-line dispute returns us to the narratives, images, and myths with which I want to conclude this meditation on the human family. In the midst of the polemics, Pat Moony of the RAFI was quoted as saying, "When a foreign government comes into a country, takes blood without explaining the real implications to local people, and then tries to patent and profit from the cell line, that's wrong. Life should not be subject to patent monopolies" (RAFI 1994:7). The patent monopoly part is true enough, but penetration by a foreign power to take blood evokes much more than intellectual property issues. Indeed, some of the indigenous organizations critical of the HGDP called it the "vampire project" (Kahn 1994:721). I cannot help but hear Moony's quote in the context of periodically surfacing stories in Latin America about white North Americans stealing body parts, sucking blood, and kidnapping children to be organ donors. The factual accuracy of the accounts is not the point, even though the dubious standards of evidence to which commentators have been held when the stories appear in U.S. news articles and radio talk shows appall me. What matters in this chapter is the stories themselves, that is, the ready association of technoscience with realms of the undead, tales of vampires, and transgressive traffic in the bloody tissues of life. Sampling blood is never an innocent symbolic act. The red fluid is too potent, and blood debts are too current. Stories lie in wait even for the most carefully literal-minded. Blood's translations into the sticky threads of DNA, even in the aseptic databases of cyberspace, have inherited the precious fluid's double-edged power. The genome lives in the realm of the undead in myriad ways that cannot be contained by rational intentions, explicit explanations, and literal behavior. The stories get at structures of power and fantasy that must be faced in all their displaced, uncanny truth.

Table 8.1, Night Births and Vampire Progeny, is a rough guide through a tiny region of the mine-strewn territory. My chart is indebted to three

Table 8.1   Night Births and Vampire Progeny

Image	OncoMouse™	Gorilla-suited bride	SimEve
Source	*Science* magazine	*American Medical News*	*Time* magazine
Kin category	species	family	race
Reproductive practice	genetic engineering	professional investment	cybergenesis by morphing
Narratives and myths	night births in the laboratory	Bad investments yield polluted offspring.	masculine parthenogenesis
	scientific enlightenment	Reverse alchemy turns gold into base metal.	mind children
	Plato's allegory of the cave		Orestian Trilogy
	heroic quest	racialized heterosexuality vampire-toothed bride	Pygmalion and Galatea
Slogan	"where better things for better living come to life"	"If you've made an unholy alliance … "	"love that will forever remain unrequited."

mainline publications within technoscientific professional and popular culture. Pursuing the symptomatic logic of this chapter, my technique is resolute over-reading. I know no better strategy to deal with the vermin-infested normality of rational discourse. Just state the obvious. Say what should not have to be said.

Running several times in *Science* magazine in 1989–1990, Du Pont's wonderful advertisement for OncoMouse™, the first patented animal in the world, provides my first text (Figure 8.2, Stalking Cancer).[26] OncoMouse™ contains a cancer-causing bit of DNA, called an onco-gene, derived from the genome of another creature and implanted by means of genetic engineering techniques. A model for breast cancer research, the redesigned rodent is like a machine tool in the workshops for the production of knowledge. OncoMouse™ is a transgenic animal whose scene of evolution is the laboratory. Inhabiting the nature of no nature, OncoMouse™'s natural habitat is the fully artifactual space of technoscience. Symbolically and materially, OncoMouse™ is where the categories of nature and culture implode for members of technoscientific cultures. For that very reason, the mouse has been at the center of

Fig. 8.2. Du Pont advertisement from *Science* magazine for OncoMouse™, April 27, 1990. Courtesy of Du Pont NEN products. On May 19, 1995 Du Pont announced its intent to divest its medical products businesses. The former Du Pont NEN products business will become NEN life science products.

controversy since its production. Defined by a spliced genome, identified with a spliced name, patented, and trademarked, OncoMouse™ is paradigmatic of nature enterprised up. What interests me here, however, are the stories that are crusted like barnacles onto the striking advertising image.

Du Pont's white mouse is in the midst of a heroic travel or quest narrative and part of a noble hunt in which the cancer enemy is stalked. Epistemophilia, the lusty search for knowledge of origins, is everywhere. The mouse climbs out of a womblike, geometric cave toward the light of knowledge, evoking the narrative elements of the Western Enlightenment and of Plato's allegory of the cave. OncoMouse™ is "available to researchers only from Du Pont, where better things for better living come to life." Like it or not, we are catapulted into the narrative fields that contain Frankenstein and his monster and all the other alluring scenes of night births in the mythological culture of science. The laboratory repeatedly figures as an uncanny place, where entities that do not fit, do not belong, cannot be normal—that transgress previously important categories—come into being. I am drawn to the laboratory for this essential narrative of epistemological and material power. How could feminists and antiracists in this culture do without the power of the laboratory to make the normal dubious? Raking ambivalence and strong visitations from a culturally specific unconscious, however, are the price of this alliance with the creatures of technoscience. Reproduction is afoot here, with all of its power to reconfigure kinship. In the proliferating zones of the undead, the kin categories of species are undone and redone, all too often by force. Consciously or unconsciously, whoever designed this ad knew all the right stories. Enlightenment has never been more pregnant with consequences—semiological, financial, and technological—for the human family.

Family imagery is much more explicit and far more ominous in my next text, an ad for Prepaid Medical Management, Inc. (PreMed), which was published in *American Medical News* on August 7, 1987 (Figure 8.3. If you've made an unholy HMO alliance, perhaps we can help). PreMed tells physicians that it can help get them out of unprofitable contracts with health maintenance organizations (HMOs) that had promised a financially sound patient base and quality care but instead delivered profits for distant shareholders and high administrative fees for doctors. PreMed claims to have aided physicians in establishing locally controlled and fiscally sound HMOs in which doctors could determine whom they treated and how they practiced medicine. There is little question that these are pressing concerns in the context of a medicine-for-profit system, in which many patients are uninsured, underinsured, or covered

If You've Made An Unholy HMO Alliance, Perhaps We Can Help.

All across the country, physicians who once had visions of a beautiful marriage to an HMO have discovered that the honeymoon is over.

Instead of quality care and a fiscally sound patient-base, they end up accepting reduced fees and increased risks. Plus a lot of new rules that make it more like administrating than practicing medicine.

And while you're doing a lot of the administration, the HMO is charging administrative fees in the neighborhood of 17% to 20%. Small wonder, these HMOs continue to reward distant shareholders with record returns while participating physicians get nothing but grief.

Most doctors have felt there's little or no hope—that "we're trapped" with no way out. But that's not true. There are alternatives. And we're in the business of providing them.

At Prepaid Medical Management Inc., we help physicians develop their own HMOs, negotiate with hostile HMOs or leave contractual situations that have turned sour. And we've been doing it for seven years. In the process, we've helped a number of physician groups profitably leave contracts with national HMOs and establish locally controlled plans with solid fiscal track records.

If you'd like to discuss the alternatives available to your group or IPA, give PreMed President Ed Petras a call. It's not too late to do something about an unholy alliance.

PREMED®

8400 Normandale Lake Blvd.
Suite 1180
Minneapolis, MN 55437
1-800/833-7612

Fig. 8.3. Courtesy of Premed. Advertisement from *American Medical News,* August 7, 1987.

by public plans that pay much less for services than private insurers. Although not referring directly to the larger context, the ad appeared in the midst of an epidemic of national publicity about high Medicare and Medicaid patient loads in urban HMOs, African American crack-addicted and AIDS-infected mothers and babies in inner cities, and astronomical malpractice insurance costs, particularly for urban obstetricians.

The PreMed verbal text makes no reference to race, gender, or class, but I think these codes structure the ad. "Accepting reduced fees and increased risks" is a code for accepting too many poor patients who do not have private insurance. The code, if not a more complicated reality, biases readers to see those high-risk, poor patients as overwhelmingly people of color, especially African American. The visual scene of a wedding and the verbal text about an unholy alliance propel the reader to see the patient as female and black and the doctor as male and white. An unholy alliance is "miscegenation," the bloodsucking monster at the heart of racist and misogynistic terror.[27]

Finally, it is the double disguise, the twice-done veiling of the bride that makes the ad so flagrantly about what it literally covers up with a joke: the class-structured, racialized, sexual politics of U.S. reproductive health and the further withdrawal of medical services from already underserved populations. A white-medical-coat-clad, stethoscope-wearing, prosperous-looking white man with just the right amount of graying hair is putting a gold wedding band on the ring finger of a black gorilla-suited bride in a white wedding gown and veil. The bride is doubly not there. Present are only two disguises: the wedding dress and gorilla suit. The implied infected or addicted pregnant Black woman who is always, in the code, on welfare, is denied in advance.[28] The surface of the ad insists that it is I, not PreMed, who am both making the connection of the gorilla-suited bride with African American women and putting the wedding scene into the context of reproductive health care. Can't I take a joke? But my power to be amused is vitiated by the searing memory of just where African American women fit historically into systems of marriage and kinship in white heterosexual patriarchy in the United States. *Miscegenation* is still a national racist synonym for infection, counterfeit issue unfit to carry the name of the father, and a spoiled future. The bitter history of the scientific and medical animalization of people of African descent, especially in the narratives of the great chain of being that associated apes and Black people, further accounts for my poor sense of humor. The gorilla suit cannot be an innocent joke here, and good intentions are no excuse. The lying disguises cannot hide what they deny.

But this bride is less a living—or a reconstructed—gorilla than an undead monster. She is not a creature in an Akeley diorama, whose natural types always glowed with health. The gorilla-suited bride is the type of no type. Her lips are parted just enough to show the gleam of a bright white tooth. The bride is a vampire, equipped with the tool for sucking the blood of the husband and polluting his lineage. The shining tooth echoes the brilliant gold of the wedding ring. The wedding night bodes ill. The conventional trope of the scientist-husband of nature generating the legitimate, sacred fruit of true knowledge in the womb of the wife's body is engaged here with chilling modifications. A metaphor for the magical power of science, alchemy is about the generative sexual practices of the craft, which are a kind of marriage that yields gold from base metal. Alchemy is about holy alliances, true marriages with gleaming children. In the PreMed advertisement, the narrative is reversed, and an "unholy alliance" threatens to mutate the promised gold of a medical-career investment into the base metal of a nonproductive practice. "If you've made an unholy alliance, perhaps we can help." Call upon PreMed and enjoy the fruit of a productive union. Be flexible; make the required structural adjustments to stimulate the production of wealth—and its flow upward to the deserving professional classes. Leave that unnatural and unprofitable alliance with infected bodies. A healthy family life demands no less.

The PreMed ad almost seems out of its time. It shouldn't still be possible to publish such an image in a scientific medical magazine. But it is possible. The fierce resurgence of explicit racist, sexist, and class-biased discourse of many kinds all over the world, and exuberantly in the United States, give all too much permission for this merely implicit and latent joke.

My third text, by contrast, wants to be firmly on the side of the antiracist angels. All the signs of liberal multiculturalism pervade *Time* magazine's cover image for its special fall 1993 issue on immigration (Figure 8.4, The New Face of America). These angels, however, turn out to exist in cyberspace. The *Time* cover is a morphed portrait of a being I call SimEve. In the background is a matrix of her mixed cybergenetic kin, all resulting from different "racial" crosses effected by a computer program. "Take a good look at this woman. She was created by a computer from a mix of several races. What you see is a remarkable preview of . . . The New Face of America." Indeed. We are abruptly returned to the ontology of databases and the marriage of genomics and informatics in the artificial life worlds that reconstitute what it means to be human. Here, the category so ethereally and technically reconfigured is race. In an odd computerized updating of the typological categories of the nineteenth and early twentieth centuries, the programmer who gave birth to SimEve and her many siblings generated the ideal racial

Fig. 8.4. Time magazine's morphed simeve. Backed by the racial-ethnic, computer generated matrix for Time's "Rebirthing America" special issue, Fall 1993. Photograph by Ted Thai. Reprinted with permission.

synthesis, whose only possible existence is in the matrices of cyberspace. Genetic engineering is not yet up to the task, so it falls to the computer sciences alone for now. Full of new information, the First Family reconstructed by Jay Matternes has had a transgenic change of form, to reemerge from *Time*'s computer womb as morphed ideal citizens, fit for the "Rebirthing of America." If the biotechnological genetics laboratory was the natural habitat and evolutionary scene fusing nature and culture for OncoMouse™'s version of the origin of life, SimEve's primal story takes place in the first morphing program for the personal computer, called Morph 2.0, produced by Gryphon Software Corporation.[29]

This technology has proved irresistible in the United States for 1990s mass cultural racialized kinship discourse on human unity and diversity. Never has there been a better toy for playing out sexualized racial fantasies, anxieties, and dreams. The reverie begins in cross-specific morphing, with the compelling computer-generated composite of human and chimpanzee faces on the cover of the 1992 *Cambridge Encyclopedia of Human Evolution*.[30] Like all portraits, this photograph records and shapes social identity. Soberly looking straight at the reader, the mature face is intelligent and beautiful. Like Carl Akeley's taxidermic reconstructions, this morphed face feeds a deep fantasy of touch across the ethnospecific categories of nature and culture. Unframed by any such specificity, the face seems to bring word about an original transformation in universal natural history.

On the contemporary human register, Gillette's shaving ads on television show the transformation of men's faces into each other across a racial spectrum, producing a utopic multiethnic male bonding. In the September 1994 Great American Fashion issue of the feminism-lite magazine *Mirabella*, the prominent photographer Hiro produced the computer-generated cover image from many photos of exquisitely beautiful multiracial, multiethnic women. Asked by the editors to give them a photo to represent "the diversity of America," Hiro did a simulated (and very light-skinned) woman.[31] A tiny microchip floats through space next to her gorgeous face. I read the chip as a sign of insemination, of the seminal creative power of Hiro, a modern Pygmalion/Henry Higgins creating his Galatea/Eliza Doolittle.[32] But the seminal power is not just Hiro's; it is the generative power of technology. Pygmalion himself has been morphed; he has become a computer program. Internationally, Benetton's ads, including its morphed racial transforms and its magazine *The United Colours of Benetton*, are the most famous. As Celia Lury put it, eschewing the distinction between cloth and skin, Benetton deals with the color of skin as a fashion palette (Lury 1994). Benetton produces a stunningly beautiful, young, stylish panhumanity composed by mix-and-match techniques. Diversity, like DNA, is the code of codes. Race, in Sarah Franklin's words, becomes a fashion accessory (Franklin 1994).

Pop star Michael Jackson brings this last point to its highest perfection. Spanning the range of chosen and imposed bodily "technologies" from cosmetic surgery, genetic skin disease, erotic performance in "private" and "public" life, clothing, costume, music videos—and mortal aging in spite of it all—Jackson's morphing practices have reshaped him by race, sexuality, gender, species, and generation. In the music video "Black and White," Jackson racially morphed himself by computer. In "real life," while a skin disease blanched his skin, he altered his facial features through cosmetic surgery, which produced race, generation, and gender effects. His childlike persona and his alleged transgressive relations with young boys morphed him into a permanent, if not altogether popular or safe, Peter Pan figure. His performance in Walt Disney's Epcot Center in the 3-D, 15-minute science-fiction production of *Captain E/O* caps the picture. As Ramona Fernandez put it, Jackson "tropes his body constantly.... In [*Captain E/O*] Jackson is both Mickey and a postmodern Peter Pan accompanied by bodies created by Lucas.... His transmuting body enacts and re-enacts the multiple problematics of race, generation, and gender" (1995:245). Analyzing Jackson's transmogrifications of himself and others through computer video technology into Cleopatra, ghoul, panther, machine, and superhero. Fernandez locates Jackson's socially significant shape-shifting within the traditions of African American tricksterism. The difference between human and machine, as well as the differences among species, are all fair game for Jackson's antiorigin narratives. As biologist Scott Gilbert writes, "If one wanted to find the intermediate morph of race, gender, and class, Michael Jackson may well be it. This makes him a science fiction 'representative' of humanity: and this is exactly how he depicts himself in *Captain E/O*."[33] This is humanity according to Epcot, where a potent trickster slipped into the monument to a clean and healthy America.[34]

Beginning unambiguously as an African American boy with striking talent, Jackson became neither black nor white, male nor female, man nor woman, old nor young, human nor animal, historical person nor mythological figure, homosexual nor heterosexual. These shape changes were effected through his art, the medical and computer technology of his culture, and the quirks of his body. Surely not even his brief marriage, least of all to Elvis Presley's daughter, could save him from the oxymoronically ineradicable stigmata of morphing. Science and fiction implode with special force in Jackson's iconic body, which is a national treasure of the first order. Jackson, however, is a much less safe representative for rebirthing the nation than the smoothly homogenized SimEve of *Time* magazine.

Not limited to specialists working for transnational corporations, weekly news magazines, official encyclopedias, or world-class entertainers, morphing is a participant reproductive sport too. In Las Vegas, in the Luxor, at the entrance to the gambling casino's reconstructed tomb of the eighteenth-dynasty Egyptian king Tutankhamen, there is a morphing machine that looks like the ordinary photomats in which one can get a quick snapshot. For five dollars a picture, one can enter the box, select the "gene machine" option, indicate whether one will be reproducing with a live partner or with a video model (human or animal), and then make further choices to determine the race and sex of the resulting child. The morphing machine is not choosy about the biological sex of the parent material. The racial menu for the child is African American, Hispanic, Asian, and Caucasian. Only if one chooses Caucasian are there any further choices, not an unfamiliar belief, but the choices are limited to hair and eye color. Then the machine photographs the parents-to-be, digitally combines them, and shoots out a child of the desired specifications. The child comes out at various ages, from toddler to adolescent. The gene machine is just another way of playing the combinations in Las Vegas at the end of the millennium.[35]

All this is surely not the naturalized typologies of Teddy Bear Patriarchy's early-twentieth-century racial discourse. Nor, in these popular cultural examples, including *Time*'s SimEve, are we subjected to PreMed's version of racial-sexual crossing. So why do I feel so uncomfortable? Shouldn't I be happy that the patently constructed nature of racial and gender categories is so obvious? In the face of resurgent racial hatred all around, what's wrong with a little obvious ideology for butterbrickle multiculturalism? Do we always have to order rocky road? Am I just having a dyspeptic attack of political correctness inevitably brought on by indulging in the pleasures of high-technology commodification within multinational capitalism? Why shouldn't the United Nations' *Family of Man* be morphed into the New World Order's *United Colours of Benetton*? Certainly the photography has advanced, and the human family seems naturally to be the story of the progress of technology.

To address the discomfort, let us look more closely at the *Time* special issue on immigration. In the note from the managing editor on page 2, we learn that *Time* imaging specialist Kin Wah Lam created the matrix of progeny in Figure 8.4 out of photographs of seven male and seven female models, each assigned to a racial-ethnic category. The top (female) and side (male) photos were electronically "mated" to produce the cyber-genetic offspring. Each figure is a pleasant-faced but undramatic nude bust, a "natural" man or woman, enhanced modestly by the understated makeup and minimal hairstyling. All the figures are young adults, and all the unions are chastely heterosexual, although presumably the computer could do a bit better than the technology of eggs and sperm on that score. In their defense, the editors' purpose was "to dramatize the impact of multiethnic marriage, which has increased dramatically in the United States during the latest wave of immigration." Still, the trope of reproductive heterosexual marriage is as firmly ensconced here as in the worlds of *The Fossil Footprint Makers of Laetoli* or *The Family of Man*. The mixing of immigration could be dramatized by many other practices. The sense of utter homogeneity that emanates from *Time's* matrix of diversity is numbing. The blacks are not very black; the blonds are not very blond; the range of skin color would require the best chromatography to distinguish one promising golden hue from another. These figures of the new humanity look like I imagine a catalog of replicants for sale off-world in *Blade Runner* might look—young, beautiful, talented, diverse, and programmed to fulfill the buyer's wishes and then self-destruct. Unlike the terrible white-supremacist scenes of *Birth of a Nation* in 1915, nothing about race and ethnicity in *Time's* "Rebirthing of a Nation" speaks about racial domination, guilt, and hatred. Nothing here is scary, so why am I trembling?

As Claudia Castañeda put it in her argument about "morphing the global U.S. family," "the racism here does not consist in the establishment of a hierarchy for domination based on biologized or even culturized racial difference. Its violence consists in the evacuation of histories of domination and resistance (and of all those events and ways of living that cannot be captured in those two terms) *through* technological (but still decidedly heterosexual) reproduction" (Castañeda 1994).[36] The denials and evasions in this liberal, antiracist, technophilic exercise are at least as thick as they are in the PreMed ad. All the bloody history caught by the ugly word *miscegenation* is missing in the sanitized term *morphing*. Multiculturalism and racial mixing in *Time* magazine are less achievements against the odds of so much pain than a recipe for being innocently raptured out of mundane into redeemed time. It is the resolute absence of history, of the fleshy body that bleeds, that scares me. It is the reconfirmation of the Sacred Image of the Same, once again under the sign of difference, that threatens national rebirth. I want something much messier, more dangerous, thicker, and more satisfying from the hope for multiculturalism. To get that kind of national reproductive health delivery is going to take addressing past and present sexualized racial power, privilege, exclusion, and exploitation. I suspect the nation will have to swallow the castor oil of sober accountability about such racialized sex before morphing looks like much fun to most of its citizens.[37]

Alongside a photo of the imaging specialist, labeled with a classically orientalist caption, "Lam creates a mysterious image," *Time*'s managing editor tells us still more about the cybergenesis of the woman on the cover: "A combination of the racial and ethnic features of the women used to produce the chart, she is: 15% Anglo-Saxon, 17.5% Middle Eastern, 17.5% African, 7.5% Asian, 35% Southern European and 7.5% Hispanic. Little did we know what we had wrought. As onlookers watched the image of our new Eve begin to appear on the computer screen, several staff members fell in love. Said one: 'It really breaks my heart that she doesn't exist.' We sympathize with our own lovelorn colleagues, but even technology has its limits. This is a love that must forever remain unrequited."

Themes running throughout the essay implode in this unlikely black hole. Early-century racialized ethnic categories reappear as entries in an electronic database for a truly odd statistical population analysis. A virtual woman is the result, fathered like Galatea, Pygmalion's creature, with which he fell in love. The curious erotics of single-parent, masculine, technophilic reproduction cannot be missed. SimEve is like Zeus's Athena, child only of the seminal mind—of man and of a computer program. The law of the nation, like that laid down by Athena for Athens in the Orestian trilogy, will be the Law of the Father. The Furies in cyberspace will not be pleased. In the narrative of romantic love, SimEve forever excites a desire that cannot be fulfilled. This is precisely the myth infusing dreams of technological transcendence of the body. In these odd, but conventional, technoscientific erotics, the actual limits of technology only spur the desire to love that which cannot and does not exist. SimEve is the new universal human, mother of the new race, figure of the nation; and she is a computer-generated composite, like the human genome itself. She is the second- and third-order offspring of the ramifying code of codes. She ensures the difference of no difference in the human family.

POSTSCRIPT

Throughout this chapter, racial discourse has persistently pivoted on sexual hygiene, and the therapeutic scene has been the theater of nature in the city of science. I am sick to death of bonding through kinship and "the family," and I long for models of solidarity and human unity and difference rooted in friendship, work, partially shared purposes, intractable collective pain, inescapable mortality, and persistent hope.

It is time to theorize an "unfamiliar" unconscious, a different primal scene, where everything does not stem from the dramas of identity and reproduction. Ties through blood—including blood recast in the coin of genes and information—have been bloody enough already. I believe that there will be no racial or sexual peace, no livable nature, until we learn to produce humanity through something more and less than kinship. I think I am on the side of the vampires, or at least some of them. But, then, since when does one get to choose which vampire will trouble one's dreams?

NOTES

1. Eugenics is race-hygiene or race-improvement discourse. For the history of eugenics, the classics include Haller 1963; Kevles 1985; Chorover 1979; and Cravens 1978. The development of Mendelian genetics after 1900, in the context of the dominant interpretation of the writing of the late-nineteenth-century German biologist August Weismann, which separated the passage of acquired characteristics from the genetic continuity of the germinal plasm, gradually eroded much of the racial and eugenic discourse I am discussing here. But many U.S. life scientists did not consistently rely on that distinction in their approach to evolution and race until near midcentury, and they certainly did not use Mendelian genetics to develop an antiracist scientific position. If they did insist on the separation of nature and culture, the effect was likely to harden into a genetic, trait-based eugenic doctrine even less open to "liberal," environmentalist contestation. For meanings of "race," see Stocking 1968; Stepan 1982; Barkan 1992; Harding 1993; Gould 1981; and Goldberg 1990.

2. For African American women's configurations of racial discourse, including scientific doctrines, in the late nineteenth and early twentieth centuries, see Carby 1987.

3. Charlotte Perkins Gilman's *Herland* (1979), serialized in *The Forerunner* in 1915, is full of the unself-critical white racialism that wounded so much of American feminism. Grant's writing (1916) is replete with unadulterated Nordic superiority and condemnation of race-crossing. A corporation lawyer, Madison Grant was a leader in eugenics, immigration restriction, and nature conservation politics—all preservationist, nativist, white-supremist activities. See Haraway 1989:57.

4. The full story of the Akeley African Hall is told in Haraway 1989:26–58, 385–88.

5. The discipline of population genetics—as opposed to the more ecologically minded population biology—has tended to exclude the development of organisms from their explanatory hypotheses and to rely almost exclusively on mutation and other ways to alter the frequency and products of individual genes to account for evolutionary change at all levels. Working against this severely limited focus, Scott Gilbert argues that for evolution above the subspecies or population level, changes in developmental pattern are key. Drawing on the molecular analysis of genes critical to homologous developmental pathways in a wide range of organisms—analytical procedures only possible since the late 1980s—Gilbert, Optiz, and Raff (1996) discuss the idea of homologies of process, as well as of older homologies of structure, in the context of a new evolutionary synthesis that emphasizes, unlike population genetics, embryology, macroevolution, and homology. In this new synthesis, the developmental or morphogenetic field is "proposed to mediate between genotype and phenotype. Just as the cell (and not its genome) functions as the unit of organic structure and function, so the morphogenetic field (and not the genes or the cells) is seen as a major unit of ontogeny,

whose changes bring about changes in evolution" (Gilbert, Optiz, and Raff 1996:357). I think this kind of evolutionary synthesis, in the context of the much more common "gene individualist" arguments in 1990s genomic and biotechnological discourse, is both refreshing and scientifically exciting. In the kind of work Gilbert signals and contributes to, neither the dominant gene/population nor genome/database formulations take one to the center of evolutionary questions. Gilbert, Optiz, and Raff's proposals should remind the reader that my chart seriously oversimplifies the debates going on today in molecular biology, development, and evolution.

6. For an overview of these complex developments, see Mayr and Provine 1980; Kaye 1986; Simpson 1967; Dobzhansky 1962; and Keller 1992.

7. The African American physical anthropologist Ashley Montagu Cobb at Howard University, one of the very few doctoral Black experts in the field, was not asked to sign the document. In the context of constitutively self-invisible, international, white scientific hegemony, his signature seemed to imply racial favoritism, not universalist, culture-free, scientific authority. In a spirit of peace, I won't even mention the gendering of the new plastic universal man—until he starts hunting in a species-making adaptation that will defeat my present restraint.

8. This account is an illustrative caricature of much more contradictory processes and practices within which the UNESCO documents lived. For a fuller but still inadequate account, see Haraway 1989:197–203. The cartoon version of the sharing way of life in the following section of this essay is argued in sober detail in Haraway 1989:186–230, 405–8.

9. The infamous gem of Man-the-Hunter theorizing was Washburn and Lancaster 1968. Woman the Gatherer made her debut in Linton 1971. She was fleshed out in Tanner and Zihlman 1976.

10. If one is weary of narrative drama and its unmarked psychoanalytic, political, and scientific universalist plots, feminist theory is the place to turn. See de Lauretis 1984:103–57; LeGuin 1988:1–12; Kim and Alarcón 1994; Sandoval 1991; V. Smith 1994.

11. Mr. Matternes refused permission to publish his painting in this chapter. *Fossil Footprint Makers of Laetoli* can be seen in *National Geographic Magazine,* April 1979, pp. 448–49.

12. Ongoing debate over the origin of modern *Homo sapiens* is another effort to track humanity's travels, with Africa again at the center of controversy. Since the late 1980s, the main alternative hypotheses are the multiregional origin account, founded on comparative anatomical studies, and the out-of-Africa theory, grounded in mitochondrial-DNA (mtDNA) analyses that are interpreted to mean that the most recent common ancestor of all living humans is a female who lived in Africa perhaps as recently as 112,000 years ago (Gibbons 1995:1271). The sperm contribute no mitochondria (a kind of cell organelle) to the fertilized egg cell, so mtDNA is inherited only through the female line. Providing a kind of clock, genetic changes accumulate over time. The mtDNA from the sampled populations living in Africa, itself an immense continent, shows the most variation compared to all other studied mtDNA taken from modern people living in different major geographical areas. This fact ought to give giant pause in the face of any generalizing genetic arguments about people of African descent, including the idea that modern races have much, if any, genetic meaning—if any such reminder is needed to maintain a skeptical attitude about claims that genetic bases justify contemporary racial classifications. This issue should be kept firmly in mind in addressing resurgent claims about heritability of IQ and association of IQ differences with ethnic/racial groups. The flap surrounding publication of *The Bell Curve* is the most important recent controversy. See Herrnstein and Murray 1994; Jacoby and Glauberman 1995. The fact that the greatest reservoir of human variation exists in Africa ought also to make organizers of genetic databases of human nuclear DNA think harder about how to develop reference composite standards for the species. Showing how deeply embedded the idea

of race still is in physical anthropology, Brendan Brisker (1995), a graduate student in the Anthropology Board at the University of California at Santa Cruz, analyzed the inadvertent use of racial typologies in the geographical sampling procedures and central arguments in the first mtDNA paleoanthropological studies. A special issue of *Discover* in November 1994 sketches the renewed debate in the 1990s about the scientific reality of race, and Lawrence Wright (1994) describes the controversy in the United States about racial typologies built into the U.S. census, which do not reflect the current multiplying racial/geographical categories and mixes claimed by people.

13. The special pull-out section of this *Science* magazine annual issue on the genome was dedicated to databases. See also Nowak 1993:1967.

14. Making life into a force of production and reorganizing biology for corporate convenience can be followed in Yoxen 1981; Wright 1986; and Shiva 1993.

15. The incisive critique of human sociobiology is Kitcher 1987. On unit-of-selection debates, see Brandon and Burian 1984. Defying classification as technical or popular, Dawkins 1976 and 1982 are the best expositions of the logic of the fierce competitive struggle to stay in the game of life, relying on strategies of flexible accumulation that strangely seem so basic to postmodern capitalism as well. For the theory of flexible accumulation in political economy, see Harvey 1989. For multilevel feminist working of the theme of flexibility in the American biomedical body, see Martin 1994.

16. The idea of nature and culture "enterprised up" is borrowed from Marilyn Strathern 1992, a treatment of assisted conception and English kinship in the period of British Thatcherism.

17. In eugenics thinking, the good of the "race" is the central ideological value. The collective aspect is hard to overstress. In 1990s genetic biomedical discourse, the "race"—either humanity as a whole or a particular racial category such as "white people"—plays little or no role, but individual reproductive investment decisions and individual genetic health are central.

18. A good place to start reading on the subject is Kevles and Hood 1992. Flower and Heath (1993) show how the semiotic-material definition of the human species in the world's genetic databases works through the multiple and heterogeneous processes that construct a reference sequence, or "consensus DNA sequence," as "the" human genome.

19. On "agency" in Internet habitats, see Waldrop 1994.

20. The supplement to the *Oxford English Dictionary* puts the first uses of the term *genom* [*sic*] in the 1930s, but the word did not then mean a database structure. That sense emerged from the consolidation of genetics as an information science, and especially since the 1970s.

21. I am indebted to an unpublished manuscript by the UCSC anthropology graduate student Cori Hayden (1994a). See Cavalli-Sforza et al. 1991; RAFI 1993:13; Spiwak 1993; and RAFI 1994.

22. I have been instructed by Giovanna DiChiro (1995a and 1995b) on what and who will count as science and as scientists. I draw also on Tsing 1993a and Cussins 1994. All three analysts trouble inherited categories of body and technology, nature and culture, wilderness and city, center and margin—all of which are part of producing the ideological distinction between modern and traditional that makes it seem odd for indigenous peoples to be savvy users and producers of genome discourse. For excellent analysis of problematic discourses of racial difference in ecofeminism, partly rooted in continuing separation of nature and culture and turning to "native" women as resources against the violations of industrial culture, see Sturgeon (1997).

23. See Star and Griesemer 1989 for development of the concept of boundary objects.

24. The scramble for the control of "biodiversity," itself quite a recent discursive object, is complex, global, and fraught with consequences for ways of life. Hayden 1994b discusses the 1991 "biodiversity prospecting" agreement between INBio, a Costa Rican

nonprofit environmental institute, and Merck, Sharpe and Dohme, the world's biggest pharmaceutical firm. The agreement is a controversial effort to control biopiracy and turn biodiversity resources in "gene rich" developing countries to their advantage. Biodiversity prospecting arrangements, the Human Genome Diversity Project, debt-for-nature swaps, the Biodiversity Convention, and GATT are just a few examples of the emerging institutional structure shaping human relations to nature in a world where the relations of technoscience to wealth and well-being have never been tighter. See World Resources Institute et al. 1993; Juma 1989; Shiva 1993.

25. See Tsing 1993b for a subtle ethnographic treatment of the complexities of what counts as marginal/central and local/global in an area of Indonesia that is also at the heart of environmental controversies.

26. Reprinted with permission of Du Pont NEN Products. On May 19, 1995, Du Pont announced its intent to divest its Medical Products businesses. The former Du Pont NEN Products business will become NEN Life Science Products.

27. According to the *Oxford English Dictionary,* the term *miscegenation* was coined in the United States in 1864.

28. A $3-million National Pregnancy and Health Survey of 2,613 women who gave birth at 52 hospitals around the nation in 1992 suggests how many and which U.S. pregnant women actually use substances that could harm the fetus (and the bottom line for an HMO). Conducted for the National Institute on Drug Abuse and released in September 1994, the study concludes that more than 5 percent of the four million U.S. women who gave birth in 1992 used illegal drugs, while about 20 percent used cigarettes and/or alcohol. Smokers and drinkers were more likely to use illegal drugs than were ethanol and nicotine abstainers. White women were more likely to drink or smoke during pregnancy than women of color (23 percent of white women drank, compared to 16 percent African American and 9 percent Hispanic; 24 percent of white women smoked, compared to 20 percent Black and 6 percent Hispanic). The racial categories here are crude and partial, but they still have limited utility. Poor, less-educated, unemployed, and unmarried women were more likely to use illegal drugs than more privileged women. About 11 percent of pregnant African American women used such drugs, compared to 5 percent of white and 4 percent of Hispanic mothers-to-be. That still means that more than half of the 221,000 pregnant women who used illegal drugs were white, 75,000 were Black, and 28,000 Hispanic. Alcohol and tobacco can harm a developing fetus as much as or more than illegal drugs but with less social and financial stigma. Overall, about 820,000 babies were born to smokers and 757,000 to imbibers. The same baby can show up in all the user categories. The study showed that most women tried to avoid illegal drugs, alcohol, and smoking during pregnancy, but few who used these powerful substances succeeded entirely. See Connell 1994:A7. The need for supportive, nonpunitive treatment for women trying to have a healthy pregnancy could hardly be clearer. Along with readily available, pro-woman, substance-treatment programs for those with any of these addictions, raising the incomes and improving the educations of women would likely be the most successful public health measures. Such measures would far outstrip the benefit to child and maternal health from intensive neonatal care units in high-tech hospitals, not to mention the dubious health results from criminalizing users. There is an unholy alliance between medicine as a system and millions of pregnant women in the United States, and it is reflected in the incomes of physicians compared to the incomes of at-risk mothers-to-be. The direction of flow of precious bodily fluids is the reverse of that suggested by the gleaming tooth and gold wedding band of the PreMed ad.

29. Selling in early 1994 for $239 for Macintoshes and $169 for Windows-using machines, Morph was widely used by scientists, teachers, special-effects designers for Hollywood movies, businesspeople making presentations, and law enforcement personnel, for

example, for aging missing children. A competitor in the market, PhotoMorph, came with graphics for practicing—"women turning into men, a girl turning into an English sheepdog, a frog turning into a chicken" (Finley 1994:F1–2). Finley illustrated his article with a series of morphed transformations between the competing personal computer giants, Apple Computer cofounder Steve Jobs and Microsoft founder Bill Gates. Mergers in the New World Order can be effected by many means. Needless to say, anyone still believing in the documentary status of photographs had better not get a copy of Morph, go to the movies, or look at the missing children on milk cartons.

30. Morphed photograph by Nancy Burson in Jones, Martin, and Pilbeam 1992. Thanks to Ramona Fernandez of the University of California at Santa Cruz for sending me this example.

31. Thanks to Giovanna DiChiro, University of California at Santa Cruz, for the tip on this image and for Hiro's comments from the *Today Show* of August 17, 1994.

32. The computer chip "impresses" its form on the morphed woman; the chip "informs" its electronic progeny in enduring Aristotelian doctrines of masculine self-reproduction that have "impressed" thinkers in the West for many centuries. The perfecting of the copy of the father in the child could be marred by the lack of transparency in the medium of the mother. Mutations on this theme proliferate in cyberspace, as in many other technoscientific wombs at the end of the Second Christian Millennium. For a discussion, which informs my chapter, of doctrines of impression, reproduction, and sanctity in medieval women saints, see Park 1995.

33. Scott Gilbert, personal e-mail communication, September 26, 1995, in response to a previous version of "Universal Donors." Thanks to Gilbert for insisting that I include "Black and White."

34. Fernandez (1995) emphasizes the trickster theme in her essay on traveling through Disney's many worlds, reading with the mixed cultural literacies required in the turn-of-the-century United States.

35. Thanks to Rosi Braidotti and Anneke Smelik, new parents of two lovely morphed off-spring, for this description of what they found possible in America in 1995. These sober European feminist theorists testified that they bonded instantly with their cyberchildren when they saw the compelling photographs of offspring so like and unlike themselves. The emotions were quite potent, even if the children were a little ethereal. I think there is potential here for population-reducing ways of having one's own children after all, in as great a number as one's willingness to put $5 in the machine will allow.

36. Castañeda's and my interpretations of the figures in this issue of *Time* evolved together in conversation, her hearing of my talk for a History of Consciousness colloquium Feb. 9, 1994, and my reading of her paper. I also draw on undergraduate students' readings of these images in a final exam in my fall 1993 course Science and Politics.

37. Meanwhile, fitting the analysis found in Emily Martin's *Flexible Bodies*, U.S. corporations attempt to capitalize on a particular version of multiculturalism. For an unembarrassed argument, see J. P. Fernandez 1993. See also Kaufman 1993.

REFERENCES

Appiah, Kwame Anthony. 1985. "The Uncompleted Argument: Du Bois and the Illusion of Race." *Critical Inquiry* 12:21–35.

———. 1990. "Racisms." In *Anatomy of Racism*, edited by D. T. Goldberg. Minneapolis: University of Minnesota Press.

Barkan, Elazar. 1992. *The Retreat from Scientific Racism: Changing Concepts of Race in Britain and the United States Between the World Wars*. New York: Cambridge University Press.

Bereano, Philip. 1994. "Broad Coalition Challenges Patents on Life." Press release, June 6.

Brandon, Robert N., and Richard M. Burian, eds. 1984. *Genes, Organisms, Populations: Controversies over the Units of Selection.* Cambridge: MIT Press.

Brisker, Brendan R. 1995. "Rooting the Tree of Race: mtDNA and the Origin of *Homo sapiens.*" Unpublished manuscript, Anthropology Board, University of Santa Cruz.

Carby, Hazel V. 1987. *Reconstructing Womanhood: The Emergence of the Afro-American Woman Novelist.* New York: Oxford University Press.

Castañeda, Claudia. 1994. "Transnational Adoption as U.S. Racist Complicity?" Paper read at American Ethnological Society Meeting, April.

Cavalli-Sforza, L. L., A. C. Wilson, C. R. Cantor, R. M. Cook-Deegan, and M. C. King. 1991. "Call for a Worldwide Survey of Human Genetic Diversity: A Vanishing Opportunity for the Human Genome Project." *Genomics* 11:490–91.

Chorover, Stephen L. 1979. *From Genesis to Genocide: The Meaning of Human Nature and the Power of Behavior Control.* Cambridge, MA: M.I.T. Press.

Connell, Christopher. 1994. "Pregnancy Study Shows Bad Habits." *The Santa Rosa Press Democrat,* September 13, A7.

Corteau, Jacqueline. 1991. "Genome Databases: Special Pull-out Section." *Science* (Genome Issue, Maps and Databases) 254:201–207.

Cravens, Hamilton. 1978. *The Triumph of Evolution: American Scientists and the Heredity Environment Controversy, 1900–47.* Philadelphia: University of Pennsylvania Press.

Cussins, Charis. 1994. "Ontological Choreography: Agency through Objectification in Infertility Clinics." *Social Studies of Science* 26: 575–610.

Dawkins, Richard. 1976, 1989. *The Self Gene.* New York: Oxford University Press.

———. 1982. *The Extended Phenotype: The Gene as a Unit of Selection.* London: Oxford University Press.

de Lauretis, Teresa. 1984. *Alice Doesn't: Feminism, Semiotics, and Cinema.* Bloomington: Indiana University Press.

DiChiro, Giovanna. 1995a. "Local Actions, Global Visions: Women Trnasforming Science, Environment, and Health in the U.S. and India." Ph.D. diss., History of Consciousness Board, University of California at Santa Cruz.

———. 1995b. "Nature as Community: The Convergence of Environment and Social Justice." In *Uncommon Ground: Towards Reinventing Nature,* edited by W. Cronon. New York: Norton, 298–320.

Dobzhansky, Theodosius. 1962. *Mankind Evolving: The Evolution of the Human Species.* New Haven: Yale University Press.

Du Bois, W. E. B. 1971. "The Conservation of Races." In *A W. E. B. Du Bois Reader,* edited by A. G. Paschal. New York: Macmillan, 19–31.

———. 1989. (orig. 1903). *The Souls of Black Folks.* New York: Bantam Books.

Erickson, Deborah. 1992. "Hacking the Genome." *Scientific American* (April):128–37.

Fernandez, John P. 1993. *The Diversity Advantage: How American Business Can Out-perform Japanese and European Companies in the Global Marketplace.* New York: Lexington Books, Macmillan.

Fernandez, Ramona. 1995. "Pachuco Mickey." In *From Mouse to Mermaid: The Politics of Film, Gender, and Culture,* edited by E. Bell, L. Haas, and L. Sells. Bloomington: Indiana University Press.

Finley, Michael. 1994. "The Electronic Alchemist." *San Jose Mercury News,* March 6, F1–2.

Flower, Michael, and Deborah Heath. 1993. "Anatamo-politics: Mapping the Human Genome Project." *Culture, Medicine, and Psychiatry* 17:27–41.

Franklin, Sarah. 1994. "Comments on Paper by Celia Lury." Center for Cultural Studies, University of California at Santa Cruz.

Gibbons, Ann. 1995. "Out of Africa—at Last?" *Science* 267 (March 3): 1272–73.

Gilbert, Scott F., John M. Optiz, and Rudy Raff. 1996. "Resynthesizing Evolutionary and Developmental Biology." *Developmental Biology* 173:357–12.

Gilman, Charlotte Perkins. 1979 [1915]. *Herland.* Serialized in *The Forerunner,* 1915. London: The Women's Press.

Goldberg, David Theo, ed. 1990. *Anatomy of Racism.* Minneapolis: University of Minnesota Press.

Gould, Stephen Jay. 1981. *The Mismeasure of Man.* New York: W. W. Norton.

Grant, Madison. 1916. *The Passing of the Great Race, or the Racial Basis of European History.* New York: C. Scribner.

Haller, Mark. 1963. *Eugenics, Hereditarian Attitudes in American Thought.* New Brunswick, NJ: Rutgers University Press.

Haraway, Donna. 1989. *Pimate Visions: Gender, Race, and Nature in the World of Modern Science.* New York: Routledge.

Harding, Sandra. 1993. *The "Racial" Economy of Science: Toward a Democratic Future.* Bloomington: Indiana University Press.

Harvey, David. 1989. *The Condition of Postmodernity: An Enquiry into the Origins of Cultural Change.* Oxford: Basil Blackwell.

Hayden, Cori. 1994a. "The Genome Goes Multicultural." Unpublished manuscript, Anthropology Board, University of California at Santa Cruz.

———. 1994b. "Purity, Property, and Preservation." Paper read at American Anthropological Association Meetings, November 27–December 4, at Atlanta, GA.

Herrnstein, Richard J., and Charles Murray. 1994. *The Bell Curve: Intelligence and Class Structure in American Life.* New York: The Free Press.

Jacoby, Russell, and Naomi Glauberman, eds. 1995. *The Bell Curve Debate: History, Documents, Opinions.* New York: Times Books/Random House.

Jones, Steve, Robert Martin, and David Pilbeam, eds. 1992. *The Cambridge Encyclopedia of Human Evolution.* New York: Cambridge University Press.

Juma, Calestous. 1989. *The Gene Hunters: Biotechnology and the Scramble for Seeds.* Princeton: Princeton University Press.

Kahn, Patricia. 1994. "Genetic Diversity Project tries Again." *Science* 266 (November 4): 720–22.

Kauffman, L. A. 1993. "The Diversity Game: Corporate America Toys with Identity Politics." *Village Voice* (August 31): 29–33.

Kaye, Howard L. 1986. *The Social Meaning of Biology: From Darwinism to Sociobiology.* New Haven: Yale University Press.

Keller, Evelyn Fox. 1992. *Secrets of Life, Secrets of Death.* New York: Routledge.

Kevles, Daniel J. 1985. *In the Name of Eugenics: Genetics and the Uses of Human Heredity.* New York: Knopf.

Kevles, Daniel J., and Leroy Hood, eds. 1992. *The Code of Codes: Scientific and Social Issues in the Human Genome Project.* Cambridge: Harvard University Press.

Kim, Elaine H., and Norma Alarcón, eds. 1994. *Writing Self, Writing Nation: Essays on Theresa Hak Kyung Cha's Dictée.* Berkeley: Third Woman Press.

Kitcher, Philip. 1987. *Vaulting Ambition: Sociobiology and the Quest for Human Nature.* Cambridge: MIT Press.

LeGuin, Ursula. 1988. "The Carrier-Bag Theory of Fiction." In *Women of Vision,* edited by D. D. Pont. New York: St Martin's Press, 1–12.

Linton, Sally. 1971. "Woman the Gatherer. Male Bias in Anthropology." In *Women in Perspective: A Guide for Cross-Cultural Studies,* edited by S.-E. Jacobs. Urbana: University of Illinois Press, 9–21.

Lury, Celia. 1994. "United Colors of Diversity: Benetton's Advertising Campaign and the New Universalisms of Global Culture, a Feminist Analysis." Paper delivered at the Center for Cultural Studies, University of California at Santa Cruz.

Martin, Emily. 1994. *Flexible Bodies: Tracking Immunity in American Culture from the Days of Polio to the Days of AIDS.* Boston: Beacon Press.

Marx, Karl. 1976. *Capital.* Translated by Ben Fowkes. Vol. 1. New York: Random House.

Mayr, Ernst, and William B. Provine. 1980. *The Evolutionary Synthesis: Perspectives on the Unification of Biology.* Cambridge: Harvard University Press.

Nowak, Rachel. 1993. "Draft Genome Map Debuts on Internet." *Science* 262 (December 24): 1967.

Park, Katherine. 1995. "Impressed Images: Reproducing Wonders." Paper read at Histories of Science/ Histories of Art, November 3–5, at Harvard and Boston Universities.

RAFI (Rural Advancement Foundation International). 1993. "Patenting Indigenous Peoples." *Earth Island Journal* (Fall):13.

———. 1994. "Following Protest, Claim Withdrawn on Guaymi Indian Cell Line." *GeneWatch: A Bulletin of the Council for Responsible Genetics* 9(3–4): 6–7.

Sandoval, Chéla. 1991. "U.S. Third World Feminism: The Theory and Method of Oppositional Consciousness in the Postmodern World." *Genders* 10:1–24.

Shiva, Vandana. 1993. *Monocultures of the Mind: Perspectives on Biodiversity and Biotechnology.* London: Zed Books.

Simpson, George Gaylord. 1967, revised (1st ed. 1949). *The Meaning of Evolution.* New Haven: Yale University Press.

Smith, Victoria. 1994. "Loss and Narration in Modern Women's Fiction." Ph.D. diss., History of Consciousness Board, University of California at Santa Cruz.

Spiwak, Daniela. 1993. "Gene, Genie, and Science's Thirst for Information with Indigenous Blood." *Abya Yala News* 7 (3–4): 12–14.

Star, Susan Leigh, and James R. Griesemer. 1989. "Institutional Ecology, 'Translations,' and Boundary Objects: Amateurs and Professionals in Berkeley's Museum of Vertebrate Zoology, 1907–39." *Social Studies of Science* 19:387–420.

Steichen, Edward. 1955. *The Family of Man.* New York: Maco Magazine Corporation for the Museum of Modern Art.

Stepan, Nancy Leys, and Sander L. Gilman. 1993. "Appropriating the Idioms of Science: The Rejection of Scientific Racism." In *"Racial" Economy of Science,* edited by S. Harding. Bloomington: Indiana University Press, 170–93.

Stepan, Nancy. 1982. *The Idea of Race in Science: Great Britain, 1800–1960.* Hampden, CT: Archon Books.

Stocking, George, Jr. 1968. *Race, Culture, and Evolution: Essays in the History of Anthropology.* New York: The Free Press.

———. 1993. "The Turn-of the Century Concept of Race." *Modernism/Modernity* 1:4–16.

Strathern, Marilyn. 1992. *Reproducing the Future: Anthropology, Kinship, and the New Reproductive Technologies.* New York: Routledge.

Sturgeon, Noël. 1996. "The Nature of Race: Discourses of Racial Difference in Ecofeminism." In *Ecofeminism: Multidisciplinary Perspectives.* Bloomington and Indianapolis: University of Indiana Press.

———. 1997. *Ecofeminist Natures.* New York: Routledge.

Tanner, Nancy, and Adrienne Zihlman. 1976. "Women in Evolution, Part I: Innovation and Selection in Human Origins." *Signs* 1:585–608.

Tsing, Anna Lowenhaupt. 1993a. "Forest Collisions: The Construction of Nature in Indonesian Rainforest Politics." Unpublished manuscript.

———. 1993b. *In the Realm of the Diamond Queen: Marginality in an Out-of-the-Way Place.* Princeton: Princeton University Press.

UNESCO. 1952. *The Race Concept: Results of an Inquiry.* Paris: UNESCO.

Waldrop, M. Mitchell. 1994. "Software Agents Prepare to Sift the Riches of Cyberspace." *Science* 265 (August 12): 882–83.

Washburn, S. L., and C. S. Lancaster. 1968. "The Evolution of Hunting." In *Man the Hunter,* edited by R. Lee and I. DeVore. Chicago: Aldine, 293–303.

World Resources Institute, INBio (Costa Rica), Rainforest Alliance (U.S.), and the African Centre for Technology Studies (Kenya), eds. 1993. *Biodiversity Prospecting: Using Generic Resources for Sustainable Development.* A contribution to the WRI/IUCN/UNEP G*lobal Biodiversity Strategy,* WRI/IUCN/UNEP, May 1993.

Wright, Lawrence. 1994. "One Drop of Blood." *The New Yorker* (July 25): 46–55.

Wright, Susan. 1986. "Recombinant DNA Technology and Its Social Transformation." *Osiris* 2 (second series): 303–60.

Yoxen, Edward. 1981. "Life as a Productive Force: Capitalizing the Science and Technology of Molecular Biology." In *Science, Technology, and the Labour Process,* edited by L. Levidow and B. Young. London: CSE Books, 66–122.

9

CYBORGS TO COMPANION SPECIES: RECONFIGURING KINSHIP IN TECHNOSCIENCE

for my father, 50 years a sportswriter at the *Denver Post*

EXCERPTS FROM "NOTES OF A SPORTSWRITER'S DAUGHTER," SPRING–FALL, 2000

(1) Cayenne, our year-old Australian Shepherd bitch, is in full "teenage mode," popping like drops of Leyden frost on a hot stove. Things she did on cue yesterday without question, today fail to engage her roving mind. Back to basics! I have written "shut up and train" across my forehead. Peace reigns in her lusty soul if she gets at least five miles a day of running and a few other bouts of vigorous activity. Cheap to a good home . . .

(2) Ms. Cayenne Pepper continues to colonize all my cells—a sure case of what the biologist Lynn Margulis calls symbiogenesis. I bet if you checked our DNA, you'd find some odd transfections between us. Her saliva must have the viral vectors; her darter-tongue kisses are irresistible. Co-evolution in the naturecultures of companion species land has as many punctuated equilibria as Stephen J. Gould could ever have wished. Margulis and Gould, opponents in evolutionary theory in their lives, are fused in Cayenne and me.

(3) Roland Dog, our Aussie-Chow six-year-old, was beautiful at the agility trials Saturday. He had speed, drive, heart, and he was paying attention. We would have gotten two legs out of our three runs if I hadn't

literally gotten a mental white-out at a jump choice in mid-course each time. He was great; I was middle-aged and unused to even novice high functioning after getting up at 4 A.M. to drive a hundred miles to spend time with my dogs!

(4) Dear Vicki,[1]

Watching Roland with you lurking inside my head over the last week made me remember that such things are multidimensional and situational, and describing a dog's temperament takes more precision than I achieved. We go to an off-leash, large, cliff-enclosed beach in Santa Cruz almost every day. There are two main classes of dogs there: retrievers and meta-retrievers. Roland is a meta-retriever. (My partner points out there is really a third class of dogs too—the "nons"—not in the game at issue here.) Roland'll play ball with us once in a while (or anytime we couple the sport with a liver cookie or two), but his heart's not in it. The activity is not really self-rewarding to him, and his lack of style there shows it. But meta-retrieving is another matter entirely. The retrievers watch whoever is about to throw a ball or stick as if their lives depended on the next few seconds. The meta-retrievers watch the retrievers with an exquisite sensitivity to directional cues and microsecond of spring. These meta-dogs do not watch the ball or the human; they watch the ruminant-surrogates-in-dog's-clothing. Roland in meta-mode looks like an Aussie-Border Collie mockup for a lesson in Platonism. His forequarters are lowered, forelegs slightly apart with one in front of the other in hair-trigger balance, his hackles in mid-rise, his eyes focused, his whole body ready to spring into hard, directed action. When the retrievers sail out after the projectile, the meta-retrievers move out of their intense eye and stalk into heading, heeling, bunching, and cutting their charges with joy and skill. The good meta-retrievers can even handle more than one retriever at a time. The good retrievers can dodge the metas and still make their catch in eye-amazing leaps—or surges into the waves, if things have gone to sea. Since we have no ducks or other surrogate sheep or cattle on the beach, the retrievers have to do duty for the metas. Some retriever people take exception to this multitasking of their dogs (I can hardly blame them), so those of us with metas try to distract our dogs once in a while with some game they inevitably find much less satisfying. I drew a mental Larson cartoon on Thursday watching Roland, an ancient and arthritic Old English Sheepdog, a lovely red tricolor Aussie, and a Border Collie mix of some kind form an intense ring around a shepherd-lab mix, a plethora of motley Goldens, and a game pointer who hovered around a human who—liberal individualist in Amerika to the end—was trying to throw his stick to his dog only. Meanwhile, in the distance, a rescue whippet was eating up sand in roadrunner fashion, pursued by a ponderous, slope-hipped German Shepherd Dog.

Why do I feel all of this is about the extended, cross-species family of a sportswriter's daughter?

PREAMBLE

This is a chapter of fragments, of work-in-progress, of dog-eaten props and half-trained arguments. But I offer this set of notes toward a future work as a training diary for reshaping some stories I care about a great deal, as a scholar and as a person in my time and place. Telling a story of co-habitation, co-evolution, Whiteheadian concrescence, and embodied cross-species sociality, "Kinship in Technoscience" compares two cobbled together figures—cyborgs and companion species—to ask which might more fruitfully inform livable politics and ontologies in current life worlds. These figures are hardly polar opposites. Cyborgs and companion species each brings together the human and non-human, the organic and technological, carbon and silicon, freedom and structure, history and myth, the rich and the poor, the state and the subject, diversity and depletion, modernity and postmodernity, and nature and culture in unexpected ways. Besides all that, neither a cyborg nor a companion animal pleases the pure of heart who long for better protected species boundaries and sterilization of category deviants. Nonetheless, the differences between even the most politically correct cyborg and an ordinary dog matter.

I begin with stories, histories, ecologies, and technologies of the space-faring NASA machine-organism hybrids named cyborg in 1960. Those cyborgs were appropriated to do feminist work in Reagan's Star Wars times of the mid-1980s. By the end of the millennium, however, cyborgs could no longer do the work of a proper herding dog to gather up the threads needed for serious critical inquiry. So I go happily to the dogs to explore the birth of the kennel in order to help craft tools for science studies in the present time, when secondary Bushes threaten to replace the old growth of more livable naturecultures in the carbon budget politics of all water-based life on earth. Having worn the scarlet letters, "Cyborgs for earthly survival!" long enough, I now sport a slogan only Schutzhund women could have come up with, when even a first nip can result in a death sentence: "Run fast; bite hard!"[2]

This is a story of biopower and biosociality, as well as of technoscience. Like any good Darwinian, I tell a story of evolution. In the mode of (nucleic) acidic millennialism, I tell a tale of molecular differences, but one less rooted in Mitochondrial Eve in a neocolonial Out of Africa and more rooted in those first mitochondrial canine bitches who got in the way of man making himself yet again in the Greatest Story Ever Told. Instead, those bitches insisted on the history of companion species, a very mundane and ongoing sort of tale, one full of misunderstandings, achievements, crimes, and renewable hopes. And so, mine is a story told by a student of the sciences and a feminist of a certain generation who has gone utterly to the dogs, literally. Dogs, in all their historical complexity, matter here. Dogs are not just an alibi for other themes; dogs are fleshly material-semiotic presences in the body of technoscience. Dogs are not surrogates for theory here; they are not here just to think with. They are here to live with. Partners in the crime of human evolution, they are in the garden from the get-go, wily as Coyote.

Whitehead (1948, 1969) talks about "the concrete" as "an actual entity as a concrescence of prehensions"—there are no pre-constituted subjects and objects in his world. Stressing the processual character of reality, he called actual entities "actual occasions." His philosophy is one among many resources for figuring "aliberal" subjects and objects, which/who are constituted in relational process. Subjects and objects (and kinds, genres, genders) are the products of their own relating, through many kinds of "emergent ontologies" (Verran 2001), or "ontological chore-ographies" (Cussins 1996); or "scale-making" in space and time (Tsing 2000). For Whitehead, "objectifications" had to do with the way "the potentiality of one actual entity is realized in another actual entity." This is very promising philosophical bait for training science studies folk to understand companion species in both storied deep time, which is chemically etched in the DNA of every cell, and in very recent doings, which leave more odoriferous traces.

And like a decadent gardener who can't keep good distinctions between natures and cultures straight, the shape of my kin networks looks more like a trellis, an esplanade, than a tree. You can't tell up from down, and everything seems to go sidewise. Such snake-like, sidewinding traffic is one of my themes. My garden is full of snakes, full of trellises, full of indirection. Instructed by evolutionary population biologists and bioanthropologists, I know that multidirectional gene flow—multidirectional flows of bodies and values—is and has always been the name of the game of life on earth. In that spirit, it is certainly the way into the kennel. Unfairly, I will risk alienating my old doppelganger, the cyborg, in order to try to convince my colleagues and comrades that dogs might be better guides through the thickets of technobiopolitics in the Third Millennium of the Current Era.

I. CYBORGS

An Evolutionary Cartoon of Enhanced Man in Space

Most Western narratives of humanism and technology require each other constitutively: how else could man make himself? Man births himself through the realization of his intentions in his objects; that is the quest story of masculinist, single-parent, self-birthing. Those objects—those realized intentions—return in the form of the threat of the instrument's surpassing the maker; thus emerges the dialectic of technophilic, technophobic apocalypse. The myth system is simple and old; cyborg practices are much less simple and much more recent.[3]

The term "cyborg" was coined by Manfred Clynes and Nathan Kline in 1960 to refer to the enhanced man who could survive in extraterrestrial environments. They imagined the cyborgian man-machine hybrid would be needed in the next great technohumanist challenge—space flight. The travel tale is a birth narrative. A designer of physiological instrumentation and electronic data-processing systems, The Australian-Austrian Clynes was the chief research scientist in the Dynamic Simulation Laboratory at Rockland State Hospital in New York. Director of research at Rockland State, Kline was a clinical psychiatrist. Their article was based on a paper the authors gave at the Psychophysiological Aspects of Space Flight Symposium sponsored by the U.S. Air Force School of Aviation Medicine in San Antonio, Texas. Enraptured with cybernetics, Clynes and Kline (1960, 27) thought of cyborgs as "self-regulating man-machine systems." That paper featured a white lab rat implanted with an osmotic pump under its skin to permit the continuous injection of chemicals to regulate basic physiological parameters. The join of pump/machine and organism, effected through the engineering of feedback-controlled communication circuits, produced an ontologically new, historically specific entity: the cyborg, the enhanced command-control-communication-intelligence system (C^3I). Here, the machine is not other to the organism, nor is it a simple instrument for effecting the purposes of the organism. Rather the machine and the organism are each communication systems joined in a symbiosis that transforms both.

This cyborg is a technohumanist figure of the Cold War and the heyday of the space race. Escape from the earth, from the body, from the limits of merely biological evolution is the message and the plot. Man is his own invention; biological evolution fulfills itself in the evolution of technology. Any emergent ethics of care for the hybrid machine-organism resolves into blissed-out, jacked-in terror of the communications-machinic self. A plethora of actors and a motley of agencies reduce to One, at least in the myth. Co-evolution and mutual co-constitution in this story re-solve into the figure of transcendent self-surpassing, not into a tale of mundane and mortal co-inhabiting, where the struggle for a practice of co-flourishing across categories might be sought. And, naturally, rats go first where no man has gone before.

Plainly, not all cyborgs have agreed to abide by this birth contract. In my own "Cyborg Manifesto" in the mid-1980s, I tried to write another surrogacy agreement, another trope, another figure for living within and honoring the skills and practices of contemporary technoculture without losing touch with the permanent war apparatus of a non-optional post-nuclear world and its transcendent, very material lies. Cyborgs can be figures for living within contradictions, attentive to the naturecultures of mundane practices, in opposition to the dire myths of self-birthing, embracing mortality as the condition for life, and alert to the emergent historical hybridities actually populating the world at all its contingent scales.[4]

However, cyborg refigurations hardly exhaust the tropic work required for ontological choreography in technoscience. Indeed, I have come to see cyborgs as junior siblings in the much bigger, queer family of companion species, in which reproductive biotechno-politics are generally a surprise, sometimes even a nice surprise. I know perfectly well that a U.S. middle-aged white woman with a dog playing the sport of agility is no match for Man in Space or Bladerunner and their transgenic kin in the annals of philosophical inquiry or the ethnography of naturecultures. Besides, (1) self-figuration is not my task; (2) transgenics are not the enemy; and (3) contrary to lots of dangerous and unethical projection in the Western world that makes domestic canines into furry children, dogs are not about oneself. Indeed, that is the beauty of dogs. They are not a projection, nor the realization of an intention, nor the telos of anything. They are dogs; i.e...., a species in obligatory, constitutive, protean relationship with human beings.

There cannot be just one companion species; there have to be at least two to make one. It is in the syntax; it is in the flesh. Dogs are about the inescapable, contradictory story of relationships—co-constitutive relationships in which none of the partners pre-exist the relating, and the relating is never done once and for all. Historical specificity and contingent mutability rule all the way down, into nature and culture, into naturecultures. There is no foundation; there are only elephants supporting elephants supporting elephants all the way down. Dogs might be better guides to what Karen Barad (1995) calls intra-action than Niels Bohr's troubling quantum phenomena at the scale of wave forms and elementary particles. Inter-action implies that already existing actors get together and act. Intra-action implies something much messier, much less determinate, ontologically speaking. Whitehead knew that; he must have had a dog. Famously, Freud certainly did. No wonder he knew something about subject making.

I am certain that, in addition to composing "Notes of a Sportswriter's Daughter," I will soon write a "Companion Species Manifesto" (Haraway 2003c). For now my task is more modest. It has three parts; to wit, (1) establishing that companion animals are only one kind of companion species and that neither category is very old in American English, (2) appropriating molecular biologists to affirm an origin story good enough for dogs and humans to get on together, and (3) turning to cats, in the guise of tigers, to suggest how the technocultural apparatus of biodiversity practices and discourses in dogland torques the origin story toward a more salubrious complexity. I'll finish with a tangled cat's cradle figure for doing technoscience studies among companion species.

II. COMPANION SPECIES
Dramatis Personae

In United States English, "companion animal" is a recent category, linked to the medical and psycho-sociological work done in veterinary schools and related sites from the middle 1970s (Beck and Katcher 1996). This is the research that told us that, except for non-dog loving New Yorkers who worry to excess about unscooped dog shit in the streets, having a dog (or, *in extremis,* a cat or even a hamster) lowers one's blood pressure and ups one's chances of surviving childhood, surgery, and divorce. Certainly, written references in European languages to animals serving as companions, rather than as working or sporting dogs, for example, predates this biomedical, technoscientific literature by centuries. However, "companion animal" enters technoculture through the land-grant academic institutions housing the vet schools. That is, "companion animal" has the pedigree of the mating between technoscientific expertise and late industrial pet-keeping practices, with their democratic masses in love with their domestic partners, or at least with the non-human ones. Companion animals can be horses, dogs, cats, or a range of other beings willing to make the leap from pet or lab beast to the biosociality of service dogs, family members, or team members in cross-species sports. Generally speaking, one does not eat one's companion animals (nor get eaten by them); and one has a hard time shaking colonialist, ethnocentric, ahistorical attitudes to those who do.

"Companion species" is a much bigger and more heterogeneous category than companion animal, and not just because one must start including such organic beings as rice, bees, tulips, and intestinal flora, all of whom make life for humans what it is—and vice versa. I want to rewrite the keyword entry for "companion species" to insist on four tones simultaneously resonating in the linguistic, historical voice box that makes uttering this term possible. First, as a dutiful daughter of Darwin, I insist on the tones of the history of evolutionary biology, with its key categories of populations, rates of gene flow, variation, selection, and biological species. All of the debates in the last 150 years about whether the category denotes a real biological entity or merely figures a convenient taxonomic box provide the over- and undertones. Species is about biological kind, and scientific expertise is necessary to that kind of reality. Post-cyborg, what counts as biological kind troubles any previous category of organism. The machinic is internal to the organic and vice versa in irreversible ways. Second, schooled by Thomas Aquinas and other Aristotelians, I remain alert to species as generic philosophical kind and category. Species is about defining difference, rooted in polyvocal fugues of doctrines of cause. Third, with an indelible mark on my soul from a Catholic formation, I hear in species the doctrine of the Real Presence under both species, bread and wine, the transubstantiated signs of the flesh. Species is about the corporeal join of the material and the semiotic in ways unacceptable to the secular Protestant sensibilities of the American academy and to most versions of the human sciences of semiotics. Fourth, converted by Marx and Freud, I hear in species filthy lucre, specie, gold, shit, filth, wealth. In *Love's Body*, Norman O. Brown taught me about the join of Marx and Freud in shit and gold, in specie. I met this join again in modern U.S. dog culture, with its exuberant commodity culture, its vibrant practices of love and desire, its mongrel technologies of purebred subject and object making. Pooper scoopers for me is quite a joke. In sum, "companion species" is about a four-part composition, in which co-constitution, finitude, impurity, and complexity are what is.

A. Who's on First? An Account of Co-evolution[5]

Pleasures and anxieties over beginnings and endings abound in contemporary dog worlds. This should not be surprising when we are awash in millennial discourses. Why shouldn't dogs get in an apocalyptic bark or two? Dog tales demand a serious hearing; they concern the basic *dramatis personae* in the ecological theater and the evolutionary play of rescripted naturecultures in technonatural, biosocial modernity (Hutchinson 1965, Rabinow 1992). This modernity is a living fictional territory; it is always here and now, in the technopresent. With reference to anthropology's late and little-lamented "ethnographic present," the technopresent names the kind of time I experience inside the *New York Times* Science Tuesday section and on the front pages and business pages so attuned to the animation and cessation of NASDAQ. History in the technopresent is Whig time enterprised up (Strathern 1992); i.e., this history is reduced to the vehicle for getting to the technopresent. In the technopresent, beginnings and endings implode, such that the eternal here and now energetically emerges as a gravity well to warp all subjects and objects in its domain. I write this paper suspended in this odd, millennial, American chronicity; but in this dimensionally challenged medium, I sense some code fusions promising another and better story about animals, machines, and people. I sense the emergence of companion species after the departure of possessive individuals and hermetically sealed objects, who will have finally succumbed to their own alien invasion of the earth. In this paper, I want to tell the story of companion species in the context of diversity discourses in U.S. dog worlds.

Evolutionary origin stories are always a good place in U.S. technoscientific worlds to check for the moves of nature and culture on the board game of widely disseminated Western metaphysics and for the players in the current versions of the game. In recent years, the long-running dog-wolf romance has a stirring new series. The origin of dogs might be a humbling chapter in the story of *Homo sapiens,* one that allows for a deeper sense of co-evolution and co-habitation and a reduced exercise of hominid hubris in shaping canine natureculture.

Accounts of the relations of dogs and wolves proliferate, and molecular biologists tell some of the most convincing versions. Robert Wayne and his colleagues at UCLA studied mitochondrial DNA (mtDNA) from 162 North American, European, Asian, and Arabian wolves and from 140 dogs representing 67 breeds, plus a few jackals and coyotes (Vilá, et al. 1997). Their analysis of mtDNA control regions concluded that dogs emerged uniquely from wolves—and did so much earlier than scenarios based on archaeological data permit. The amount of sequence divergence and the organization of the data into clades support the emergence of dogs more than 100,000 years ago, with very few separate domestication events. Three-quarters of modern dogs belong to one clade; i.e., they belong to a single maternal lineage. The early dates give *Canis familiaris*[6] and *Homo sapiens* roughly the same calendar, so folks walking out of Africa soon met a wolf bitch who would give birth to man's best friends. And, building a genetic trellis—not a tree—as they went, both dogs and people walked back into Africa (Templeton 1999). These have been species more given to multidirectional traveling and consorting than to conquering and replacing, never to return to their old haunts again. No wonder dogs and people share the distinction of being the most well-mixed and globally geographically distributed large-bodied mammals. They shaped each over a long time. Their pedigrees are a proper mess.

Further—in a story familiar from the post–World War II studies of human population gene frequencies that were so important to the early 1950s anti-racist UNESCO statements and to subsequent reforms of physical anthropology and genetics teaching—dog mtDNA haplotypes do not sort out by breed, indicating that breeds have diverse doggish ancestries. "Pure" breeds are an institutional fiction, if one that threatens the health of animals regulated by the story. Variations of many genes and markers within breed exceed variations between populations of dogs and wolves. And, in another lab's study, "greater mtDNA differences appeared within the single breeds of Doberman pinscher or poodle than between dogs and wolves," even while "there is less mtDNA difference between dogs, wolves and coyotes than there is between various ethnic groups of human beings" (Coppinger and Schneider 1995, 33). Genetic difference studies are a high stakes game, and emphases on similarity or divergence shift with the theoretical bets laid.

Findings from Wayne's lab have been controversial, partly because the mtDNA clock doesn't measure up to the accuracy demanded by Swiss watchmakers. At an International Council for Archaeozoology symposium in 1998 at the University of Victoria, controversy waxed over Wayne's arguments. Relevant to this paper are implications for thinking about agency in dog-human interactions. Wayne argued that to domesticate dogs took a lot of skill, or it would have happened more often. His story bears the scent of the anatomically wolfish hunting dog, and this dog is a man-made hunting tool/weapon. In this version, morphologically differentiated dogs did not show up in the fossil or archaeology record until 12,000–14,000 years ago because their jobs in settled post–hunter-gatherer, paleoagricultural communities did not develop until then; so they got physically reshaped late in the relationship. People call the shots in both chapters of a story that makes "domestication" a one-sided human "social invention." But archaeozoological expert Susan Crockford, who organized the Victoria symposium, disagreed. She argued that human settlements provided a species-making resource for would-be dogs in the form of garbage middens and—my addition—concentrations of human bodily waste. If wolves could just calm their well-justified fear of *Homo sapiens*, they could feast in ways all too familiar to modern dog people. "Crockford theorizes that in a sense, wild canids domesticated themselves" (Weidensaul 1999, 57; Crockford 2000).

Crockford's argument turns on genes that control rates in early development and on consequent paedomorphogenesis. Both the anatomical and psychological changes in domesticated animals compared to their wild relatives can be tied to a single potent molecule with stunning effects in early development and in adult life—thyroxine. Those wolves with lower rates of thyroxine production, and so lower titres of the fright/flight adrenaline cocktail regulated by thyroid secretions, could get a good meal near human habitations. If they were really calm, they might even den nearby. The resulting pups who were the most tolerant of their two-legged neighbors might themselves make use of the caloric bonanza and have their own puppies nearby as well. A few generations of this could produce a being remarkably like current dogs, complete with curled tails, a range of jaw types, considerable size variation, doggish coat patterns, floppy ears, and—above all—the capacity to stick around people and forgive almost anything. People would surely figure out how to relate to these handy sanitary engineers and encourage them to join in useful tasks, like herding, hunting, watching kids, and comforting people. In a few decades, wolves-become-dogs would have changed, and that interval is too short for archaeologists to find intermediate forms.

Crockford made use of the 40-year continuing studies of Russian fur foxes, beginning in the 1950s, which have been much in the recent popular science news (Weidensaul 1999; Trut 1999; Belyaev 1969). Unlike domesticated animals, wild farmed foxes vigorously object to their captivity, including their slaughter. In what were originally experiments designed to select tamer foxes for the convenience of the Soviet fur industry, geneticists at the Siberian Institute of Cytology and Genetics found that by breeding the tamest kits from each fox generation—and selecting for nothing else—they quickly got very dog-like animals, complete with non-fox attitudes like preferential affectional bonding with human beings and phenotypes like those of Border Collies.[7] By analogy, wolves on their way to becoming dogs might have selected themselves for tameness. People got in the act when they saw a good thing.

With a wink and a nod to problems with my argument, I think it is possible to hybridize Wayne's and Crockford's evolutionary accounts and so shamelessly save my favorite parts of each—an early co-evolution, human-canine accommodation at more than one point in the story, and lots of dog agency in the drama of genetics and co-habitation. First, I imagine that many domestication sequences left no progeny, or offspring blended back into wolf populations outside the range of current scientific sensors. Marginally fearless wolfish dogs could have accompanied hunter-gatherers on their rounds and gotten more than one good meal for their troubles. Denning near seasonally moving humans who follow regular food-getting migration routes seems no odder than denning near year-round settlements. People might have gotten their own fear/aggression endocrine systems to quell murderous impulses toward the nearby canine predators who did garbage detail and refrained from threatening. Paleolithic people stayed in one place longer than wolf litters need to mature, and both humans and wolves reuse their seasonal sites. People might have learned to take things further than the canines bargained for and bring wolf-dog reproduction under considerable human sway. This radical switch in the biopolitics of reproduction might have been in the interests of raising some lineages to accompany humans on group hunts or perform useful tasks for hunter-gatherers besides eating the shit. Paleoagricultural settlement could have been the occasion for much more radical accommodation between the canids and hominids on the questions of tameness, mutual trust, and trainability.

And, above all, on the question of reproduction. It's on this matter that the distinction between dogs and wolves really hinges; molecular genetics may never show enough species-defining DNA differences. Rather, the subtle genetic and developmental biobehavioral changes through which dogs got people to provision their pups might be the heart of the drama of co-habitation. Human baby sitters, not Man-the-Hunter, are the heroes from doggish points of view. Wolves can reproduce independently of humans; dogs cannot. Even Italian feral dogs still need at least a garbage dump (Boitani et al. 1995). As Coppinger and Schneider summarized the case: "In canids with a long maturation period, growth and development are limited by the provisioning capacity of the mother. . . . Wolves and African hunting dogs solved the pup-feeding problem with packing behavior, in coyotes the male helps, and jackal pairs are assisted by the 'maiden aunt.' The tremendous success of the domestic dog is based on its ability to get people to raise its pups" (1995, 36). People are part of dogs' extended phenotype in their Darwinian, behavioral ecological, reproductive strategies. Pacé Richard Dawkins.

Two points emerge from this evolutionary origin story: (1) co-evolution makes humans and dogs companion species from "the beginning," but with historically changing and specific sets of interspecies biotechnosocial relations and with agency a mobile and distributed matter; and (2) the fine arts of molecular genetics and hormone biochemistry are indispensable for this account of the agency of nature in the person of dog-wannabe wolves. The latest in sequencing machinery, sampling protocols, and DNA comparison software are crucial to the tale of "nature" making the first moves in a "social" invention. But this nature does not have the shape of the specters from the recent U.S. science and culture wars, where unruly science studies people were accused of arguing that scientists invented nature rather than reported on her in a mood of humble truth-telling. Here, with those worried realist warriors, I am also arguing that hominids did not "invent" nature or culture (wolves become dogs), then or now, but that all of the players emerge in a kind of White-headean concrescence, where none of the actors precede, finished, their interaction. Companion species take shape in interaction. They more than change each other; they co-constitute each other, at least partly. That's the nature of this cat's cradle game. And the ontology of companion species makes room for odd bedfellows—machines; molecules; scientists; hunter-gatherers; garbage dumps; puppies; fox farmers; and randy bitches of all breeds, genders, and species.

I want to use the figure of companion species to do a lot of analytical and associative work. Figures are powerful attractors that collect up the hopes, fears, and interests of collectives. Figures promise to fulfill hopes in a sense related to Christian realism (Auerbach 1953). Companion species are figures of a relational ontology, in which histories matter; i.e., are material, meaningful, processual, emergent, and constitutive. In the past, I have written about cyborgs, and cyborgs are a kind of companion species congeries of organisms and machines located firmly in the Cold War and its offspring. Equally on my mind have been genetically engineered laboratory organisms like OncoMouse™, also companion species tying together many kinds of actors and practices. Dogs and humans as companion species suggest quite different histories and lives, compared to cyborgs and engineered mice, emergent over the whole time of species being for the participants. In much of my own work, I have tried to figure out the consequences for biology and for cultural theory and politics of the implosion of biologics and informatics in post–World War II life sciences. In this implosion, organisms lost their ontological privilege to genomes, those wonderful generators of new wealth, new knowledge, and mutated ways of living and dying. While I take for granted many of the consequences of the implosion of biologics and informatics in shaping ways of being and knowing in the technopresent, I am here attending to a related but different sort of implosion—that of the utterly "natural" and the wholly "technical," where, for example, endangered species in the necessarily managed wilderness wear electronic sensors and live in habitats monitored by satellites as a crucial part of their biological reproductive apparatus. It remains to be seen if this arrangement will be an Evolutionary Stable Strategy (Dawkins 1982), but it has surely become a figure of biosocial modernity. Simply put, biodiversity has become dependent upon high technology in many parts of the world. The physical implosion of the "natural" and the "technical," materially-semiotically speaking, is a normal, everyday, earthly fact in the most biophillic, diversity-committed communities every bit as much as in the most technophillic worlds. And none of it is innocent—or guilty.

Is there a moral to this story? Dogs invented themselves; they are not an invention of humans? Or dogs and people shaped each other in a long and complicated history, where the story of the wannabe dogs taking the first steps is more convincing than its opposite? If dogs are a human technology, so also is the reverse true, as part of an extended phenotype in a canine sociobiological tale. I like the co-evolution story better than either the version that the dogs did it, or the people did it. It redoes the story of the human place in nature in homely ways that also impact on fortifications between categories of nature and culture.

There are stakes here beyond what we think about dog evolution. The stakes are how we think about liveliness and agency in different worlds. We require a multi-species and a multi-expertise way of doing/thinking worlds and ways of life, and that requires muting the command/communication/control/intelligence idiom of cyborgs.

Companion species are, among other things, a serious feminist matter, right at the heart of the ongoing Western feminist effort to do better than recycling idioms of liberalism and their benefits-maximizing, bounded, and independent selves as archetypes of freedom. Companion species offer a kind of bypass surgery for liberal idioms of both individuals and of diversity. Companion species do this right in the belly of the monster—inside biotechnology and the New World Order, Inc. Genders, breeds, races, lines, species—all the kinds are in play in these narrative morphings, with material-semiotic consequences. This is concrescence from the point of view of the birth of the kennel, in ongoing, relentlessly historical layers of practice, where all the actors and agencies are not human.

B. Biodiversity Goes to the Dogs[8]

Genetic disease is not news to dog people, and perhaps especially pure-bred dog people. Many breeders and owners—some willingly, some not—have become used to thinking about the genetic difficulties common to their breeds and even about polygenic traits with unknown modes of inheritance and strong environmental and developmental components affecting expression, like canine hip dysplasia. In myriad ways, genetic disease discourse shapes communities of practice for owners, breeders, researchers, dog rescue activists, breed clubs, kennel clubs, journalists, shelter workers, veterinarians, dog sports competitors, and trainers. There is much to say about the fascinating cultures of genetic disease in dogs, but in this paper I want to focus on a much more unsettling topic in purebred dog land: genetic diversity in small populations. First, let us look at why genetic diversity concerns are news—and hard to digest news—for most dog people, in spite of the long history of pop-ulation genetics and its importance for the modern theory of natural selection and the neo-Darwinian synthesis and its offspring.

Genetic culture for both professionals and non-professionals, espe-cially but not only in the United States, has been strongly shaped by medical genetics. Human genetic disease is the moral, technoscientific, ideological, and financial center of the medical genetic universe. Typo-logical thinking reigns almost unchecked in this universe; and nuanced views of developmental biology, behavioral ecology, and genes as nodes in dynamic and multi-vectorial fields of vital interactions are only some of the crash victims of high-octane medical genetic fuels and gene-jockey racing careers. For my taste, genomes are too much made up of invest-ment opportunities of the "one region–one product" sort, a kind of enterprised-up descendant of the "one gene–one enzyme" principle that proved so fruitful in research. Taken one at a time, genes, especially disease-related genes, induce brain damage in those trying to come to grips with genetic diversity issues and their consequences.

Evolutionary biology, bio-social ecology, population biology, and population genetics (not to mention history of science, political econ-omy, and cultural anthropology) have played a woefully small role in shaping public and professional genetic imaginations, and all too small a role in drawing the big money for genetic research. Considering only dog worlds, my preliminary research turns up millions of dollars in grants going into genetic disease research (even though peanuts compared to dollars for genetic research in organisms like mice who are convention-ally models for human disease; dog genetics gets more money as it is shown that genome homologies across taxonomic divisions make ca-nines ideal for understanding lots of human conditions, e.g., narcolepsy, bleeding disorders, and retinal degeneration), and only a few thousand dollars (and lots of volunteer time from both professionals and lay col-laborators) going into canine genetic diversity research.

The emergence since the 1980s of biodiversity discourses, environ-mentalisms, and sustainability doctrines of every political color on the agendas of myriad NGOs and of First World institutions like the World Bank, the International Union for the Conservation of Nature and Natu-ral Resources (IUCN), and the Organization for Economic Cooperation and Development (OECD), as well as in the Third World, has made a difference in this situation.[9] The notoriously problematic politics and also the compelling naturalcultural complexity of diversity discourses requires a shelf of books, some of which have been written. I think the emergence of genetic diversity concerns in dog worlds only makes sense historically as a wavelet in the set of breakers constituting transnational, globalizing, biological and cultural diversity discourses, in which genes and genomes (and immune systems) are major players. Noticing some of the conditions of emergence of a discourse is not the same thing as reducing its value to ideological stepchild status. Quite the opposite: I am compelled by the irreducible complexity—morally, politically, culturally, and scientifically—of diversity discourses, including those leashed to the genomes and gene pools of purebred dogs and their canine relatives in and out of "nature."

The last few paragraphs are preparation for logging onto the Canine Diversity Project website, www.magma.ca/~kaitlin/diverse.html, owned by Dr. John Armstrong, a lover of Standard and Miniature Poodles and a faculty member in the Department of Biology at the University of Ottawa until his death in the summer of 2001. Armstrong wrote and distributed as widely as possible his analyses of the effects of a popular sire and a particular kennel in Standard Poodles. Also the owner of CANGEN-L, Armstrong conducted collaborative research with dog health and genetics activists to study whether longevity is correlated to the degree of inbreeding. Aiming in its introductory sentence to draw the attention of dog breeders to "the dangers of inbreeding and the overuse of popular sires," the Diversity Project website started in 1997. Used by at least several hundred dog people of several nationalities, in the first three and half months of 2000, the site registered 4,500 logons. I have myself learned a tremendous amount from this website; I appreciate the quality of information, the controversies engaged, the evident care for dogs and people, the range of material, and the commitments to issues I am concerned with. I am also professionally acutely alert to the semiotics—the meaning-making machinery—of the Canine Diversity Project website.

Animated by a mission, the site draws its users into its reform agenda at every turn.[10] Some of the rhetorical devices are classical American tropes rooted in old popular self-help practices and evangelical Protestant witness, devices so ingrained in U.S. culture that few users would be conscious of their history. For example, right after the introductory paragraph with the initial link terms, the Diversity Project website leads its users into a section called, "How You Can Help." The question the visitor confronts is like that used in advertising and in preaching—Have you been saved? Have you taken the Immune Power pledge? (slogan from an ad for a vitamin formulation in the 1980s). Or, as the Diversity Project puts the query, "Ask the Question—Do you need a '*Breed Survival Plan*'?" This is the stuff of subject-reconstituting, conversion and conviction discourse (Harding 1999).

The first four highlighted linkage words in the opening paragraphs of the website are "popular sires," a common term for many years in purebred dog talk about the overuse of certain stud dogs and the consequent spreading of genetic disease; "Species Survival Plans," a term that makes a new link for dog breeders to zoos and the preservation of endangered species; "wild cousins," which places dogs with their taxonomic kin and reinforces considering purebreds within the family of natural (in the sense of "wild"), and frequently endangered, species; and "inherited disease," firmly in last place on the list and of concern primarily because a high incidence of double autosomal recessives for particular diseases is an index of lots of homozygosity in purebred dog genomes. Such high incidences of double recessives are certainly related to excessive in- and line breeding, which are diversity-depleting practices. But, as I read it, the soul of the website is the value of diversity for itself in the semiotic framework of evolutionary biology, biodiversity, and biophilia (Wilson 1988, 1992), not diversity as an instrument for solving the problem of genetic disease. Of course, these two values are not mutually exclusive; indeed, they are complementary. But priority matters. In that sense, "breeds" become like endangered species, inviting all the wonderful apparatus we have become familiar with in wildlife biology at the turn of the millennium.

The web site is constructed as a teaching instrument; it constructs its principal audience as engaged lay breeders and other committed dog people. These are the subjects invited to declare for a breed survival plan. Secondarily, scientists of whatever specialty might learn from using the site, but scientists are more teachers here than they are researchers or students. Nonetheless, there are plenty of trading zones and boundary objects linking lay and professional communities of practice in this very inviting site. Further, the nature of a website, as opposed to many other writing technologies (King forthcoming), resists reduction to single purposes and dominating tropes. Links lead many places, and these paths are explored by users, within the webs initially spun by designers, to be sure, but rapidly spiraling out of the control of any designer, no matter how broad minded. The Internet is hardly infinitely open, but its degrees of semiotic freedom are many.

"Popular sires" is sufficiently recognized that the linking term would appeal to the tender-footed neophyte thinking about genetic diversity. For one thing, the link stays with dogs as the principal focus of attention, and does not launch the user into a universe of marvelous creatures in exotic habitats whose utility as models for dogs is hard to swallow for many breeders, even those interested in such non-dog organisms and ecologies in other contexts. "Species Survival Plans," on the other hand, open up controversial metaphoric and practical universes for breeders of purebred dogs and, if taken seriously, would require major changes in ways of thinking and acting.[11] The first obvious point is that "survival plans" connote that something is endangered. The line between a secular crisis and a sacred apocalypse is a very thin one in U.S. American discourse, where millennial matters are written into the fabric of the national imagination from the first Puritan City on a Hill to Star Trek and its sequelae. The second obvious point is that the prominent role given to species survival plans on the Canine Diversity Project website invites a reproductive tie between natural species and purebred dogs. This is one of those ties where the natural and the technical keep close company, semitotically and materially.

To illustrate this point, I will dwell on the material on my screen after I click on "Species Survival Plan" and follow up with a click on "Introduction to a Species Survival Plan." I am teleported to the website for the Tiger Information Center; and, appreciating a face-front photo of two imposing tigers crossing a stream, I am presented with a paper on "Regional and Global Management of Tigers," by R. Tilson, K. Taylor-Holzer and G. Brady. Now, I know lots of dog people love cats, contrary to popular stereotypes about folks being either canine or feline in their affections. But tigers in zoos around the world and in shrunken "forest patches spread from India across China to the Russian Far East and south to Indonesia" is a leap out of the kennel and the show ring or herding trials. I learn that three of the eight recognized subspecies of tigers are already extinct, a fourth on the brink, and all the wild populations stressed. Ideally, the goal of a SSP masterplan for an endangered species is, out of existing animals in zoos and some new "founders" brought in from "nature," to create viable, managed, captive populations to maintain as much of the genetic diversity for all the extant taxa of the species as possible. The purpose is to provide a genetic reservoir for the reinforcement or reconstitution of wild populations where necessary and possible. A practical SSP, "because of space limitations generally targets 90% of genetic diversity of the wild populations for 100–200 years as a reasonable goal." I am in love with the hopefulness of that kind of reasonableness. The "Zoo Ark" for the tigers, lamentably, has to be more modest because the resources are too few and the needs too great. An SSP is a complex, cooperative management program of the American Zoo and Aquarium Association (AZA).

What does developing and implementing a SSP involve? The short answer is—a long list of companion species of organic, organizational, and technological kinds. A minimum account of such companion species must include: the World Conservation Union's specialist groups who make assessments of endangerment; member zoos, with their scientists, keepers, and boards of governors; a small Management Group under the

AZA; a database maintained as a Regional Studbook, using specialized software like SPARKS (Single Population and Records Keeping System) and its companion programs for demographic and genetic analysis, produced by the International Species Information System; funders; national governments; international bodies; and, hardly least, the animals in danger. Crucial operations within a SSP are measurements of diversity and relatedness. One wants to know Founder Importance Coefficients (FIC) as a tool for equalizing relative founder contributions and minimizing inbreeding. Full and accurate pedigrees are precious objects for an SSP. Mean Kinship (MK) and Kinship Values (KV) rule mate choice in this sociobiological system. "Reinforcing" wild species requires a global apparatus of technoscientific production, where the natural and the technical have very high coefficients of semiotic and practical inbreeding.[12]

Purebred dog breeders also value deep pedigrees, and they are accustomed to evaluating matings with regard to breed standards, which is a complex, non-formulaic art. Inbreeding is not a new concern. So what is so challenging about a SSP as a universe of reference? The definition of populations and founders is perhaps first. Discussions among engaged breeders on CANGEN—i.e., people sufficiently interested in questions of genetic diversity to sign on and post to a specialized Internet mailing list—show that dog people's "lines" and "breeds" are not equivalent terms to wildlife biologists' and geneticists' "populations." The behavior properly associated with these different words is quite different. A dog breeder educated in the traditional mentoring practices of the fancy will attempt through line breeding, with variable frequencies of outcrosses, to maximize the genetic/blood contribution of the truly "great dogs," who are rare and special. The great dogs are the individuals who best embody the type of the breed. The type is not a fixed thing, but a living, imaginative hope and memory. Kennels often are recognized for the distinctiveness of their dogs, and breeders point proudly to their kennel's own founders, and breed club documents to the breed's founders. The notion of working to equalize the contribution of all of the founders in the population geneticists' sense is truly odd in traditional dog breeders' discourse. Of course, a SSP, unlike nature and unlike dog breeders, is not operating with adaptational selectional criteria; the point of a SSP is to preserve diversity as such as a reservoir. Small populations are subject to intense extinction pressures—loss of habitat, fragmented subpopulations no longer able to exchange genetic material, loss of genes through the random process called genetic drift, crisis events causing population crashes like famines or diseases, and on and on.

The SSP is a conservation management plan, not nature, however conceptualized, and not a breed's written standard or an individual breeder's interpretation of that standard. Like a SSP, a breed standard is also a kind of large-scale action blueprint, but for other purposes. Some breeders talk of those purposes in capital letters, as the Original Purpose of a given breed. Others are not typological in that sense and are attuned to dynamic histories and evolving goals within a partly shared sense of breed history, structure, and function. These breeders are keenly aware of the need for selection on the basis of many criteria as holistically as possible to maintain and improve a breed's overall quality and to achieve the rare special dogs. They take these responsibilities very seriously; and they are not virgins to controversy, contradiction, and failure. They are not against learning about genetic diversity in the context of the problems they know or suspect their dogs face. Some breeders—a very few, I think—embrace genetic diversity discourse and population genetics. They worry that the foundation of their breeds might be too narrow and getting narrower. But the breeder's art does not easily entertain adopting the heavily mathematical and software-driven mating systems of a SSP. I witness in my research several courageous breeders insisting on deeper pedigrees and regular calculations of coefficients of inbreeding, with efforts to hold them down where possible. But the breeders I overhear are loathe to cede decisions to anything like a master plan. In my judgment, they do not see their own dogs or their breed primarily as biological populations. The dominance of specialists over local and lay communities in the SSP world does not escape dog breeders' attention. Most of the breeders whom I overhear squirm if the discussion stays on a theoretical population genetics level and if few if any of the data come from dogs, rather than, say, a Malagasy lemur population or lab-bound mouse strain. In short, breeders' discourse and genetic diversity discourse do not hybridize smoothly, at least in the F1 generation. This mating is what I hear breeders call a "cold outcross" that they worry risks importing as many problems as it solves.

There is much more to the Canine Diversity Project website than the SSP links, and if I had the space to examine the rich texture of the whole website, many more sorts of openings, repulsions, inclusions, attractions, and possibilities would be evident for seeing the ways dog breeders, health activists, veterinarians, and geneticists relate to the question of diversity. At the very least, the serious visitor to the website could get a decent elementary education in genetics, including Mendelian, medical, and population genetics. Fascinating collaborations between individual scientists and breed club health and genetics activists would emerge. The differences within dog people's ways of thinking about genetic diversity and inbreeding would be inescapable, as the apocalyptic and controversial Jeffrey Bragg's "evolving breeds" and Seppala Siberian

Sled Dogs meet John Armstrong's more modest Standard Poodles (and his more moderate action plan, "Genetics for Breeders: How to Produce Healthier Dogs") or Leos Kraal's and C. A. Sharp's ways of working in Australian Shepherd worlds. Links would take the visitor to the extraordinary Code of Ethics of the Coton de Tulear Club of America and this breed's alpha-male geneticist activist, Robert Jay Russell, as well as to the online documents with which the Border Collie web site teaches genetics relevant to that fascinating breed. The visitor could follow links to the molecular evolution of the dog family, updated lists of current DNA gene tests in dogs, discussions of wolf conservation and wolf taxonomic debates, accounts of a cross-breeding (to a Pointer) and backcross project in Dalmatians to eliminate a common genetic disease and of importing new African stock in Basenjis to deal with genetic dilemmas. One could click one's way to discussions of infertility, stress, and herpes infections, or follow links to endocrine disrupter discourse for thinking about how environmental degradation might be affecting dogs, as well as frogs and people, globally. Right in the middle of the Diversity Project website is a bold-type invitation to join the mailing list Armstrong ran until his death, the Canine Genetics Discussion Group (CANGEN-L), where a sometimes rough and tumble exchange among heterogeneous lay and professional dog people stirred up the pedagogical order of the website.

So dogs, not tigers—and breeds, not endangered species—actually dominate on the Canine Diversity Project website. But the metaphoric, political, and practical possibilities of those first links to Species Survival Plans attach themselves like well positioned ticks on a nice blade of grass, waiting for a passing visitor from purebred dog land. Frontline defenses are not always enough.[13] We are in the fiercely local and linked global zones of technobiopolitics, where few species are more than a click away. Naturalcultural survival is the prize.

III. CAT'S CRADLE: A CONCLUSION IN TANGLES
What I Like about the Material-Semiotic Knot, the Literalized Figure, of Companion Species

1. Networks of co-constitution, co-evolution, communication, collaboration abound to help us rethink issues of communication and control at the heart of the cyborg figure.
2. Humans, other organisms, artifacts, and technologies are all players, a requirement of an aliberal approach. The relationship is the smallest possible unit of analysis.
3. Likewise, scientists, lay people, and dogs are all on the play bill in this evolutionary drama.

4. Companion species are not involved in another Hegelian confrontation of self-other, culture-nature, or similar dualisms.[14]
5. Companion species are not another version of a Marxist humanist dialectic of nature remade by labor. The making goes in too many directions.
6. The story is more Whiteheadean, full of his kind of objectifications.
7. This story provokes finding non-anthropomorphic ways to figure agencies and actors.
8. Companion species throw comparative methods into crisis because the norm stabilizing comparison wobbles; e.g., consciousness will not do for considering animal well being. Companion species discourse does not produce an animal rights or human rights agenda, but does insist on complex ethical discourse.
9. Here we have situated co-constitution, with inherited pasts of many kinds, rather than dialectical unity; i.e., situatedness displaces teleology analytically and morally. This is all about origin stories. How we might live and flourish is a permanent, finite, mundane question; there is no way out, especially in terms of extraterrestrial projects of man (species) evolving toward bodylessness.
10. Companion species worlds are at home with the non-heroic dailiness of epistemological/ontological/ethical action; the birth of the kennel is a homely story.
11. Leigh Star's question (1991) cannot be evaded: *cui bono?* Who lives, and lives well, and who dies and why, in companion species relations?
12. We have real histories of dogs and people; not The Dog and The Human; in co-constitution there are layers of practices and many chronicities. Scale is made, not given (Tsing 2000).
13. There is always a necessary weave of narrative and other material-semiotic practices.
14. Full of cross talk and questions about locations of expertise and authority, this story is not cynical about telling the truth. Practices must be relentlessly situated inside truth telling, and vice versa.
15. The story requires considering seriously "companion animals" and complex moral-scientific action outside the straight-jackets of much animal rights discourse, feminist and otherwise. "Companion species" is not a very friendly notion for those "animal rights" perspectives that rely on a scale of similarity to human mentality for assigning value. Both people and their partners are co-constructed in the history of companion species, and the

issues of hierarchy and cruelty, as well as colleagueship and responsibility, are open and polyvalent, both historically and morally. Also, "companion species" does not prejudge the category of the "species"; they could be artifacts, organisms, technologies, other humans, etc. The simple and obvious point is that nothing is self-made, autocthonous, or self-sufficient. Origin stories have to be about fraught histories of consequential relationships. The point is to engage "ontological choreography" in the yearning for more livable and lively relationships across kinds, human and non-human.

16. Dog worlds become a place to work through idioms and practices around diversity, including in environmental politics and human-animal relationship politics, both of which are areas of major feminist concern.

17. The literalized figure of companion species addresses the long history of feminist critique of possessive individualism. By "literalized," I mean materially semiotically engaged, fleshly and significant all the way down.

18. The literalized figure of companion species does semiotic-material work on idioms of breed, species, sex, reproduction, behavior, genome.

19. The literalized figure of companion species foregrounds relations of communities of practice in relation to intersectionally (Crenshaw 1993; gender, race, nation, and species are only a start). Attention to differentiated expertise and differentiated literacies—whether called "lay" or "professional"—is required.

20. The literalized figure of companion species invites "intersectional analysis" of key themes: breeds and the history of eugenics; technology and the organic body; histories of class, race, gender, and nation.

21. A key question is: who cleans up the shit in a companion species relationship?

NOTES

1. Vicki Hearne is a dog writer, philosopher, and email correspondent, who died in the summer of 2001. See Hearne (1986).
2. This slogan can be found on T-shirts and windshield stickers among enthusiasts of the dog sport called Schutzhund, which involves competition in tracking, protection work, and obedience.
3. See Gray (1995) for a rich set of documents and accounts of cyborg worlds. See Tofts, Jonson, and Cavallaro (2002) for a striking intellectual history of cyberculture.
4. See Haraway (1985) for my effort to live in the naturecultures of Marxist feminism conjugated with technoscience studies in the time of Reagan's Star Wars. See Sofoulis

(2002) for an incisive account of the fate of that cyborg material-semiotic doppleganger. See Goodeve (2000) for an extended interview on my cyborg and her associates.

5. The dog-human co-evolution story below is slightly revised from "For the Love of a Good Dog: Webs of Action in the World of Dog Genetics," Haraway 2003a.

6. Recent taxonomic revisions make dogs into a subspecies of wolves, *Canis lupens familiaris*, rather than into a species of their own, *Canis familiaris*. This technical issue has multiple consequences beyond the scope of this paper. See Coppinger and Coppinger 2001, pp. 273–282. For a critique of Vilá, et al's dating of dog evolution from mtDNA data, see Coppinger and Coppinger 2001, pp. 283–294.

7. Like much in the former USSR, this trickster drama of worker safety, industrial efficiency, and evolutionary theory and genetics in the far north devolved in the post–Cold War economic order. Since the salaries of the scientists at the Genetics Institute have not been paid, much of the breeding stock of tame foxes has been destroyed. The scientists scramble to save the rest—and fund their research—by marketing them in the West as pets with characteristics between dogs and cats. A sad irony is that if the geneticists and their foxes succeed in surviving in this enterprise culture, the population of remaining animals bred for the international pet trade will have been genetically depleted by the slaughter necessitated by the rigors of post-Soviet capitalism and commercializing the animals not for fur coats but as pets. Note also the tones of the Lysenko affair in the story of the evolution of tame Soviet foxes.

8. The section on biodiversity in dogland below is drawn from an earlier version of parts of "Cloning Mutts, Saving Tigers: Ethical Emergents in Technocultural Dogland," in Franklin and Lock, 2003.

9. See, for example, *World Conservation Strategy,* IUCN, 1980; the Bruntland Report, *Our Common Future,* WECD, 1987; Agenda 21; Convention on Biodiversity, 1992; Guiding Principles on Forests; *Valuing Nature's Services,* WorldWatch Institute Report of Progress toward a Sustainable Society, 1997; *Investing in Biological Diversity,* Cairns Conference, OECD, 1997; *Saving Biological Diversity: Economic Incentives,* OECD, 1996.

10. I am using a version of the website online in 2000.

11. The Rare Breed Survival Trust in the UK (mainly for poultry, sheep, pigs, cattle, and other "farm livestock heritage" animals not usually thought of as either companion animals—especially not as "pets"—or wild animals, including the working collie dogs that the Trust attends to), and its journal *The Ark,* would repay close attention in relation to action in dog worlds. Thanks to Sarah Franklin and Thelma Rowell for handing me into *The Ark.*

12. SSP is a North American term. Europeans have European Endangered Species Programs (EESPs); Australasians have Australasian Species Management Programs, and China, Japan, India, Thailand, Malaysia, and Indonesia all have their own equivalents. This is global science of indigenous species.

13. Information for those whose lives are not ruled by real ticks and real dogs: Frontline[TM] is a new-generation tick and flea control product that has made dogs' and dog people's lives much less irritable.

14. I am in obvious and deep alliance with Bruno Latour on these matters. See. for example, Latour 1993.

REFERENCES

Auerbach, Eric. 1953. *Mimesis: The Representation of Reality in Western Literature.* Princeton, NJ: Princeton University Press.

Barad, Karen. 1995. "Meeting the Universe Halfway: Ambiguities, Discontinuities, Quantum Subjects, and Multiple Positionings in Feminism and Physics." In *Feminism, Science, and the Philosophy of Science: A Dialog,* edited by L. H. Nelson and J. Nelson. Norwell: Kluwer Press.

Beck, Alan, and Aaron Katcher. 1996. *Between Pets and People: The Importance of Animal Companionship.* 2nd ed. West Lafayette, Ind.: Purdue University Press.

Belyaev, D.K. 1969. "Domestication of Animals." *Science Journal,* UK 5:47–52.

Boitani, L., F. Francisci, P. Ciucci, and G. Andreoli. 1995. "Population Biology and Ecology of Feral Dogs in Central Italy." In Serpell, ed., pp. 217–244.

Brown, Norman O. 1966. *Love's Body.* New York: Random House.

Clynes, Manfred E., and Nathan S. Kline. 1960. "Cyborgs and Space." *Astraunatics,* September 26–27, 1960, pp. 5–76.

Coppinger, Raymond, and Lorna Coppinger. 2001. *Dogs: A Startling New Understanding of Canine Origin, Behavior, and Evolution.* New York: Scribner.

Coppinger, Raymond, and Richard Schneider. 1995. "Evolution of Working Dogs." In Serpell, pp. 21–47.

Crenshaw, Kimberle. 1993. "Demarginalizing the Intersection of Race and Sex: A Black Feminist Critique of Antidiscrimination Doctrine, Feminist Theory, and Antiracist Politics." In *Feminist Legal Theory: Foundations,* edited by D. Kelly Weisberg, pp. 383–395. Philadelphia: Temple University Press.

Crockford, Susan J. 2000. "Dog Evolution: A Role for Thyroid Hormone Physiology in Domestication Changes." In *Dogs through Time: An Archaeological Perspective,* edited by S. Crockford, pp. 11–20. Oxford, England: BAR International Series 889.

Cussins, Charis Thompson. 1996. "Ontological Choreography: Agency through Objectification in Infertility Clinics," *Social Studies of Science,* 26:575–610.

Dawkins, Richard. 1982. *The Extended Phenotype: The Gene as a Unit of Selection.* London: Oxford University Press.

Franklin, Sarah and Margaret Lock, eds. 2003. *Remaking Life and Death.* Santa Fe, NM: SAR Press.

Goodeve, Thyrza. 2000. *How Like a Leaf: Donna Haraway, an Interview with Thyrza Goodeve.* New York: Routledge.

Gray, Chris Hables, ed. 1995. *The Cyborg Handbook.* New York: Routledge.

Haraway, Donna. 1985. "A Manifesto for Cyborgs," *Socialist Review,* no. 80, pp. 65–108.

———. 2003a. "For the Love of a Good Dog: Webs of Action in the World of Dog Genetics." In *Race, Nature, and the Politics of Difference,* edited by Donald Moore. Durham, NC: Duke University Press.

———. 2003b. "Cloning Mutts, Saving Tigers: Ethical Emergents in Technocultural Dogland." In Franklin and Lock.

———. 2003c. *The Companion Species Manifesto.* Chicago: Prickly Paradigm Press.

Harding, Susan. 1999. *The Book of Jerry Falwell.* Princeton, NJ: Princeton University Press.

Hearne, Vicki. 1986. *Adam's Task: Calling Animals by Name.* New York: Knopf.

Hutchinson, George Evelyn. 1965. *The Ecological Theater and the Evolutionary Play.* New Haven: Yale University Press.

King, Katie. forthcoming. *Feminism and Writing Technologies.* Manuscript, University of Maryland at College Park.

Latour, Bruno. 1993. *We Have Never Been Modern,* translated by Catherine Porter. Cambridge: Harvard University Press.

Rabinow, Paul. 1992. "Artificiality and Enlightenment: From Sociobiology to Biosociality." In *Incorporations,* edited by J. Crary and S. Qwinter, pp. 234–252. New York: Zone Books.

Serpell, James, ed. 1995. *The Domestic Dog: Its Evolution, Behaviour, and Interactions with People.* Cambridge: Cambridge University Press.

Sofoulis, Zoe. 2002. "Cyberquake: Haraway's Manifesto." In Tofts et al.

Star, Susan Leigh. 1991. "Power, Technology and the Phenomenology of Conventions: On

Being Allergic to Onions." In *A Sociology of Monsters: Power, Technology, and the Modern World*, edited by John Law, pp. 26–56. Oxford: Basil Blackwell.

Strathern, Marilyn. 1992. *Reproducing the Future: Anthropology, Kinship, and the New Reproductive Technologies*. New York: Routledge.

Templeton, Alan. 1999. "Human Race in the Context of Human Evolution: A Molecular Perspective." Paper for the Wenner Gren Foundation Conference on Anthropology in the Age of Genetics, June 11–19, Teresopolis, Brazil.

Tofts, Darren, Annemarie Jonson, and Alessio Cavallaro, eds. 2002. *Prefiguring Cyberculture: An Intellectual History*. Cambridge: MIT Press.

Trut, Lyudamila N. 1999. March–April. "Early Canid Domestication: The Fox-Farm Experiment." *American Scientist* 87 (March–April): 160–169.

Tsing, Anna. 2000. "Inside the Economy of Appearances." *Public Culture* 12, no. 1: 115–144.

Verran, Helen. 2001. *Science and an African Logic*. Chicago: University of Chicago Press.

Vilá, Carles, Peter Savolainen, Jesús E. Maldonado, Isabel R. Amorim, John E. Rice, Rodney L. Honeycutt, Keith A. Crandall, Joakim Lundeberg, and Robert K. Wayne. 1997. "Multiple and Ancient Origins of the Domestic Dog." *Science* 276 (June 13):1687–1689.

Weidensaul, Scott. 1999. "Tracking America's First Dogs," *Smithsonian Magazine*, March 1, pp. 44–57.

Whitehead, Alfred North. 1948. *Science and the Modern World*. New York: Mentor; orig. 1925.

———. 1969. *Process and Reality*. New York: Free Press; orig. 1929.

Wilson, E. O. 1992. *The Diversity of Life*. New York and London: W. W. Norton.

———. ed. 1988. *Biodiversity*. Washington, DC: National Academy Press.

10

CYBORGS, COYOTES, AND DOGS: A KINSHIP OF FEMINIST FIGURATIONS
and
THERE ARE ALWAYS MORE THINGS GOING ON THAN YOU THOUGHT! METHODOLOGIES AS THINKING TECHNOLOGIES

An interview with Donna Haraway conducted in two parts by Nina Lykke, Randi Markussen, and Finn Olesen

PART I: CYBORGS, COYOTES AND DOGS: A KINSHIP OF FEMINIST FIGURATIONS

Interviewer: Let us start with the Cyborg Manifesto.[1] Many women have been fascinated by the idea that the cyborg could be a woman. Why did you insist on the femaleness of the cyborg?

Donna Haraway: For me the notion of the cyborg was female, and a woman, in complex ways. It was an act of resistance, an oppositional move of a pretty straightforward kind. The cyborg was, of course, part of a military project, part of an extraterrestrial man-in-space project. It was also a science fictional figure out of a largely male-defined science fiction. Then there was another dimension in which cyborgs were female: in popular culture, and in certain kinds of medical culture. Here cyborgs appeared as patients, or as objects of pornography, as "fem-bots"—the iron maiden, the robotisized machinic, pornographic female. But the whole figure of the cyborg seemed to me potentially much more interesting than that. Moreover, an act of taking over a territory seemed like a fairly straightforward, political, symbolic technoscientific project.

From my point of view, the cyborg was a figure that collected up many things, among them the way that post–World War II technoscientific cultures were deeply shaped by information sciences and biological sciences, by the implosion of informatics and biologics that was already well under way by the end of World War II, and that has only deepened in the last fifty years that transformed conditions of life very deeply. These are not matters of choice, neither are they matters of determinism. These are deep materializations of very complex sociotechnical relations. What interested me was the way of conceiving of us all as communication systems, whether we are animate or inanimate, whether we are animals or plants, human beings or the planet herself, Gaia, or machines of various kinds. This common coin of theorizing existence, this common ontology of everything as communication-control-system was what interested me. It made me very angry and anxious, but interested me in more positive ways, too. Among other things I was attracted by an unconscious and dreamlike quality, and I was interested in affirming not simply the human-machine aspect of cyborgs, but also the degree to which human beings and other organisms have a kind of commonality to them in cyborg worlds. It was the joint implosion of human and machine, on the one hand, and human and other organisms, on the other, within a kind of problematic of communication that interested me about the cyborg. There were many levels in this, for example labor process issues: the particular ways that women—working-class women, women of color, women in Third World countries with export processing zones that would attract international capital for micro-electronics manufacture—were implicated in the labor process of cyborg production, as scientists, too, although in relative minorities. Women occupied many kinds of places in these worlds, in biomedicine, in information sciences, but also as a preferred workforce for transnational capitals. Strategies of flexible accumulation involved the productions of various kinds of gender, for men and for women, that were historically specific. The cyborg became a figure for trying to understand women's place in the "integrated circuit"[2]—a phrase produced by feminist socialists.

Moreover, the cyborg was a place to excavate and examine popular culture including Science Fiction, and, in particular, feminist Science Fiction. A novel like *Superluminal* by Vonda McIntyre[3] made a strong use of cyborg imagery in complex, interesting ways that were quasi-feminist. Joanna Russ' clone sister fiction of the mid-1970s[4] and, certainly, Octavia

Butler's work[5] intrigued me a lot. There was a great deal of feminist cultural production, which was working with the cyborg in fascinating ways.

Also, the cyborg seemed to me a figuration that was specifically hard for psychoanalysis to account for. But in contrast to what a lot of people have argued, I do not think of the cyborg as without an unconscious. However, it is not a Freudian unconscious. There is a different kind of dreamwork going on here; it is not ethical, it is not edenic, it is not about origin stories in the garden. It is a different set of narrations, figurations, dreamwork, subject formations, and unconscious work. These sorts of figurations do not exclude many kinds of psychoanalytic work, but they are not the same thing. It was important to me to have a way of dealing with figurations in technoscience that were not quite so hegemonized by psychoanalysis as I found it developed around me in really lively places of feminist cultural work such as film theory. Some marvellous work has been done with Freudian or post-Freudian tools here, but they did not seem right for the analysis of technoscience. So I turned to literature as well as biology and philosophy, and questions of figurations interested me a lot.

Cyborgs are also places where the ambiguity between the literal and the figurative is always working. You are never sure whether to take something literally or figuratively. It is always both/and. It is this undecidability between the literal and the figurative that interests me about technoscience. It seems like a good place to inhabit. Moreover, the cyborg involves a physicality that is undeniable and deeply historically specific. It is possible to extend the cyborg image into other historical configurations, allegorically or analogically, but it seems to me that it had a particular historical emergence. You can use it to inquire into other historical formations, but it has a specificity.

In a way, you know, I am doing this analysis of the meanings attached to the cyborg retrospectively. I cannot imagine that I thought all these things in 1983 [laughter]. It is a funny thing to look back at something I actually began writing seventeen years ago . . .

Interviewer: Please, tell us about the intriguing history of the Cyborg Manifesto, which has taken on a life of its own in a way that academic papers seldom do.

Donna Haraway: I began writing the manifesto in 1983. *Socialist Review* in the United States wanted socialist feminists to write about the future of socialist feminism in the context of the early Reagan era and the retrenchment of the left that the 1980s was witnessing. Barbara Ehrenreich and I, and many other American socialist feminists, were invited to contribute. Moreover, Frigga Haug and the feminist collective of the West German socialist journal *Das Argument* wanted me to write about reproductive technologies, and the cyborg is an obvious place for making reflections on the technologification of reproduction. Almost at the same time, a left democratic group in the former Yugoslavia was holding a conference and I was designated as one of the American representatives from *Socialist Review*. I wrote a version of the Cyborg Manifesto for this occasion, although I actually did not deliver my paper at the conference. Instead, a small group of us made a demonstration about the division of labor at the conference, where the women were invisibly doing all the work, while the men were not so invisibly doing all the propounding! So in the beginning the Cyborg Manifesto had a very strong socialist and European connection.

Interviewer: Where did you read the word, cyborg, the first time? Do you remember that?

Donna Haraway: I do not remember. I tried to remember it, and it felt like I made the word up, but I cannot have made it up. I read Norbert Wiener, but I do not think I got it there. I did not read Clynes and Kline[6] until way after I had written the Cyborg Manifesto. I did not know about Clynes and Kline and that fabulous connection of the psychiatrist, the systems engineer, and the mental hospital. It was a graduate student of mine, Chris Gray,[7] who told me about the cyborg article of Clynes and Kline from 1960.

Interviewer: How do you yourself look upon the remarkable history of the Cyborg Manifesto? How do you evaluate the reception, in terms both of positive and negative responses?

Donna Haraway: I am astonished ... But to answer your question, I can tell you that the reactions, right from the beginning, were very mixed. At *Socialist Review* the manifesto was considered very controversial. The *Socialist Review* East Coast Collective truly disapproved of it politically and did not want it published. But the Berkeley *Socialist Review* Collective did, and it was Jeff Escoffier, a very interesting gay theorist and historian, who was my editor at the Berkeley Collective, and he was very enthusiastic about the paper. So from the beginning the manifesto was very controversial. There were some who regarded it as tremendously anti-feminist, promoting a kind of blissed-out, techno-sublime euphoria. Those readers completely failed to see all the critique. They would read things that for me are highly ironic and angry, a kind of contained ironic fury—they would read these things as my literal position, as if I was embracing and affirming what I am describing with barely restrained fury.

The reading practices of the Cyborg Manifesto took me aback from the very beginning, and I learned that irony is a dangerous rhetorical strategy. Moreover, I found out that it is not a very kind rhetoric, because it does things to your audience that are not fair. When you use irony, you assume that your audience is reading out of much the same sort of experiences as you yourself, and they are not. You assume reading practices that you have to finally admit are highly privileged and often private. The manifesto put together literacies that are the result of literary studies, biology, information sciences, political economy and a very privileged and expensive travel and education. It was a paper that was built on privilege, and the reading practices that it asks from people are hard. I learned something about that from certain receptions of the manifesto. On the other hand, most of my readers shared the same privileges [laughter].

There were also readers who would take the Cyborg Manifesto for its technological analysis, but drop the feminism. Many science studies people, who still seem tone-deaf to feminism, have done this. It is generally my experience that very few people are taking what I consider all of its parts. I have had people, like *Wired Magazine* readers, interviewing and writing about the Cyborg Manifesto from what I see as a very blissed-out, techno-sublime position.

But I have also had this really interesting reception from young feminists—a reception which I love. They embrace and use the cyborg of the manifesto to do what they want for their own purposes. They have completely different histories from mine, from this particular moment of democratic socialism and socialist feminism, the transition of the 1980s which I just narrated. This is not their history at all. They have a totally different relationship to cultural production, to access to media, to use of computers for performance art and other purposes, to technomusic. They have, to my pleasure and astonishment, found the Cyborg Manifesto useful for queer sexuality work, and for certain kinds of queer theory that take in technoscience. I found myself to be an audience here. In this context, I am one of the readers of the manifesto, not one of the writers. I did not write that manifesto, but I love reading it [laughter]. These young feminists have truly rewritten the manifesto in ways that were not part of my intention, but I can see what they are doing. I think it is a legitimate reading, and I like it, but it really wasn't what I wrote. So sometimes people read the manifesto in ways that are very pleasant surprises to me, and sometimes it is really distressing to be confronted with the reading practices. But, anyway, it is a hard paper to read. Difficulty is an issue. On the other hand, I swear, I meet people without academic training who read the manifesto and who do not give up. They read it for what they want, and they just do not care about the difficulty issue.

Interviewer: I have been teaching gender and technoculture to registered nurses, and for many of them, the manifesto was a revelation. It helped them to see their practice as nurses in a new light and to avoid being caught in the dilemma between a humanistic and partly technophobic concept of care, on the one hand, and, on the other, the powerful and un-critically self-glorifying visions of progress, embedded in the discourses of medical science. Your cyborg was for them a critical tool, a position from which they could think their professional identity differently.

Donna Haraway: This is very interesting. I think that part of the fem-inist argument of the manifesto is exactly in line with this. It is neither technophobic, nor technophilic, but about trying to inquire critically into the worldliness of technoscience. It is about exploring where real people are in the material–semiotic systems of technoscience and what kinds of accountability, responsibility, pleasure, work, play, are engaged, and should be engaged.

Interviewer: Another aspect of the cyborg that I would like to ask you about is, how you evaluate the danger that it might lose its critical po-tential and become a mainstream figure, closed within a certain main-stream narrative, since it today—much more than when you started writing about it in 1983—has become a so obvious and inescapable part of society and culture.

Donna Haraway: I think that as an oppositional figure the cyborg has a rather short half-life [laughter], and indeed for the most part, cyborg figurations, both in technical and popular culture, are not, and have never been oppositional, or liberatory, or had a critical dimension in the sense that I use critique, i.e., in the sense that things might be otherwise.

It is a sense of critique that is not negative, necessarily, except in the particular way that the Frankfurt School understood negativity—a way which I think is really worth remembering and holding on to. It is critique in the deep sense that things might be otherwise. There is much of the Frankfurt School that I have never embraced, but that sense of critique as a freedom project is important. There was a certain amount of work, and there even still is a certain amount of work in that freedom project that oppositional, or critical cyborgs can do, but I agree that it is much less true now than it was in 1983. Precisely because of the kind of tightening of the Internet around us all; precisely because we are now in the matrix in such a relentlessly literal way that there is some really new tropic work that has to be done in this figure.

I take figurations and the question, how they work, very seriously, as a practice for trying to understand what collects up the life and death concerns of people. It seems to me that we need a whole kinship system of figurations as critical figures, and in that sense I think cyborg figurations can continue to do critical work. But it can quickly become banal, and mainstream, and comforting. The cyborg may be an alibi that makes the technoscientific bourgois figure comfortable, or it may be a critical figure.

Interviewer: You pointed out that a whole kinship of figurations is needed...

Donna Haraway: Yes, [laughter] littermates, a kennel, a breed...

Interviewer: So I would like to leave the cyborg and look at another figuration that has emerged in your work: the coyote. I read the coyote figure in your texts[8] as a figuration that becomes necessary because your complex approach to the deconstruction of the dichotomy between "nature" and "culture" implies a refusal to consider non-human "nature" as nothing but stupid, soulless matter. To me your coyote figure is a figuration in which the search for alternative understandings of the phenomena we used to call "nature" is embedded. But why did you choose this particular figuration?

Donna Haraway: It is partly a regional issue. You know, I am a Westerner, not just in the sense of inheriting Western traditions, but I am from the western United States. Coyote figures are important to Native Americans in many places in North America, including various groups in the southwestern United States. When I use the coyote figure, a double issue is at stake. First of all, my use of the coyote is marked by the middle-class, white feminist appropriation of Native American symbols, about which one must be very suspicious. There is a particular way in which feminist spirituality has operated in a rather colonial way to Native American practices. I have a certain criticism of my own use of the coyote figuration on this background. However, saying that I do not mean to dismiss or to forbid what I and others have been doing in terms of using Native American symbols. What I want is to add a certain caution, because figures do travel, and they travel outside of their places of emergence in various ways, and certain figures like the raven and the coyote do work in Anglo culture, as well as in Native culture. We do live in a world that is made up of complexly webbed layers of locals and globals, and who is to say that Native American symbols are to be less global than those produced by Anglo-Americans? Or who is to say that one set of symbols has got to stay local, while all the other ones get to figure so-called globalization? So I think there is a way in which this cross-talk between figurations is politically interesting, although certainly not innocent.

Thus, the coyote is a specific figuration. It is not nature in a Euro-American sense and not about resources to the makings of culture. Moreover, coyote is not a very nice figure. It is a trickster figure, and, particularly in Navaho figurations, the coyote is often associated with quite distressing kinds of trickster work. Coyote is about the world as a place that is active in terms that are not particularly under human control, but it is not about the human, on the one side, and the natural, on the other. There is a communication between what we would call "nature" and "culture," but in a world where "coyote" is a relevant category, "nature" and "culture" are not the relevant categories. Coyote disturbs nature/culture ontologies.

I chose coyote and not, for example, Spiderwoman, because of the already overdetermined feminist appropriations of the latter, and for one thing the coyote is not female, particularly...

Interviewer: Is it post-gender?

Donna Haraway: No! I have no patience with the term "post-gender." I have never liked it.

Interviewer: But you used it in the manifesto...[9]

Donna Haraway: Yes, I did. But I had no idea that it would become this "ism"! [Laughter] You know, I have never used it since! Because post-gender ends up meaning a very strange array of things. Gender is a verb, not a noun. Gender is always about the production of subjects in relation to other subjects, and in relation to artifacts. Gender is about material-semiotic production of these assemblages, these human-artifact assemblages that are people. People are always already in assemblage with worlds. Humans are congeries of things that are not us. We are not self-identical. Gender is specifically a production of men and women. It is an obligatory distribution of subjects in unequal relationships, where some have property in others. Gender is a specific production of subjects in sexualized forms where some have rights in others to reproductivity, and sexuality, and other modes of being in the world. So gender is specifically a system of that kind, but not continuous across history. Things need not be this way, and in this particular sense that puts focus on a critical relationship to gender along the lines of critical theory's "things need not be this way"—in this sense of blasting gender I approve of the term "post-gender." But this is not "post-gender" in a utopian, beyond-masculine-and-feminine sense, which it often is taken to mean. It is the blasting of necessity, the non-necessity of this way of doing the world.

Interviewer: Going back to the coyote and your choice to include that in your kinship of potentially critical figurations instead of such explicitly female figures as Spiderwoman or the goddess—did that have something to do with coyote being post-gender in the sense that you just defined?

Donna Haraway: Oh yes! It has much to do with "post-gender" in the sense of blasting the scandal of gender and with a feminism that does not embrace Woman, but is for women. This kind of "post-gender" involves the powerful theories of intersection that came out of post–colonial theory, and women of color feminist theory, and that came overwhelmingly, though not only, from people who had been oppressed in colonial and racial ways. They insisted on a kind of relentless intersectionality, that refused any gender analysis standing on its own, and in this context, I find that the term "post-gender" makes sense. Here it can be understood as a kind of intensified critical understanding of these many threads of production of inequality.

Interviewer: To go a bit further into your deconstruction of the nature/culture dichotomy, I will ask you to comment on your concept of the "apparatus of bodily production."[10] Like the cyborg and coyote figurations. This concept is a useful tool, when you want to shift the traditional nature/culture boundaries and create new ways of understanding bodies, as well as the sex/gender dichotomy. How do you yourself look upon the link between the concept of "apparatus of bodily production" and the breaking down of the "sex/gender" dichotomy?

Donna Haraway: Sex and gender theory is an analytical device that is clearly indebted to a way of doing the world that works through matter/form categories. It is a deeply Aristotelian dichotomy. It works on the cultural appropriation of nature for the teleologial ends of mind. It has terribly contaminated roots. None the less it has been a useful tool for analysing the sex/gender system. In that sense, it was a radical achievement at a certain moment. But the analytical work was mistaken for the thing itself, and people truly believed, and believe, in sex and gender as things. It is the mistake of misplaced concreteness. Instead it is important to remember the contaminated philosophical tradition which gives us tools of that kind. In order to do the world in other than Platonist and Aristotelian ways, in order to do ontology otherwise, in order to get out of a world that is done by notions of matter/form, or production/raw material, I feel aligned with ways of getting at the world as a verb, which throws us into worlds in the making and apparatuses of bodily production—without the categories of form and matter, and sex and gender, . . .

Interviewer: And without reducing everything either to purely social constructions or purely natural things?

Donna Haraway: Absolutely. I am neither a naturalist, nor a social con-structionist. Neither–nor. This is not social constructionism, and it is not technoscientific, or biological determinism. It is not nature. It is not culture. It is truly about a serious historical effort to get elsewhere.

Interviewer: You have recently included a new member in your kinship of potential critical figurations: the dog. Why?

Donna Haraway: Dogs are many things. They occupy many kinds of categories: breeds, populations, vermin, figures, research animals, pets, workers, sources of rabies, the New Guinea singing dog, the Dingoes, etc. Dogs are very many kinds of entities. The ontology of dogs turns out to be quite big, and there are all those names for dogs that are about various kinds of relationalities. Dogs engage many kinds of relationality, but one kind that is practically obligatory is with humans. It is almost part of the definition of a dog to be in relationship with humans, although not necessarily around the word "domestication." Though "domestication" is a very powerful word, it is not altogether clear. In fact, it is probably not true that humans domesticated dogs. Conversely, it is probably true from an evolutionary and historical point of view that dogs took the first steps in producing this symbiosis. There are a lot of interesting biological-behavioural stories that have a certain evidential quality. These are partly testable stories, partly not testable stories. So dogs have this large array of possible ontologies, that are all about relationship and very heavily about relationships with humans in different historical forms. For people, dogs do a tremendous amount of semiotic work. They work for us not only when they are herding sheep; they also work as figures, and dogs figure back very important kinds of human investments.

For me, there are many, many ways in which I am interested in dogs. I am interested in the fact that dogs are not us. So they figure not-us. They are not just cute projections. Dogs do not figure mirror-of-me. Dogs figure another species, but another species living in very close relationship; another species in relation to which the nature/culture divide is more of a problem than a help, when we try to understand it. Because dogs are neither nature, nor culture, not both/and, not neither/nor, but something else.

Interviewer: The notion of companionship becomes important here, I assume?

Donna Haraway: Yes, although the notion of companionship is a very modern way of seeing the dogs. The notion of the companion animal is a quite recent invention. Seeing dogs as companion animals, but not pets, is a rather recent contestation. We have necessarily to be in an ethical relationship with dogs, because they are vulnerable to human cruelty in very particular ways, or to carelessness, or stupidity. So dogs become sites of meaning-making and sites of inquiry: ethical inquiry, ontological inquiry, inquiry about the nature of sociality, inquiry about pedagogy and training and control, inquiry about sadism, about authoritarianism, about war (the relationship between the infantry and the war dog as tools in military history), etc. Dogs become good figures to think with—in all sorts of circumstances. There is the development of service dogs, for example, the seeing-eye dogs and other sorts. There are all the different ways that dogs are brought into relationship with human need, or human desire. There are dogs as toys, toy dogs, dogs as livestock guardians in charge of protecting sheep against wolves, bears, coyotes, and so on. Working dogs interest me a lot and so does the relationship of a human being and a dog in the sports world. There are also dependency issues, but dogs are not surrogate children. Dogs are adult. Adult dogs should not be infantilized! When you live with a dog you live with another adult who is not your species. I find this cross-species companionship and the questions of otherness that are involved really intersting. Dogs confront us with a particular kind of otherness that raises many questions, ethical, ontological, political, questions about pleasure, about embodiment etc.

Interviewer: How does the dog relate to the cyborg and the coyote? Is it an in-between figure in the kinship of figurations?

Donna Haraway: It is, and in that sense, you know, I feel like I have written about many sorts of entities that are neither nature, nor culture. The cyborg is such an entity, and the coyote; and the genetically engineered laboratory research animal OncoMouse[TM11] is also in this odd family—this queer family that is neither nature, nor culture, but an interface. The family includes, for me, in terms of what I have written about, personally, the cyborg, the coyote, the OncoMouse[TM], the FemaleMan,[12] the feminists, the history of women within feminist analysis, the dogs in my new project, and, of course, the non-human primates.[13] All these are entities that require one to be confused about the categories of nature and culture.

Interviewer: Are they all on the same level, or do you consider the cyborg to be a kind of meta-category?

Donna Haraway: Well, sometimes the cyborg functions as a meta-category, but I am actually much happier to demote it to one of the litter. Sometimes I do end up saying these are all cyborg figures, but I think that is a bad idea. I like to think of the cyborg as one of the litter, the one that requires an awful lot of intervention in order to survive [laughter]. . . . It has to be technically enhanced in order to survive in this world.

PART II: THERE ARE ALWAYS MORE THINGS GOING ON THAN YOU THOUGHT! METHODOLOGIES AS THINKING TECHNOLOGIES

Interviewer: I would like to start the second part of the interview with a question about your writing style. When I teach feminist theory, I often advise the students to focus not only on the line of argument of your texts, but also to read them in a literary way, i.e., to give attention to the metaphors, images, narrative strategies and to study how you make the literary moves explicit. I think that you, in a very inspiring way, practice your tenet about "scientific practice" as a "story-telling practice" (Haraway 1989: 4). Your deconstructions of the barriers between theory and literature make your texts extremely rich; theoretical content, methodology, style and epistemology go hand in hand. How did you come to this kind of theory writing?

Donna Haraway: Well, there are lots of ways of talking about this. First of all, it is not altogether intentional. Writing does things to the writer. Writing is a very particular and surprising process. When I am writing, I often try to learn something, and I may be using things that I only partly understand, because I may have only recently learned about them from a colleague, a student, a friend. This is not altogether a scholarly proper thing to do. But I do that from time to time, and it affects style. It is like a child in school learning to use a new word in a sentence.

Interviewer: Would you compare this to a literary intuitive way of writing?

Donna Haraway: Yes. My texts are full of arguments, it must be said [laughs jokingly]. But my style of writing is also intuitive. It absolutely is. And I like that. I like words. They are work, but they are also pleasurable.

Interviewer: This means that it is possible to keep going back to your texts and still find new inspiring layers of meaning like in literary texts.

Donna Haraway: Yes, in a sense, I do think that they are literary texts.

Interviewer: Your efforts to transgress the barriers between theory and literature make me think of other scholars within the feminist tradition, for example Luce Irigaray, and the ways in which she deliberately links writing strategies and epistemology. Could you tell us, how you look upon these links as far as your own work is concerned?

Donna Haraway: Well, my style is not only intuitive, but also the result of deliberate choice, of course. Sometimes people ask me "Why aren't you clear?" and I always feel puzzled, or hurt, when that happens, thinking "God, I do the best I can! It's not like I'm being deliberately unclear! I'm really trying to be clear!" But, you know, there is the tyranny of clarity and all these analyses of why clarity is politically correct. However, I like layered meanings, and I like to write a sentence in such a way that—by the time you get to the end of it—it has at some level questioned itself. There are ways of blocking the closure of a sentence, or of a whole piece, so that it becomes hard to fix its meanings. I like that, and I am committed politically and epistemologically to stylistic work that makes it relatively harder to fix the bottom line.

When you ask how I came to this, I think that it is actually something I inherited out of my theological formation. In an academic sense, I am trained in biology, in literature, and in philosophy. Those are my academic backgrounds—together with history, but that came later. But I am also deeply formed by theology, and particularly by Roman Catholic theology and practice. I learned it. I studied it. It is deep in my bones. I started reading St. Thomas when I was about twelve years old, because of the advice of a confessor. It was a way of dealing with doubts about faith. This confessor was a very young priest, a Jesuit, who was ordained with my uncle. He advised me to read St. Thomas. It was a very strange reading for a twelve-year-old girl [laughter]. It was very confusing. I did not understand a word [more laughter].

Interviewer: But it was your first meeting with layered meanings?

Donna Haraway: Not exactly. But I had this whole relationship with priests, who themselves were struggling with things. It is a very personal history.... Anyway, there was a particular theological frame, which was very powerful for me. Actually, it was not St. Thomas, but more the whole framework and, in particular, the idea that as soon as you name something and believe in a name, there is an act of idolatry involved; the idea that the names of God are always finally deeply suspect; the idea that spirituality has a much more negative quality to it; the idea that if you seriously are trying to deal with something that is infinite, you should not attach a noun to it, because then you have fixed and set limits to that which is limitless, and the whole point of God is about a kind of eternal totality that is not the totality of a system. It is not a systemic totality. It is a different kind of totality. It is an unnameableness. It is the theological tradition that focuses on unnameableness.

Interviewer: Like the Sunni tradition in Islam?

Donna Haraway: Yes, but there are many traditions that have a commitment to this kind of negativity. There is a strong current within Catholicism that has a commitment to this kind of negativity, and within Quaker practice as well. That is the theological formation that I think is strongest for me—also as regards its relationship to the proofs of the existence of God. These are not about design, and not about causality, but about the reality of infinity, about the truth of limitlessness, which, as I see it, is existent. To me, this is an existentialist idea, and not a design idea. Any proof of the existence of God is almost a kind of joke from that point of view.

Well, you know, I am, of course, a committed atheist and anti-Catholic, anyway at some level. I cannot live in Christian right-wing U.S. culture and not be an anti-Christian. But that theological tradition is a very deep inheritance for me, and I think it affects my style very deeply.

Interviewer: How do you "translate" your epistemological and political commitment to the deconstruction of fixed categories into methodologies? How do we avoid fixations? Whenever we are doing research, we use certain sets of skills, which imply certain kinds of names, classifications, categorizations, standards, etc. So isn't there a latent, or even very active danger, or risk, or possibility that we will always reduce whatever we are doing—even when we have the most ambitious intention about avoiding closure of the discursive spaces, in which we theorize, analyze, etc.?

Donna Haraway: Well, obviously there is no final answer to that, because it is a permanent paradox, or dilemma. But there are some things that we can say about that dilemma.

First of all, categories are not frozen. We are more inventive than that. The world is more lively than that, including us, and there are always more things going on than you thought; maybe, as Katie King taught me, less than there should be, but more than you thought! Second, you can use categories to trouble other categories. Marilyn Strathern formulated this very wonderful aphorism: "It matters which categories you use to think other categories with" [laughter]. You can turn up the volume on some categories, and down on others. There are foregrounding and backgrounding operations. You can make categories interrupt each other. All these operations are based on skills, on technologies, on material technologies. They are not merely ideas, but thinking technologies that have materiality and effectivity. These are ways of stabilizing meanings in some forms rather than others, and stabilizing meanings is a very material practice.

Third, I find it important to make it impossible to use philosophical categories transparently. There are many philosophers who use cognitive technologies to increase the transparency of their craft. But I want to use the technologies to increase the opacity, to thicken, to make it impossible to think of thinking technologies transparently. Rather, I will foreground the work practice that thinking is. I will stress that category-making is a labor process with its own materiality that is a different kind of materiality than making a sailboat, or raising a dog, or organizing a feminist demonstration. Thinking is involved in all these material practices, but category formation, category manipulation is a different skill. I do not want to throw away the category formation skills I have inherited, but I want to see how we can all do a little re-tooling. This is a kind of modest project, an act of modest witnessing.

Interviewer: To me your article on "The Promises of Monsters" (Haraway 1992) and the way you use the semiotic square here, is a very good example. Contrary to making your thinking technology—in casu:

the semiotic square—transparent and non-conspicuous, you make it visible as a tool, as an analytical technology, as "an artificial device that generates meanings very noisily" (Haraway 1992: 304). I think that this is a very good way of showing how thinking technologies, categories, models, research designs, etc., create the object of study.

Donna Haraway: Yes, I agree with you. I like that reading of "The Promises of Monsters." Here we have this very clutchy structuralist object, and you march through the square [Donna Haraway laughs and marks a march rhythm with her fingers on the table]. It is a kind of serious joke. I think you can actually do interesting work with these tools, but I want to hear them making noise, I want to feel the friction, I do not want to increase the transparency. Obviously you have to hold the transparency at a certain level, or you cannot get anywhere. But those are tactical decisions about tools. That is a techno-scientific way of thinking.

Interviewer: Your epistemological focus on non-closure and deconstruction of fixed categories has led you beyond the impasse of standpoint feminism, but it has also been important for you to avoid the traps of relativism and nihilism, in which the rejection of stable political grounds has left some postmodern and poststructuralist thinkers. As part of your commitment to situated knowledges, you emphasize the necessity of political accountability. I think that you here point out a very important third part, so to speak, a way of navigating in between pure standpoint feminism and pure postmodern relativism. Do you yourself see your position in this way?

Donna Haraway: Yes, that is the way I like to understand it, too. I am not looking for the stable ground of standpoint feminism, nor is my position relativist, nihilist. It is not sceptical. It is not cynical. I emphasize non-stable grounds, but at the same time I feel very strongly affiliated with standpoint theorists.

Interviewer: In which ways?

Donna Haraway: I feel that important work gets done with this very contaminated tool. There are obvious troubles with adopting the metaphors of perspectives, of locations and standpoints, of embodiment and privileged perspective, etc. I think the contaminations of these metaphors are obvious. But that should not stop us from understanding the crucial work that feminist standpoint theorists did, inheriting Lukács, on the one hand, and certain kinds of feminist work that we had been doing, on the other. People like Nancy Hartsock and others understood standpoint feminism as an achievement. This was an epistemological achievement that came out of a political practice that produced the possibility of understanding the reality beneath the appearances in a specifically Marxist sense, i.e., the possibility of understanding the system of domination that supports the appearance of equality; the appearance of normality, and comfort, and equality, and the market, and all of its sequelae; the appearance that men and women can simply have a few equal rights and all will be well; the appearance that race can simply be erased by a little bit of anti-discrimination. Standpoint theory produced the understanding of the deep materiality of oppression beneath all these appearances. The method of understanding was the metaphor of surface and depth, which is not the same as making the mistake of misplaced concreteness that mistakes analytic technology for the world. The world is not surface and depth. One analytic technology is about surface and depths. But you do not mistake the analytic technology for the world, because then you would have committed an act of idolatry and fetishism (in the bad sense of the word, fetish), an act of reification, and this is not what the standpoint theorists did.

Such a reading of the standpoint theorists acknowledges their work as an epistemological achievement within a particular intellectual tradition. I find that important, and I also think that this is the way in which the standpoint theorists read themselves. However, the standpoint theorists do not analyse the literary moves of their own texts, because they often do not see them as literary moves. I see them as literary moves, but not in a reductive sense. It is not in order to dismiss the texts, but in order to remind myself that this is a set of rhetorical possibilities. This kind of "literary" reading makes the standpoint theorists very suspicious. Actually, I cannot believe the number of people who, in the face of the word "narrative," think that all of a sudden you are "merely" in the realm of culture and entertainment—that all of a sudden you are not talking about what is serious. This is a terrible prejudice, which some standpoint theorists share with most political scientists and a vast majority of philosophers, too.

Interviewer: In a video, "Donna Haraway reads the *National Geographics of Primates*" (Paper Tiger Television, #126, 1987) you visualize your analytical method pedagogically by untangling a ball of yarn. You are pulling out the threads, metaphorically demonstrating a deconstructive move, I guess, critically going back to where things are coming from. How would you compare this to the Latour-inspired "follow-the-actors" approach that I think you are very committed to, as well?

Donna Haraway: Well, I see the "pulling-out-the-threads" on the video and the "follow-the-actors" approach as closely related. In my recent book *Modest Witness* (Haraway 1997), I have this family of entities, these imploded objects: chip, gene, cyborg, seed, foetus, brain, bomb, ecosystem, race. I think of these as balls of yarn, as gravity wells, as points of intense implosion or as knots. They lead out into worlds, you can explode them, you can untangle them, you can somehow loosen them up. They are densities that can be loosened, that can be pulled out, that can be exploded, and they lead to whole worlds, to universes without stopping points, without ends. Out of the chip you can in fact untangle the entire planet, on which the subjects and objects are sedimented. Similarly, you do not have to stay below the diaphragm of the woman's body when dealing with the foetus. It leads you into the midst of corporate investment strategies, into the midst of migration patterns in northeastern Brazil, into the midst of little girls doing caesarean sections on their dolls, into the midst of compulsory reproductivity and the question: What is it that makes everybody want a child these days? Who is this "everybody"?

Interviewer: How would you describe the relationship between the research subject and the figures that perform in the analysis? What is, for example, your relationship to the figures or imploded knots, chip, gene, cyborg, foetus, brain, bomb, ecosystem, race?

Donna Haraway: Figures are never innocent. The relationship of a subject to a figure is best described as a cathexis of some kind. There is a deep connection between the writing subject and the figure. It is not just about picking an entity in the world, some kind of interesting academic object. There is a cathexis that needs to be understood here. The analyst is always already bound in a cathectic relationship to the object of analysis, and s/he needs to excavate the implication of this bond, of her/his being in the world in this way rather than some other. Articulating the analytical object, figuring, for example, this family or kinship of entities, chip, gene, foetus, bomb, etc. (it is an indefinite list), is about location and historical specificity, and it is about a kind of assemblage, a kind of connectedness of the figure and the subject.

Interviewer: I would like to know about your relationship to science and technology studies, the STS-tradition. There are, for example, some obvious parallels between your work and the work of Bruno Latour, and he is, in a sense, leaving science studies now. What about you? How do you look upon science studies today? And which role does feminism play here?

Donna Haraway: Well, science studies is a kind of indefinite signifier, and that is what has made it a good place to locate oneself. It is professionalized in various ways, and that is useful. I will sometimes use science studies as a signifier for myself, and at other times I will not use it. It is a professional and strategic location, but it is not a life-long identity. Even though in some other ways it is, because there are institutional realities connected to it. People like Susan Leigh Star, and Bruno Latour, and Andy Pickering, and many others, read each other. So we end up being both deliberately and unconsciously in conversation. But this conversation and reading of each other's texts do not refer to a kind of shared origin story or genealogy. I have a very different genealogy in science studies than, say, Andy Pickering or Bruno Latour do. People like Susan Leigh Star and I share more of a genealogy in science studies that roots it, for example, in the women's health movement and in techno-scientific issues related to women's labor in the office, or to Lucy Suchman's work. You know, we share a genealogy of science studies that, among other things, situates it in relation to the history of the women's movement at least as much as it connects it to a history of a strong program, to a history of actor-network-theory (ANT), or to a history of a rejection of actor-network-theory. You know, all of those end up becoming interesting little events in the neighborhood, but not the main line of action. So in that sense, I have a kind of annoyed relationship with some of the canonized versions of the history of science studies which go like this: "Well, there was this in Edinburgh, there was that in Paris, and whatever." You know, in that narrative of science studies people like me and my buddies are always hard to incorporate. Even by people of great goodwill, such as Andy Pickering, whom I both admire and read with great pleasure, and like as a human being. Nonetheless, read his preface to *Science as Practice and Culture* (Pickering 1992) and watch the absolute indigestibility of Sharon Traweek and me. We are as the angels with the twelve trumpets. Literally. Every other figure in that introduction got a paragraph or so of analysis, in terms of what was contributed, and what he liked or objected to. But we were like blasts from John's Apocalypse [laughter]. Literally! That is the figure he used. Because we are not part of that other story in that way of telling it, and they do not know our story. They do not know it as an academic story, and they do not know it as a political story. It is a different history. So after I was already doing what I now call feminist techno-science studies, I read people like, for example, Bruno Latour. So Latour and other authors, which figure prominently in the canonized version of the history of STS, were not the origin in my story; they came after other events. And they do not get this! That there is a whole other serious genealogy of technoscience studies. So I remain irritated! [Laughter] Because we do know their genealogies, very well. And they do not know ours, even though they exist in writing; they are certainly not inaccessible! On the other hand, this does not mean that I would call myself an outsider. That would be silly of me. But I think it remains true in most academic locations, including science studies, that most feminists are both insiders and outsiders in the sense that Patricia Hill Collins theorized this insider/outsider location for African American women. Sometimes we are forced into this location, and sometimes we choose to inhabit it.

Interviewer: And I suppose the reason is the issue of feminism. . . .

Donna Haraway: Yes, we are a little hard to digest. And I think that is a good thing. On the one hand, we are so normalized, and disciplinized, and comfortable, you know, and to call ourselves outsiders is a kind of lie. But, you know, from another point of view, we are still outsiders.

Interviewer: I think that the term you borrow from Trinh Minh-ha in "The Promises of Monsters" [chapter 3] describes this position very well. It is the position of the inappropriate/d other.

Donna Haraway: Yes, you are necessarily inappropriate/d. . . . You know, I am surprised that so few people have used Trinh Minh-ha's term. I agree with you that it is a really good figuration.

Interviewer: When you emphasize that there are other stories about science studies than the canonized ones, I am reminded of the copy of the filmposter from *The Matrix* which Don Ihde presented at the conference yesterday as a kind of serious joke, suggesting that the three male figures on the original poster could represent Bruno Latour, Andrew Pickering, and himself, while the only female figure could refer to you. This was Don Ihde's way of jokingly creating a metaphor for the matrix of science studies. But when I saw Don Ihde's matrix, a different matrix of science studies immediately came to my mind. Here the three male figures were replaced by three female figures, you, Evelyn Fox Keller and Sandra Harding, while the female figure was replaced by Bruno Latour. This does not mean that I do not recognize the importance of Bruno Latour in the matrix of science studies, but I would simply consider the three other contributors more important in my feminist version of the story of science studies.

Donna Haraway: Yes, I agree. There are a lot of missing matrices or matricians! Moreover, I can add to the story that many of us have fought with Bruno Latour about feminism, and he has finally been willing to take it on. But it is never symmetrical. He is a friend and a person for whom I have enormous respect. But the asymmetry is a historical, structural problem, not a personal one. It is almost impossible for folks in those locations to get it, and feminist technoscience work always feels like trouble, like "now you are getting political again."

NOTES

The interview took place when Donna Haraway visited Denmark as keynote speaker at the conference "Cyborg Identities—The Humanities in Technical Light," October 21–22, 1999, arranged by Randi Markussen and Finn Olesen, Institute of Information and Media Sciences, Aarhus University, as part of the initiative "The Humanities at the Turn of the Millennium," Centre for Cultural Studies, Aarhus University. The interview was taken as part of a special event with Donna Haraway, organized by the FREJA research project "Cyborgs and cyberspace—between narration and sociotechnical reality"; the three interviewers are all members of the FREJA research group.

1. D Haraway, "A Cyborg Manifesto: Science, Technology, and Socialist-Feminism in the Late Twentieth Century," in D. Haraway, *Simians, Cyborgs and Women* (London: Free Association Books, 1991).
2. Cf. Haraway, "A Cyborg Manifesto," p. 170.
3. V. McIntyre, *Superluminal* (Boston: Houghton Mifflin, 1983).
4. J. Russ, *The Female Man* (New York: Bantam Books, 1975).
5. O. Butler, *Dawn* (1987), *Adulthood Rites* (1988), *Imago* (1989), all published by Warner Books, New York.
6. M. E. Clynes, and N. S. Kline, "Cyborgs and Space," *Astronautics* (September 1960). Reprinted in C. H. Gray et al., *The Cyborg Handbook* (London and New York: Routledge, 1995), pp. 29–33.
7. Ibid.
8. Cf. D. Haraway, "Situated Knowledges," in Haraway, *Simians, Cyborgs, and Women*, pp. 199ff.; and D. Haraway, "The Actors Are Cyborg: Nature Is Coyote, and the Geography Is Elsewhere: Postscript to 'Cyborgs at Large,'" in C. Penley and A. Ross, *Technoculture: Cultural Politics*, vol. 3 (Minneapolis: University of Minnesota Press, 1991), pp. 21–26.
9. Cf. Haraway, "A Cyborg Manifesto," p. 150; "The cyborg is a creature in a post-gender world..."
10. Cf. Haraway, *Simians, Cyborgs, and Women*, pp. 1977ff. and 208ff.
11. Cf. D. Haraway, *Modest_Witness@Second_Millennium. FemaleMan©_Meets_Onco-Mouse*™, *Feminism and Technoscience* (London and New York: Routledge, 1997).
12. Cf. n. 11.
13. Cf. D. Haraway, *Primate Visions: Gender, Race, and Nature in the World of Modern Science* (London and New York: Routledge, 1989).

REFERENCES

Haraway, Donna. 1989. *Primate Visions: Gender, Race and Nature in the World of Modern Science*. New York and London: Routledge.
———. 1991. *Simians, Cyborgs and Women. The Reinvention of Nature*. London: Free Association Books.

————. 1992. "The Promises of Monsters: A Regenerative Politics for Inappropriate/d Others." In *Cultural Studies*, L. Grossberg, C. Nelson, and P. Treichler, eds. London and New York: Routledge: 295–338.

————. 1997. *Modest_Witness@Second_Millenium. FemaleMan^{TM}_Meets_OncoMouse^{TM}: Feminism and Technoscience.* New York and London: Routledge.

————. 2000. *How Like a Leaf. An Interview with Thyrza Nichols Goodeve.* London, New York: Routledge.

Lykke, Nina. 1996. "Kyborg eller gudinde? Feministiske dilemmaer i det sene 20. århundredes øko- og teknokritik." I:Kvinder, køn og forskning 5 nr. 4 (1996). pp. 31–45.

Lykke, Nina. 1999. "Posthumane visioner: En postkønnet eller kvindelig cyberkultur?" I: Kvinder, køn og forskning 2: 43–53.

Lykke, Nina, Finn Olesen og Randi Markussen. 2000.

Pickering, Andrew, ed. 1992. *Science as Practice and Culture.* Chicago: University of Chicago Press.

NOTE ON SOURCES

All of the pieces in this reader were previously published elsewhere. "A Manifesto for Cyborgs" is reprinted from *Socialist Review,* no. 80 (1985), pp. 65–108. "Ecce Homo, Ain't (Ar'n't) I a Woman, and Inappropriate/d Others: The Human in a Post-Humanist Landscape" is reprinted from *Feminists Theorize the Political,* Judith Butler and Joan W. Scott, eds. (New York: Routledge, 1992), 86–100. "The Promises of Monsters: A Regenerative Politics for Inappropriate/d Others" is reprinted from *Cultural Studies,* Lawrence Grossberg, Cary Nelson, and Paula Treichler, eds. (New York: Routledge, 1992), 295–337. "Otherworldly Conversations; Terran Topics; Local Terms" is reprinted from *Science as Culture* (London) vol. 3, no. 1 (1992): 59–92. "Teddy Bear Patriarchy: Taxidermy in The Garden of Eden, New York City, 1908–1936" is reprinted from the revised version in Donna Haraway's *Primate Visions* (New York: Routledge, 1989) 26–58, 385–388. It originally appeared in *Social Text,* no. 11 (winter 1984–1985): 20–64. "Morphing in the Order: Flexible Strategies, Feminist Science Studies, and Primate Revisions" is reprinted from *Primate Encounters,* Shirley Strum and Linda Fedigan, eds. (Chicago: University of Chicago Press, 2000), 398–420. "Modest_ Witness@Second_Millennium" is reprinted from the revised version in Donna Haraway's *Modest_Witness@Second_Millennium. FemaleMan©_ Meets_OncoMouse*™ (New York: Routledge, 1997), 23–47, 276–279. It originally appeared as "Modest Witness: Feminist Diffractions in Science Studies" in *The Disunity of Science: Boundaries, Contents, and Power,* Peter Galison and David Stump, eds. (Stanford, Calif.: Stanford University Press, 1996). "Race: Universal Donors in a Vampire Culture: It's All in the Family. Biological Kinship Categories in the

Twentieth-Century United States" is abbreviated and reprinted from the revised version in Donna Haraway's *Modest_Witness@Second_ Millennium. FemaleMan©_Meets_OncoMouse*™ (New York: Routledge, 1997), 232–265, 310–315. It originally appeared in *Uncommon Ground,* William Cronon, ed. (New York: Norton, 1995), 321–366, 531–536. "Cyborgs to Companion Species: Reconfiguring Kinship in Technoscience" is forthcoming in *Chasing Technoscience: Matrix of Materiality,* Donald Ihde and Evan Selinger, eds. (Bloomington: Indiana University Press, in press for 2003). The dog-human co-evolution story is slightly revised from "For the Love of a Good Dog: Webs of Action in the World of Dog Genetics" in *Race, Nature, and the Politics of Difference,* Donald Moore, Jake Kosek, and Anand Pandian, eds. (Durham, NC: Duke University Press, forthcoming, 2003). The section on biodiversity in dogland is drawn from an earlier version of parts of "Cloning Mutts, Saving Tigers: Ethical Emergents in Technocultural Dogland," and is reprinted by permission from *Remaking Life and Death: Toward an Anthropology of the Biosciences,* edited by Sarah Franklin and Margaret Lock. Copyright© 2003 by the School of American Research, Santa Fe, N.M. The interview in chapter 10 was conducted in two parts by Nina Lykke, Randi Markussen, and Finn Olesen. Part I: "Cyborgs, Coyotes, and Dogs: A Kinship of Feminist Figurations" appeared in *Kvinder, Køn og Forskning* vol. 2 (2000), 6–15. Part II: "There Are Always More Things Going on Than You Thought! Methodologies as Thinking Technologies" appeared in *Kvinder, Køn og Forskning,* vol. 4 (2000), 52–60.

INDEX